MW01621202

Pollen Chemistry & Biotechnology

Nesrin Ecem Bayram
Aleksandar Ž. Kostic • Yusuf Can Gercek
Editors

Pollen Chemistry & Biotechnology

Editors
Nesrin Ecem Bayram
Bayburt University, Aydıntepe
Vocational College
Department of Food Processing
Bayburt, Türkiye

Aleksandar Ž. Kostic
University of Belgrade,
Faculty of Agriculture
Belgrade, Serbia

Yusuf Can Gercek
Istanbul University, Faculty of Science
Department of Biology
Istanbul, Türkiye

ISBN 978-3-031-47562-7 ISBN 978-3-031-47563-4 (eBook)
https://doi.org/10.1007/978-3-031-47563-4

This Springer imprint is published by the registered company Springer Nature Switzerland AG
The registered company address is: Gewerbestrasse 11, 6330 Cham, Switzerland

Paper in this product is recyclable.

Preface

Bee products have been used in medicine since ancient times. Their usage in alternative medicine continues to this day (such as propolis, bee pollen, bee bread, honey, etc.). Apart from this, it has been recognized as an excellent functional food ingredient. Scientific studies aimed at determining the phytochemical and physicochemical compositions of bee products have revealed their potential in the treatment of various diseases. Bee pollen, which is a mixture of pollen from different plant species, is a bee product of agglutination of pollen grains collected by bee workers. Bee pollen can be harvested using a trap fixed at the entrance of the hive. Bee pollen stands out for its content of proteins, carbohydrates, fatty acids, minerals, vitamins, and wide range of plant secondary metabolites such as phenolic compounds, carotenoids, phenylamides, etc. The composition of bee pollen varies depending on its biogeographic (regional) origin, botanical origin, and even the harvesting season. It also can be influenced by processing and storage conditions.

The global interest and increasing consumer awareness, especially regarding the nutritional and biological value of food, have led to a growing interest in bee products. From this point of view, bee pollen is of great interest in the food supplementation and food processing industries due to its high nutritional value and bioactivity. Its inclusion in various formulations such as pills, tablets, capsules, and powders helps to meet the needs of many customers. Also, it can improve bioactivity and shelf life of different food products, in particular meat, dairy, and bakery products.

With this book, which we have edited, we aimed to present the structure of bee pollen through a multidisciplinary approach. We prepared this book to provide scientists with a single source on the morphological and anatomical structure, primary and secondary metabolites, food safety assessment, microbiome, and biotechnological applications of bee pollen, as well as to encourage research on bee pollen. We extend our gratitude to all the authors and the staff of Springer Publishing for their contributions to this book.

Bayburt, Türkiye — Nesrin Ecem Bayram
Belgrade, Serbia — Aleksandar Ž. Kostić
Istanbul, Türkiye — Yusuf Can Gerçek

Contents

1 **Pollen Morphology and Anatomy with Botanical Preferences Made by Bees: An Introduction Data** ... 1
Deniz Canlı and Nesrin Ecem Bayram

2 **Amino Acids, Peptides, and Proteins of Pollen** ... 17
Rita Végh and Mariann Csóka

3 **Bee Pollen Carbohydrates Composition and Functionality** ... 51
Jasna Bertoncelj, Nataša Lilek, and Mojca Korošec

4 **Lipids in Pollen** ... 71
Aleksandar Ž. Kostić and Sofija Kilibarda

5 **Macro-, Micro-, Trace, and Toxic Elements of Pollen** ... 85
Pawel Pohl, Anna Dzimitrowicz, Piotr Jamroz, Anna Lesniewicz, Anna Szymczycha-Madeja, Maja Welna, and Krzysztof Greda

6 **Phenolic Acids in Pollen** ... 103
Aleksandar Ž. Kostić, Yusuf Can Gercek, and Nesrin Ecem Bayram

7 **Flavonoids in Pollen** ... 127
Milica Kalaba, Živoslav Tešić, and Stevan Blagojević

8 **Carotenoids and Vitamins of Pollen** ... 147
Rodica Mărgăoan and Mihaiela Cornea-Cipcigan

9 **Important Contaminants (Mycotoxins, Pesticide Residues, Pirolizidine Alkaloids) in Pollen** ... 179
Miroslava Kačániová, Natália Čmiková, and Vladimíra Kňazovická

10 **Other Bioactive Constituents of Pollen** ... 197
José Bernal, Silvia Valverde, Adrián Fuente-Ballesteros, Beatriz Martín-Gómez, and Ana M. Ares

11 **Microbiology of Pollen** ... 229
Vladimíra Kňazovická and Miroslava Kačániová

12 Physical and Bioprocessing Techniques for Improving Nutritional, Microbiological, and Functional Quality of Bee Pollen 251
Carlos Alberto Fuenmayor, Carlos Mario Zuluaga-Domínguez, and Martha Cecilia Quicazán

13 Good Practice of Pollen Collection-What Pollen Traps Are Better Choice 277
Nebojša M. Nedić

14 Techno-Functional Properties of Pollen 291
Danijel D. Milinčić, Aleksandar Ž. Kostić, Slađana P. Stanojević, and Mirjana B. Pešić

15 Bee Pollen as a Source of Pharmaceuticals: Where Are We Now? ... 319
Rachid Kacemi and Maria G. Campos

Index 337

Chapter 1
Pollen Morphology and Anatomy with Botanical Preferences Made by Bees: An Introduction Data

Deniz Canlı and Nesrin Ecem Bayram

1.1 Introduction

Honey bees, which are the most important pollinators of cultivated plants, prefer the flowers of some plant species to be more attractive. The structure of the flower, the movement of the flower, the volatile component content of the flower, the size of the flower, the color of the corolla, and finally the amount and/or quality of the pollen and nectar produced by the plant can affect the bees' visit to the flower [1, 2]. But the most important feature of plants that attract bees is their color. Bees have a trichromatic visual system that enables them to see many colors [1]. However, it was stated that the relationship between nectar volume and secreted nectar is probably genetic. Another important factor in the nectar preference of honey bees is the sugar concentration in the nectar, and there is a positive correlation between bee visitation and nectar production. However, even the best known floral sources for honey bees, nectar production can be variable in different years and regions depending on factors such as air temperature, humidity, groundwater, precipitation and soil yield [2].

Some plants actually produce little or no nectar, but they attract bees with their pollen production. Pollen is the main source of protein, especially for young bees, and therefore some plants attract pollinators by producing pollen [2]. The pollen content of each plant is unique to itself, and especially the pollen shape can be described as the fingerprint of the plants. At this point, the branch of science that studies pollen, namely palynology, comes into play. Thanks to this branch of

D. Canlı
Food, Agriculture and Livestock Vocational School, Bingol University, Bingol, Turkey
e-mail: dcanli@bingol.edu.tr

N. Ecem Bayram (✉)
Aydıntepe Vocational College, Department of Food Processing, Bayburt University, Bayburt, Turkey

N. Ecem Bayram et al. (eds.), *Pollen Chemistry & Biotechnology*,
https://doi.org/10.1007/978-3-031-47563-4_1

science, the detection of plants visited by bees can be determined using the distinctive characteristic morphological parameters of plant pollen.

1.2 The Pollen Grain

The pollen grain is an element that contains the male gametophyte produced by seed plants. Pollen grains are formed by the meiosis of pollen mother cells (PMC), which develops in sporogenous tissues in the anthers of the androecium [3]. In the adaptation process of seed plants to terrestrial life, dependence on water for reproduction has been eliminated and the male gametophyte producing sperm is protected in pollen grain [4, 5]. Pollen grain transports the sperm cells to the ovule containing the female gametophyte. Pollen reaches the female organ by means of various vectors (such as wind, water, or animals).

A mature pollen grain develops over two distinct phases. The first is microsporogenesis, whilst the following is microgametogenesis [6, 7].

1.2.1 Microsporogenesis

The development of microspores from pollen mother cells is known as microsporogenesis. The diploid pollen mother cells originate from the sporogenous tissue in pollen sacs are surrounded by a callose layer [3]. As a result of meiosis of pollen mother cells, 4-microspore groups (tetrads) are formed [8]. The tetrad type is determined by the direction of the spindle fibers pulled to the poles during meiosis [9]. Microspores then develop into pollen grains [6].

1.2.2 Microgametogenesis

Microgametogenesis is the process by which sperm is developed by the microspore nucleus going through two sequential mitoses. In this stage, a vacuole occurs in the microspore center and the microspore nucleus moves from central to eccentric position. The first pollen mitosis happens at this position and two unequal new cells occur [10]. The small cell is named the "generative cell" and the larger one is called the "vegetative cell (tube cell)" [3, 6]. The generative cell is surrounded by a membrane and floats in the cytoplasm of the vegetative cell. It becomes spindle-shaped and locates near the pollen wall [11]. The generative cell divides into two male gametes (sperm) in the second pollen mitosis. This is the final stage of microgametogenesis and mature pollen has developed [6]. The second pollen mitosis does not always occur in the pollen sacs of the anther. At the time of anthesis, the number of nuclei in the pollen grain may vary, and this situation is thought to be phylogenetically important. In many species, pollen grains contain one generative cell during the pollination, and it divides into two sperm cells in the pollen tube after germination. In angiosperms, binucleate pollen grains—one tube cell and one generative cell—are the most typical during the anthesis, whereas trinucleate grains—one tube cell and two sperm cells—are comparatively uncommon [11].

1.3 Pollen Morphology

Pollen grains have a 3-dimentional form and diverse greatly in shape and sizes. The observation of multiple orientations of a pollen grain may be needed before the morphological type is confirmed. The pollen wall provides various taxon-specific microscopic features, which the palynologists use for identification. For the aim of identifying and describing the microscopic features of pollen grains, a particular terminology that is generally accepted by palynologists is adopted. But nevertheless, in some cases, the terminology applied by the authors differs. The terminology used in this chapter follows the recommendations made by Erdtman (1963), Punt et al. (2007), and Halbritter et al. (2018) [6, 12, 13].

The main features of pollen grains that are used to identify them are summarized below: pollen unit, size, polarity and symmetry, shape, pollen wall, exine ornamentation, and apertures.

1.3.1 Pollen Unit

The term "pollen unit" describes the number of pollen grains that were fused at the time of anthesis [6]. If the microspores separate from one another and form a single unfused pollen grain, they disperse from the anther as monads. In some plants, meiotic microspores release together partly or without separating from each other and form dyads or tetrads [3]. Additionally, pollen grains that are arranged in groups of more than four are called polyads, and generally, polyads are composed of multiples of eight united grains. If all pollen grains in a whole theca are released together, it is called pollinium (plural pollinia) Fig. 1.1 [14].

1.3.2 Pollen Size

Although pollen size is typically in the range of 25–50 μm [15]. It varies by plant species, *Zostera marina*, an aquatic plant, has ribbon-like pollen that is about 2–3 mm long [16] and *Myosotis* pollen is smaller than 10 μm [17].

It requires several measurements and statistical processing of the results to specify the average size of a pollen species [3]. However, taxonomically, it is a useful

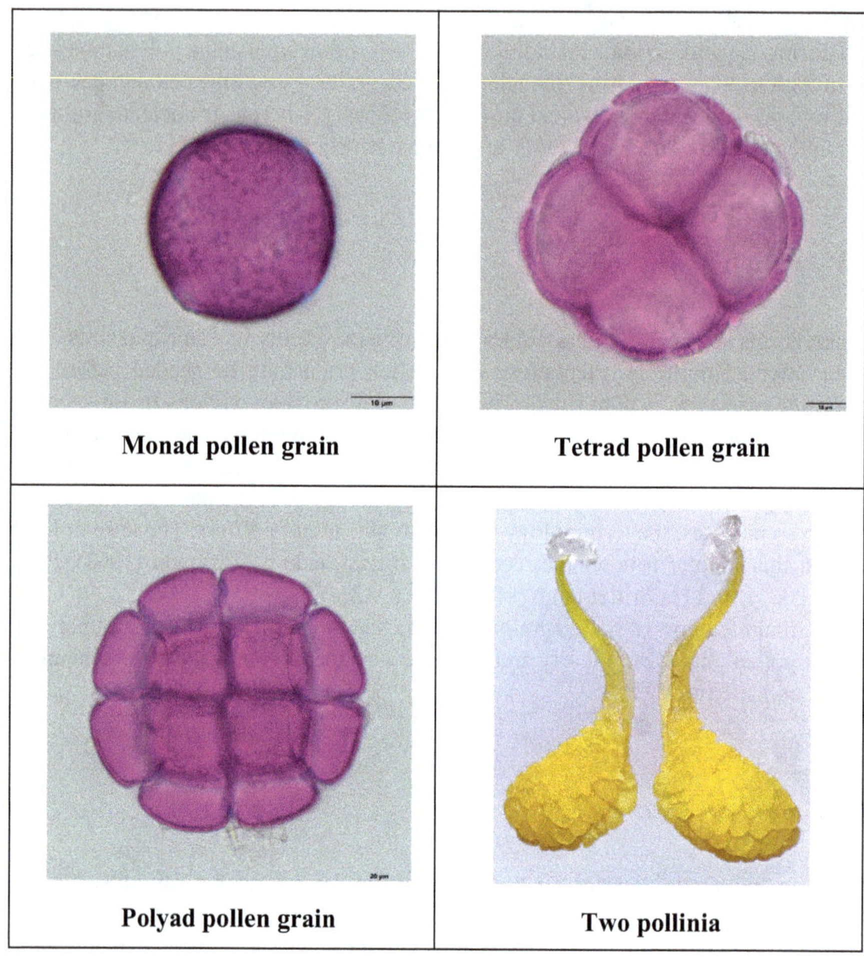

Fig. 1.1 Pollen units

feature as it is stable within taxa, and in some cases, it allows the delimitation of species that are closely related [3]. However, the preparation method(s) applied and the degree of hydration may have an impact on pollen size [6, 15, 18].

1.3.3 Polarity and Symmetry

Pollen grains may exist in several polarity and symmetry configurations since pollen grains are three-dimensional structures [19]. The polarity of the pollen is closely related to the position of the microspores within the tetrad phase. Polarity is determined when the microspores are in the tetrad stage. The direction of spindle fibers

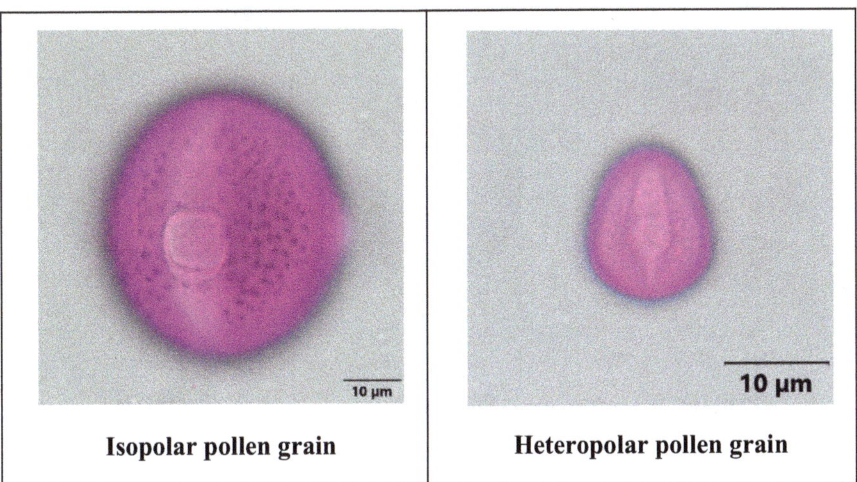

Fig. 1.2 Polarity of pollen grain

throughout meiosis and cytokinesis strongly influences where the microspores are positioned in the tetrad [6].

The zone that is closest to the tetrads center is identified as the proximal pole, while the zone that is farthest away is the distal pole. The polar axis of a pollen is the axis that extends from the proximal pole to the distal pole in the tetrad. The equatorial axis is the axis that is perpendicular to the polar axis. The proximal and distal hemispheres of isopolar pollen are similar either in shape, apertures or ornamentation and hemispheres are obviously distinct in heteropolar ones (Fig. 1.2) [3, 12].

Determining the angle of view is significant for identifying the pollen grain's morphological features. The polar view refers to a pollen grain observed from the pole, while the equatorial view is when it is observed from the equator.

The majority of pollen grains have symmetry, and this property may be used to identify the morphological type of a grain. Symmetry is defined in the polar view of the pollen. If pollen has more than one plane of symmetry, it is radially symmetrical or bilaterally symmetrical if it has a single plane of symmetry [11].

1.3.4 Shape

Pollen shape is determined by the ratio of the polar axis length of the pollen to the equatorial axis length (P/E) [20]. In angiosperms, there are three basic pollen shapes. These are oblate, spheroidal (round) and prolate (Fig. 1.3). In oblate pollen, the polar axis is shorter than the equatorial axis. The polar and equatorial axes of spheroidal pollen are almost equal in length. The polar axis of prolate pollen is longer than the equatorial axis [21].

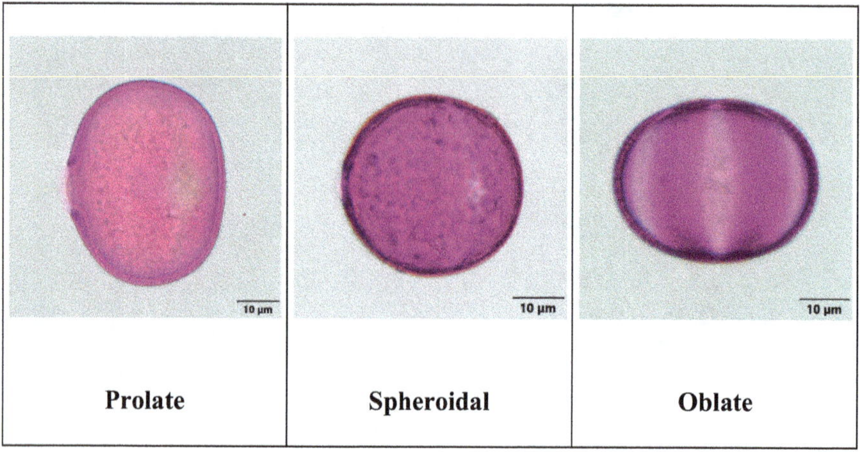

Fig. 1.3 Pollen shapes

The nomenclature for the pollen grain shapes (outline) in polar view is a bit different. Amb is another name for the outline in polar view of a pollen grain [11]. The outline in polar view can be angular (triangular, quadrangular, polygonal, rectangular, rhomboid, etc.), non-angular (circular or elliptical) or lobate [11, 12].

1.3.5 Pollen Wall

The pollen grain is surrounded by a wall called the sporoderm, and this wall basically consists of two layers, exine and intine [15, 22]. Pollen wall protects the live content of pollen grain against external dehydrating or damaging effects (temperature, pressure, etc.).

1.3.5.1 Exine Sculpturing

The exine is the outermost layer of the pollen wall. It provides protection of pollen grain from external factors. Also, the main structural support for the cytoplasm is provided by exine. Sporopollenin, a biopolymer that is resistant to acetolysis and degradation, constitutes the majority of the exine [23–25]. Exine typically consists of two sublayers: "ectexine (external exine)" and "endexine (internal exine)" [26].

Pollen grain ornamentation is related to the exine elements found on the pollen wall. The outermost layer of the ectexin "tectum" generally exhibits various sculptural properties. This feature of the exine plays an important role in the diagnosis of pollen taxa at the genus/species level. The pollen of insect-pollinated plants usually has more ornamental exines [26, 27].

Exine ornamentation type is one of the most important key characters used in the differentiation of pollen types. But ornamentation details may not be viewed by light microscopy (LM). It can be observed in greater detail with a scanning electron microscope.

Two basic categories may be used to group different ornamentation types [3]. In the first group, there are no real sculptural elements on the exine. This category includes psilate, foveolata, perforate and fossulate ornamentations that described below. The ornamentation types in the first group may not be able to be determined accurately by LM due to the low-resolution limit.

Psilate: The surface of tectum is smooth. The exine shows no particular sculpturing.
Foveolate: The tectum surface is covered with pits of 1 μm in diameter and the distance between the pores is greater than the pore diameter.
Perforate: Tectum surface covered with pores/pits smaller than 1 μm.
Fossulate: Irregularly patterned grooves cover the tectum surface.

In the second group, there are real sculptural elements on the exine surface. Six sculptural elements: spine-spinule, pilum, baculum, verruca, gemma and clava constitute different ornamentation types as described below (Fig. 1.4).

Echinate: Tectum has pointed end (spine-like) elements >1 μm long. If spine-like elements <1 μm long, ornamentation takes the micro- prefix and is named as microechinate.

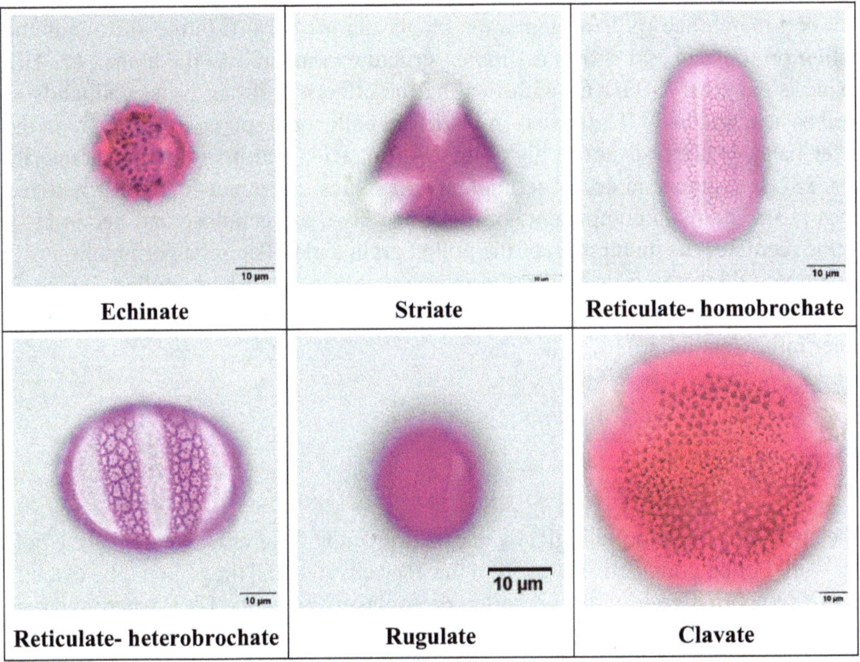

Fig. 1.4 Exine sculpturing types by LM

Baculate: There are rod-shaped not pointed elements >1 μm in length (bacula) on exine surface.

Clavate: Exine surface has club-shaped elements (clavae, sing. clava) with diameter < height. The apical part of the clave is thicker, in comparison to the bacula.

Reticulate: It is a net-like ornamentation type. Sculpturing elements are arranged in a network that meshes termed lumina (sing. lumen) and the walls are designated as muri (sing. murus). The homogeneity of reticulum is an important taxonomic feature in reticulate ornamentation. The homobrochate type has a reticulum uniform in size and the heterobrochate type has a reticulum variation in size.

Gemmate: Exine surface has non-pointed sculptural elements (gemmae, sing. gemma) with a constricted basal part, >1 μm in length, width and length are approximately close to each other.

Verrucate: Verruca is a wart-like sculptural element whose width is greater than its height, >1 μm in length, and non-constricted at the base.

Striate: Describing an ornamentation in which elongated sculptural elements (muri) are oriented parallelly and separated by grooves.

Rugulate: Elongated sculptural elements (muri) >1 μm long are arranged irregularly on the pollen wall.

1.3.5.2 Intine Properties

In a pollen grain, there is an intine layer underneath the exine and above the vegetative cell membrane (pollen membrane or plasma membran). Intine surrounds the pollen protoplasm and shows a simpler structure compared to the exine [11, 19]. Exine is extremely resistant, withstanding the effects of strong bases and acids as well as heat up to 400 °C. It exists in fossilized pollen and spores. Intin layer, on the other hand, is not resistant to high temperature, acids and microorganism activity [19, 28]. It degrades in acetolised pollen and spores and is not found in fossilized ones [13]. The main components of the intine layer are cellulose and pectin [15]. Intine regulates the maturation of the pollen grain and pollen tube germination [6].

Intine is thicker or bilayered at the apertural area in angiosperm pollen. It is basically consisting of two sub-layers. These are the outer intine (ektintine) and the inner intine (endintine) [6, 11].

1.3.6 Apertures

The aperture is the morphologically and anatomically differentiated area of the pollen wall. Pollen grain germinates from the aperture area and the pollen tube extends out of the pollen grain with the pollen protoplasm and intine [22]. Apertures perform a harmomegathy function by acting as regulators of pollen grain volume changes caused on by humidity changes [29]. In pollen grains lacking apertures,

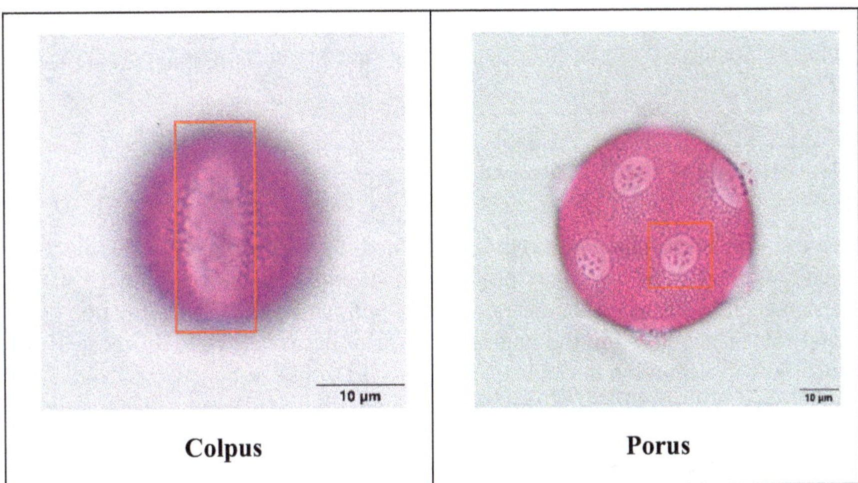

Fig. 1.5 Apertures (outlined by a red rectangle)

pollen tube formed where the exine is limited. Pollen grains with no aperture are denoted by the term inaperturate [12, 22].

Apertures could be classified into two groups depending on their morphological characteristics. The elongated aperture with a length/width ratio >2 and pointed ends is designated a colpus (pl. colpi) or furrow, while the circular aperture with a length/width ratio <2 is termed a porus (pl. pori) (Fig. 1.5) [3]. Colpus and porus are usually situated equatorially or globally; if they are located distally, they are referred to as a sulcus and an ulcus, respectively [11].

Pollen grains with porus are porate and those with colpus are colpate. It is colporate if the pore and colpus are combined in the same aperture. The prefixes mono-, di-, tri-, tetra-, penta-, hexa, and poly- (>6) with the terms porate, colpate, or colporate indicate the number of apertures [22]. For instance, a tetracolpate pollen grain is one with four elongated apertures located equatorially. The accurate aperture number of a pollen grain could be established from the polar view since the number of apertures present is not always visible in the equatorial view.

Pollen grains with apertures situated equatorially may be generally referred to as zonoaperturate (or stephanoaperturate), as in zonoporate or zonocolpate. If an aperture distrubutes globally on the pollen grain surface, it is named pantoaperturate (e.g., pantoporate, pantocolpate) [6, 11, 12].

On the aperture, there is a layer called the aperture membrane that is a constituent of the exine. The aperture membrane could be ornamented (covered with various exine elements) or smooth (psilate) [6]. The pollen of some plant taxa has a cover (partially or completely) on the aperture membrane and this is called the "operculum". The operculum essentially corresponds to a thick exine layer covering the aperture [12].

Apertures may be surrounded by a zone that varies in thickness or ornamentation; these are termed as margo (around colpus) and annulus (around porus) [6, 12].

1.4 Botanical Preferences of Bees

The most important stage in the reproduction and survival of plants is "pollination", which is defined as the transfer of pollen from the anther of a flower onto the stigma of a same (self-pollination) or different (cross-pollination) flower. Pollination plays a vital role in maintaining the natural balance of ecosystems [1]. In most of the plants (80%), pollination depends on the transport of pollen by insects. Among the insects that provide pollination, the bee plays a decisive role. Flower visiting performance of bees is directly related to pollination performance [30]. It has been reported that bees have varying degrees of effectiveness on crop quality and productivity of different plants such as apple (*Malus domestica* L.), coconut (*Cocos nucifera* L.), watermelon (*Citrullus lanatus* Thunb.), tart cherry (*Prunus cerasus* L.), cape gooseberry (*Physalis peruviana* L.), sweet cherry (*Prunus avium* L.), almond (*Prunus dulcis* (Mill.) D.A.Webb), avocado (*Persea americana* Mill.), passion fruit (*Passiflora edulis* Sims. *f. flavicarpa* Deg), citrus (*Citrus sinensis* L.), mango (*Mangifera indica* L.), guava (*Psidium guajava* L.), strawberry (*Fragaria* × *ananassa* DUCH), kiwifruit (*Actinidia deliciosa* (A.Chev.) C.F.Liang & A.R.Ferguson), pear (*Pyrus communis* L.), cranberries (*Vaccinium oxycoccos* L.), cucumbers (*Cucumis sativus* L.), sweet pepper (*Capsicum annuum* L.), tomatoes (*Solanum lycopersicum* L.), anise (*Pimpinella anisum* L.), black seed (*Nigella sativa* L.), cumin (*Cuminum cyminum* L.), sunflowers (*Helianthus annuus* L.), coriander (*Coriandrum sativum* L.), cotton (*Gossypium hirsutum* L.), pumpkins (*Cucurbita maxima* L.), soyabean (*Glycine max* L.), sesame (*Sesamum indicum* L.), cowpea (*Vigna unguiculata* L. Walp), red clover seed (*Trifolium pratense* L.) legume, pineland golden trumpet (*Angadenia berteroi* (A.DC.) Miers), mustard (*Brassica juncea* L.), green grams (*Vigna radiate* L.) and bambara groundnut (*Voandzeia subterranean* L.), coffee (*Coffea arabica* L.), acai palm (*Euterpe oleracea* Martius) and oilseed rape (*Brassica napus* L.) [28]. The fact that bees stay away from such plants causes these plants to not reach the fruit and seed development stages [30].

In addition, the botanic source of honey, which is a bee product, is determined by pollen analysis. Honey varies considerably in terms of taste, aroma, color, and bioactive properties depending on the botanical content, although the overall composition is similar [31]. Honeys with specific botanical and geographic origins have grown in popularity in recent years as customers' concerns about food quality, accurate labelling, and traceability have grown. Melissopalynological analysis is the most essential method for identifying the botanical and geographic origins of honey. Melissopalynological analyses are based on the identification of pollen in honey by microscopic observation of its morphological characteristics [19, 32]. Each pollen grain from different types of plants has a unique morphology that enables microscopic identification of them. The morphological characters such as shape and size of the pollen grain, the number, location and arrangement of apertures, the structure of the exine layer, surface ornamentations and color are used to identify pollen in honey [33]. Diagnoses could be made through several taxonomic ranks, such as family, genus, or species, depending on how well the specific characteristics of pollen grains could be seen under a light microscope. Therefore, these analyses require expertise and in-depth knowledge and are also time-consuming [32, 34–36]. The history of melissopalynological analysis dates back to 1895, when Pfister showed that pollen found in honey can be used to determine the geographic origin [37]. Although a few more researchers conducted studies after Pfister, it took 50–60 years for the microscopic analysis of honey to develop. Honey naturally contains a large amount of pollen. During the visit of the bees to the flowers to collect nectar, the pollen is carried to the hive by mixing with the nectar and sticking to the body parts of the bee. Therefore, pollens in honey are important indicators that allow us to obtain direct information about the nectar composition of honey, the flora of the region where it is produced, and the nutritional ecology of honey bees.

As a result of melissopalynological analyzes performed on honey produced in different regions in different scientific reports presented in the literature, pollen types of plants belonging to families Asteraceae, Fabaceae, Lamiaceae, Boraginaceae, Apiaceae, Brassicaceae and Rosaceae were frequently encountered in honey samples, which is an indication that plants belonging to these families are visited by bees as nectar sources [38–43]. However, plants with dominant honey production potential and pollen production potential can be counted among the species frequently visited by bees. Some of these plants are presented in Table 1.1.

Table 1.1 Some plants with dominant honey and pollen production potential [51]

Plants with dominant honey production potential	Plants with dominant pollen production potential
Arbutus andrachne L.	*Brassica napus* L.
Arbutus unedo L.	*Brassica oleracea* L.
Arctium minus (Hill) Bernh.	*Carum carvi* L.
Brassica napus L.	*Castane sativa* Miller
Calluna vulgaris (L.) Hull	*Cistus criticus* L.
Cardus nutans L.	*Cistus salvifolius* L.
Castane sativa Miller	*Cistus laurifolius* L.
Centaurea triumfetti All.	*Citrus limonia* (L.) Burn. Fil.
Ceratonia siliqua L.	*Cistrus nobilis* Lour.
Cirsium arvense (L.) Scop.	*Citrus paradisii* Macfad.
Citrus limonia (L.) Burn. Fil.	*Cornus mas* L.
Cistrus nobilis Lour.	*Crepis foetida* L.
Citrus paradisii Macfad.	*Daucus carota* L.
Citrus sinensis (L.) Osbeck	*Diospyros kaki* L. F.
Coronilla varia L.	*Diospyros lotus* L.
Daucus carota L.	*Diplotaxis tenuifolia* (L.) DC.

(continued)

Table 1.1 (continued)

Plants with dominant honey production potential	Plants with dominant pollen production potential
Diospyros kaki L. F.	*Erica arborea* L.
Diplotaxis tenuifolia (L.) DC.	*Erica manipuliflora* Salisb.
Echium italicum L.	*Eucalyptus camaldulensis* Dehn.
Echium plantagineum L.	*Hedera helix* L.
Echium vulgare L.	*Hedysarum varium* Wild.
Erica arborea L.	*Lythrum salicaria* L.
Eriobotrya japonica Lindley	*Onobrychis viciifolia* Scop.
Eucalyptus camaldulensis Dehn.	*Papaver hybridum* L.
Hedera helix L.	*Papaver somniferum* L.
Hedysarum varium Wild.	*Phacelia tanacetifolia* Bentham
Helianthus annuus L.	*Quercus robur* L.
Helianthus tuberusus L.	*Rubus canescens* DC.
Lotus corniculatus L.	*Rubus idaeus* L.
Malus sylvestris Mill.	*Salix alba* L.
Medicago sativa L.	*Salix caprea* L.
Medicago varia Martyn	*Salix fragilis* L.
Myosotis alpestris F.W.Schimidt	*Salix triandra* L.
Onobrychis viciifolia Scop.	*Taraxacum officinale* Weber
Paliurus spina-christi Miller	*Trifolium campestre* Schreb.
Phacelia tanacetifolia Bentham	*Trifolium hybridum* L.
Pyrus communis L.	*Trifolium pratense* L.
Rhododendron ponticum L.	*Trifolium repens* L.
Robinia pseudoacacia L.	*Verbascum phlomoides* L.
Rosmarinus officinalis L.	
Rubus canescens DC.	
Rubus idaeus L.	
Salix alba L.	
Salix caprea L.	
Salix fragilis L.	
Salix triandra L.	
Salvia officinalis L.	
Salvia pratensis L.	
Salvia verbenaca L.	
Salvia verticillate L.	
Solidago virgaurea L.	
Stachys annua L.	
Taraxacum officinale Weber	
Thymus longicaulis C. Presl.	
Thymus praecox Opiz	
Tilia plantyphyllos Scop.	
Trifolium campestre Schreb.	
Trifolium pratense L.	
Trifolium repens L.	
Vicia cracca L.	

In addition to melissopalynological analyzes, researches carried out for the last few years to determine the plant origin of pollen grains in bee pollen and honey have been carried out using molecular-based methods [44–47]. In this way, a method is used in which more reliable results are obtained in a shorter time compared to melissopalynological analyses [48, 49]. However, both the results obtained by the molecular method and the results obtained by the melissopalynological method support each other, and these results show that bees prefer certain plant species as nectar and pollen sources [50]. In this way, by using both melissopalynological and molecular methods, the products collected by the bee from the plant, especially bee pollen and pollen grains in honey, can be identified and the plants visited by the bees can be defined in detail.

References

1. Khalifa SA, Elshafiey EH, Shetaia AA, El-Wahed AAA, Algethami AF, Musharraf SG et al (2021) Overview of bee pollination and its economic value for crop production. Insects 12(8):688. https://doi.org/10.3390/insects12080688
2. Silici S, Özkök D (2009) Bal Arısı Biyolojisi ve Yetiştiriciliği. Silici S, Yayınevi E (eds)
3. Reyes ES, Sanchéz JS (2017) Botanical classification. In: Bee products – chemical and biological properties. Springer, pp 3–19. https://doi.org/10.1007/978-3-319-59689-1_1
4. Scott RJ, Spielman M, Dickinson H (2004) Stamen structure and function. Plant Cell Online 16:46–60. https://doi.org/10.1105/tpc.017012
5. Hafidh S, Fíla J, Honys D (2016) Male gametophyte development and function in angiosperms: a general concept. Plant Rep 29:31–51. https://doi.org/10.1007/s00497-015-0272-4
6. Halbritter H, Ulrich S, Grímsson F, Weber M, Zetter R, Hesse M, Buchner R, Svojtka M, Frosch-Radivo A (2018) Illustrated pollen terminology. Springer. https://doi.org/10.1007/978-3-319-71365-6_7
7. Keijzer CJ, Willemse MTM (1988) Tissue interactions in the developing locule of Gasteria verrucosa during microgametogenesis. Acta Botanica Neerlandica 4:475–491. https://doi.org/10.1111/j.1438-8677.1988.tb02155.x
8. Gómez JF, Talle B, Wilson ZA (2015) Anther and pollen development: a conserved developmental pathway. J Integr Plant Biol 11:876–891. https://doi.org/10.1111/jipb.12425
9. Furness CA, Rudall PJ (2001) Pollen and anther characters in monocot systematics. Grana 1–2:17–25. https://doi.org/10.1080/00173130152591840
10. Eady C, Lindsey K, Twell D (1995) The significance of microspore division and division symmetry for vegetative cell-specific transcription and generative cell differentiation. Plant Cell 1:65–74. https://doi.org/10.1105/tpc.7.1.65
11. Simpson MG (2010) Palynology. In: Plant systematics. Elsevier, pp 561–571. https://doi.org/10.1016/b978-0-12-374380-0.50012-9
12. Punt W, Hoen PP, Blackmore S, Nilsson S, Le Thomas A (2007) Glossary of pollen and spore terminology. Rev Palaeobot Palynol. https://doi.org/10.1016/j.revpalbo.2006.06.008
13. Erdtman G (1963) Palynology. In: Advances in Botanical Research, vol 1. Academic, pp 149–208. https://doi.org/10.1016/s0065-2296(08)60181-0
14. Pacini E, Hesse M (2002) Types of pollen dispersal units in orchids, and their consequences for germination and fertilization. Ann Bot 6:653–664. https://doi.org/10.1093/aob/mcf138
15. Dahl O (1970) How to know pollen and spores. Ronald O. Kapp. Q Rev Biol 45(4):406. https://doi.org/10.1086/406697

16. De Cock AWAM (1980) Flowering, pollination and fruiting in Zostera marina L. Aquat Bot 9:201–220. https://doi.org/10.1016/0304-3770(80)90023-6
17. Meudt HM (2016) Pollen morphology and its taxonomic utility in the Southern Hemisphere bracteate-prostrate forget-me-nots (Myosotis, Boraginaceae). N Z J Bot 4:475–497. https://doi.org/10.1080/0028825x.2016.1229343
18. Reitsma T (1969) Size modification of recent pollen grains under different treatments. Rev Palaeobot Palynol 3:175–202. https://doi.org/10.1016/0034-6667(69)90003-7
19. Ricciardelli d'Albore G (1997) Textbook of melissopalynology. Apimondia, Bucharest
20. Erdtman G (1943) An introduction to pollen analysis. The Ronald Press Co, New York
21. Erdtman G (1986) Pollen morphology and plant taxonomy: angiosperms. Brill Archive, Leiden
22. Erdtman G (Gunnar) (1969) Handbook of palynology- an introduction to the study of pollen grains and spores. Hafner, Copenhague
23. Li FS, Phyo P, Jacobowitz J, Hong M, Weng JK (2018) The molecular structure of plant sporopollenin. Nat Plant 5:41–46. https://doi.org/10.1038/s41477-018-0330-7
24. Wiermann R, Gubatz S (1992) Pollen Wall and Sporopollenin. Int Rev Cytol C:35–72. https://doi.org/10.1016/s0074-7696(08)61093-1
25. Steemans P, Lepot K, Marshall CP, Le Hérissé A, Javaux EJ (2010) FTIR characterisation of the chemical composition of Silurian miospores (cryptospores and trilete spores) from Gotland, Sweden. Rev Palaeobot Palynol 162(4):577–590. https://doi.org/10.1016/j.revpalbo.2010.07.006
26. Faegri K, Iversen J (1989) Textbook of pollen analysis, 4th edn. Wiley
27. Fellenberg C, Vogt T (2015) Evolutionarily conserved phenylpropanoid pattern on angiosperm pollen. Trends Plant Sci 20(4):212–218. https://doi.org/10.1016/j.tplants.2015.01.011
28. Erdtman G (1953) Pollen morphology and plant taxonomy. Soil Sci 75(3):248. https://doi.org/10.1097/00010694-195303000-00016
29. Wodehouse RP (1935) Pollen grains. Their structure, identification and significance in science and medicine. Mc Graw-Hill Publishing Co. Ltd., London
30. Savaş T (2007) Arıcılık, çeviri: Meltem Leyla Kuş,; Özgün adı: Bienen Halten, Franz Lampeitl, Bilge Kültür Sanat Yayınevi, İstanbul
31. Špánik I, Pažitná A, Šiška P, Szolcsányi P (2014) The determination of botanical origin of honeys based on enantiomer distribution of chiral volatile organic compounds. Food Chem:497–503. https://doi.org/10.1016/j.foodchem.2014.02.129
32. Von Der Ohe W, Oddo LP, Piana ML, Morlot M, Martin P (2004) Harmonized methods of melissopalynology. Apidologie 35:18–25. https://doi.org/10.1051/apido:2004050
33. Molan PC (1998) The limitations of the methods of identifying the floral source of honeys. Bee World 2:59–68. https://doi.org/10.1080/0005772x.1998.11099381
34. da Luz CFP, de Miranda Chaves SA, Cano CB (2020) Botanical and geographical origins of honey samples from Pantanal (Mato Grosso and Mato Grosso do Sul states, Brazil) certificated by melissopalynology. Grana 1:1–28. https://doi.org/10.1080/00173134.2020.1815831
35. Maurizio A (1975) Microscopy of honey. In: Crane E (ed) Honey: a comprehensive survey. Heinemann, London, pp 240–257
36. Louveaux J, Maurizio A, Vorwohl G (1978) Methods of Melissopalynology. Bee World 4:139–157. https://doi.org/10.1080/0005772x.1978.11097714
37. Pfister R (1895) Den Versuch einer Mikroskopie des Honigs. Z Anal Chem 34(1):479. https://doi.org/10.1007/bf01595875
38. Bodor Z, Kovacs Z, Benedek C, Hitka G, Behling H (2021) Origin identification of hungarian honey using melissopalynology, physicochemical analysis, and near infrared spectroscopy. Molecules 26(23):7274. https://doi.org/10.3390/molecules26237274
39. Bahloul R, Zerrouk S, Chaibi R (2022) Pollen analysis of honey from Laghouat region (Algeria). Grana 61(6):1–10. https://doi.org/10.1080/00173134.2022.2126726
40. Piroux M, Lambert O, Puyo S, Farrera I, Thorin C, L'Hostis M et al (2014) Correlating the pollens gathered by Apis mellifera with the landscape features in western France. Appl Ecol Environ Res 12(2):423–439. https://doi.org/10.15666/aeer/1202_423439

41. Mureşan CI, Cornea-Cipcigan M, Suharoschi R, Erler S, Mărgăoan R (2022) Honey botanical origin and honey-specific protein pattern: characterization of some European honeys. LWT 154:112883. https://doi.org/10.1016/j.lwt.2021.112883
42. Küçükaydın S, Tel-Çayan G, Çayan F, Taş-Küçükaydın M, Eroğlu B, Duru ME, Öztürk M (2023) Characterization of Turkish Astragalus honeys according to their phenolic profiles and biological activities with a chemometric approach. Food Biosci 53:102507. https://doi.org/10.1016/j.fbio.2023.102507
43. Bayram N, Yüzer MO, Bayram S (2019) Melissopalynology analysis, physicochemical properties, multi-element content and antimicrobial activity of honey samples collected from Bayburt, Turkey. Uludağ Arıcılık Dergisi 19(2):161–176
44. Bruni I, Galimberti A, Caridi L, Scaccabarozzi D, De Mattia F, Casiraghi M, Labra M (2015) A DNA barcoding approach to identify plant species in multiflower honey. Food Chem 170:308–315. https://doi.org/10.1016/j.foodchem.2014.08.060
45. Saravanan M, Mohanapriya G, Laha R, Sathishkumar R (2019) DNA barcoding detects floral origin of Indian honey samples. Genome 62(5):341–348. https://doi.org/10.1139/gen-2018-0058
46. Özkök A, Bilgiç HA, Kosukcu C, Arık G, Canlı D, Yet İ, Karaaslan C (2023) Comparing the melissopalynological and next generation sequencing (NGS) methods for the determining of botanical origin of honey. Food Control 148:109630. https://doi.org/10.1016/j.foodcont.2023.109630
47. Qiao J, Feng Z, Zhang Y, Xiao X, Dong J, Haubruge E, Zhang H (2023) Phenolamide and flavonoid glycoside profiles of 20 types of monofloral bee pollen. Food Chem 405:134800. https://doi.org/10.1016/j.foodchem.2022.134800
48. Casiraghi M, Labra M, Ferri E, Galimberti A, De Mattia F (2010) DNA barcoding: a six-question tour to improve users' awareness about the method. Brief Bioinform 11(4):440–453. https://doi.org/10.1093/bib/bbq003
49. Valentini A, Miquel C, Taberlet P (2010) DNA barcoding for honey biodiversity. Diversity 2(4):610–617. https://doi.org/10.3390/d2040610
50. Laha RC, Mandal S, Ralte L, Ralte L, Kumar NS, Gurusubramanian G, Satishkumar R, Mugasimangalam R, Kuravadi NA (2017) Meta-barcoding in combination with palynological inference is a potent diagnostic marker for honey floral composition. AMB Express 7(1):132. https://doi.org/10.1186/s13568-017-0429-7
51. Sorkun K (2008) Türkiye'nin Nektarlı Bitkileri, Polenleri ve Balları. Palme Yayıncılık, Ankara

Chapter 2
Amino Acids, Peptides, and Proteins of Pollen

Rita Végh and Mariann Csóka

2.1 Introduction

Floral pollen is a fine powder-like substance produced by the male reproductive organs of flowering plants. It plays a crucial role in the plant's reproduction as it carries the male gametes (sperm cells) necessary for fertilizing the female reproductive organs, resulting in the production of a seed [1]. Approximately 73% of the world's cultivated crops rely on animal pollination for successful reproduction. Honey bees (*Apis mellifera* L.) are important pollinators in several ecosystems as when visiting plants, they efficiently transfer pollen from the male reproductive organs (anthers) of one plant to the female reproductive organs (stigma) of another plant [2]. These interactions are mutually beneficial for both plants and honey bees: Plants need pollinators to produce fertile seeds, while bees rely on nutritious floral rewards that can be collected efficiently [3]. Pollen has crucial importance in the whole life cycle of honey bees like larval and adult growth and development, synthesis of tissues and enzymes, immune response, lifespan, and reproduction [4–7]. Floral pollen is the most important protein source for pollinators including honey bees, since it contains essential amino acids that bees cannot synthesise and non-essential amino acids that they can metabolically convert into each other or synthesise from certain essential amino acids [5, 7].

Honey bees are social insects and work in a division of labour within their colonies. Forager bees collect nutritional resources not only for themselves, but also for other members of the hive [8]. Pollen foragers are typically 15–17 days old worker bees, which collect pollen and transfer it to the hive. When bees flying, a positive static-electric charge is developed on their body, which facilitate them to collect and transport pollen effectively. The hair of bees allows pollen grains to stick to their bodies. Followingly, they comb pollen grains into specialized structures known as pollen baskets (corbiculae) located on their hind pair of legs. During this process, the bees moisten the pollen with nectar and glandular secretions, forming pellets

R. Végh (✉) · M. Csóka
Institute of Food Science and Technology, Department of Nutrition,
Hungarian University of Agriculture and Life Sciences, Budapest, Hungary
e-mail: Vegh.Rita@uni-mate.hu; Csoka.Mariann@uni-mate.hu

N. Ecem Bayram et al. (eds.), *Pollen Chemistry & Biotechnology*,
https://doi.org/10.1007/978-3-031-47563-4_2

[9]. Beekeepers can harvest pollen pellets (usually referred to as bee pollen) by placing a specific perforated device at the entrance of the hive. The perforations are as large that bees can pass to the hive, but a certain proportion of the collected pollen is scrapped from their legs into a tray [10].

Owing to its important nutrients and bioactive compounds (proteins, sugars, fats, minerals, vitamins, phenolics and flavonoids), **bee pollen** is considered as a human food [11]. It is widely used as a functional food ingredient or a dietary supplement throughout the world and is recognised for its healing properties as well [12, 13]. The chemical composition of pollen is species-dependent, with botanically related species producing pollen of similar composition [14, 15]. Besides, the composition of bee pollens is also influenced by their geographical origin, climatic conditions, bee species, soil type, preservation methods and storage conditions [16–19]. Based on a widely referenced study, the nutrient composition of bee pollen on a dry matter basis shows considerable variation within the following ranges: 13–55% total carbohydrates, 10–40% proteins, 1–13% lipids, 0.3–20% dietary fiber, and 2–6% ash [11]. Bee pollen is considered as a good source of dietary proteins and free amino acids [20], however, it is necessary to process it to improve the digestibility and bioavailability of these nutrients [13, 21].

2.2 General Description of Amino Acids, Peptides, and Proteins

Amino acids are the basic structural units of proteins determining their chemical, physical and biological properties [22]. Owing to their chemical nature and molecular structure, amino acids have high melting and boiling points, are soluble in water, and can readily ionize their carboxylic and amino groups in aqueous solutions. Amino acids exhibit amphoteric properties, meaning they can react with both acids and bases. Except for glycine, they possess at least one asymmetric carbon atom, allowing for isomerization. The majority of naturally occurring amino acids adopt the L configuration, which plays a pivotal role in the structure and functionality of proteins [23].

In nature, amino acids are primarily found bounded, while relatively small quantities occur in free forms. Twenty different amino acids participate in the formation of peptides and proteins, which are connected to each other through peptide bonds formed between their amino and carboxyl groups, resulting in the release of water. Depending on the number of amino acids involved, the resulting products can be classified as dipeptides (2 amino acids), tripeptides (3 amino acids), oligopeptides (4–10 amino acids), or polypeptides (>10 amino acids) [24]. Dietary **peptides** exhibit a broad spectrum of biological activities, and many of them are active even in microgram quantities. Based on the results of *in vivo* and *in vitro* studies, several peptides obtained from dietary sources have antioxidant and antibacterial potententials [25]. In addition, certain peptides may modulate the cellular, neurologic, endocrine, and immune functions [26].

Proteins generally refer to polypeptides consisting of more than 100 amino acids. Proteins account for more than 50% of the dry weight of living cells and perform an enormous number of biological functions [24]. These functions encompass a wide range of roles including being enzymes or catalytic proteins (e.g. trypsin, DNA polymerases and ligases), contractile proteins (e.g. actin, myosin, tubulin), structural or cytoskeletal proteins (e.g. tropocollagen, keratin), transport proteins (e.g. haemoglobin, myoglobin, serum albumin), effector proteins (e.g. insulin, epidermal growth factor, thyroid stimulating hormone) defence proteins (e.g. ricin, immunoglobulins, venoms and toxins, thrombin), electron transfer proteins (e.g. cytochrome oxidase, plastocyanin, ferredoxin, acetylcholine receptor), repressor proteins (e.g. Jun, Fos, Cro), and storage proteins (e.g. ferritin, gliadin) [22].

2.3 Dietary Requirements of Proteins and Amino Acids

2.3.1 Requirements for Humans

The quantity and quality of food proteins is a criterion for adequate nutrition. Protein quality refers to the amino acids profile, bioavailability and digestibility allowing the absorption of the amino acids [27]. The dietary protein requirements are commonly expressed using the Recommended Dietary Allowance (**RDA**), which represents the minimum amount of protein necessary to prevent deficiencies in 97.5% of the healthy adult population. The RDA established for protein was stated as a minimum of 10% of dietary energy or 0.8 g/kg body weight (bw). However, protein requirements may be higher in certain populations, depending on age, physical activity, and health status [28]. As the body utilizes amino acids for the synthesis of its own proteins, the protein requirement refers to the requirement for amino acids. However, only small amounts of free amino acids are consumed, and their primary sources are proteins [24]. Phenylalanine, valine, threonine, tryptophan, methionine, isoleucine, leucine, lysine, and histidine are **indispensable (essential)** for humans and, therefore, must be obtained by the food sources. Certain amino acids (Cys, Tyr, taurine, Gly, Arg, Gln, and Pro) are considered as **semi-essential** which become indispensable under specific physiological or health conditions [29].

The efficacy that a certain protein can satisfy amino acid requirements for humans is a critical determinant of its nutritive value. Amino acid requirements for adults reported by WHO/FAO/UNU (2007) are presented in Table 2.1. This dataset can be considered as a reference protein, to which the amino acid profiles of different dietary proteins can be compared [30]. The **biological value (BV)** of a protein expresses the extent to which the protein meets the amino acid requirements of the human body. A higher biological value indicates that a protein source provides a

Table 2.1 Amino acid requirements of adults [30]

Amino acid	mg/kg/day	mg/g/protein
Histidine	10	15
Isoleucine	20	30
Leucine	39	59
Lysine	30	45
Methionine + cysteine	15	22
Phenylalanine + tyrosine	25	38
Threonine	15	23
Tryptophan	4	6
Valine	26	39
Total indispensable amino acids	**184**	**277**

greater proportion of essential amino acids and is more efficiently utilized by the body for protein synthesis [27]. Sources of protein with high biological value include animal-based proteins such as eggs, milk, fish, and meat. Plant-based proteins typically have a lower biological value, but combining different plant protein sources can help improve their overall biological value. A limiting amino acid refers to an essential amino acid that is present in the lowest quantity relative to the body's requirement in a particular protein or diet. When the intake of a specific amino acid is insufficient to meet the body's needs for protein synthesis, it becomes the limiting factor for protein production [24]. Cereals are often characterized by a deficiency in lysine, while legumes may have low concentrations of sulfur-containing amino acids (methionine and cysteine) [27], so combining grains and legumes in a diet can help to complement their amino acid profiles and improve overall protein quality.

For the evaluation of the nutritional quality of proteins, the calculation of **amino acid score (AAS)** is a common procedure. The amino acid score is designed to predict protein quality in terms of the potential ability of food protein to provide the appropriate pattern of dietary essential amino acids [31]. The amino acid score of a protein or mixture of proteins can be calculated according to the following formula (Eq. 2.1) [32]:

$$AAS = \frac{\text{mg of amino acid in 1 g of test protein}}{\text{mg of amino acid in reference protein}} \times 100 \tag{2.1}$$

The **protein digestibility-corrected amino acid score (PDCAAS)**, recommended by the World Health Organization, Food and Agricultural Organization and United Nations University [30], is another important index used to evaluate the nutritional value and quality of protein sources. It provides a more reliable assessment compared to other indexes as it takes into account both the amino acid profile and protein digestibility. PDCAAS is calculated by multiplying the amino acid score (AAS) by protein digestibility (Eq. 2.2), thus, its value ranges from 0.0 (poor quality) to 1.0 (high quality) [32].

$$PDCAAS = AAS \times digestability \tag{2.2}$$

While the PDCAAS is a useful tool for routine protein quality assessment, it has limitations and disadvantages that researchers have highlighted. It can sometimes underestimate the value of high-quality proteins and overestimate the value of others. It may not be suitable for predicting the quality of plant proteins that may contain antinutritional factors, such as phytates, tannins, trypsin inhibitors, and lectins [27]. Furthermore, the availability of digestibility scores is limited to a specific set of foods. The PDCAAS also only allows for direct comparison between two proteins and is not scalable [32].

In 2013, the FAO introduced a novel scoring system, termed the **digestible indispensable amino acid score (DIAAS)**, to quantify the quality of dietary proteins [31]. The DIAAS shares a similar conceptual goal with the PDCAAS, however, the DIAAS is a more advanced and accurate method for assessing protein quality. Unlike the PDCAAS, the DIAAS does not truncate for a single-source protein, enabling a comprehensive ranking of all dietary proteins based on their quality. The concept behind the DIAAS is based on the relative digestible content of the indispensable amino acids (IAAs) and the amino acid requirement pattern. The formula used for calculating the DIAAS is presented below (Eq. 2.3) [33].

$$DIAAS = \frac{\text{mg of digestible dietary IAA in 1 g of the dietary protein}}{\text{mg of the same amino acid in 1 g of the reference protein}} \times 100 \tag{2.3}$$

2.3.2 *Requirements for Bees*

A balanced diet containing sufficient amount and quality of proteins is crucial for the proper development of honey bees. The protein requirement of worker bees and drones vary between 25–38 mg and 60–98 mg, respectively [34]. Paoli and co-workers (2014) demonstrated that young bees demand more essential amino acids compared to foragers, while foragers require a diet rich in carbohydrates [35]. Amino acids are involved in several physiological processes including growth, flight ability and immunity of bees [36]. Ten amino acids (histidine, isoleucine, leucine, lysine, methionine, phenylalanine, threonine, tryptophan, valine and arginine) cannot be synthetised by the bees and are therefore considered as essential [37]. Pollen of different plants is the primary source of proteins for bees. Pollens rich in protein generally contribute more to colony growth and bee development compared to pollens containing low amounts of protein [38]. The total pollen-need to raise one larva is between 125–188 mg for workers and 325–488 mg for drones [34]. To ensure optimal nutrient intake and avoid nutrient deficiencies, bees usually collect pollen from several botanical families at the same time [6, 15]. Whereas oligolectic species confine foraging to pollen from a single host plant family rather than to an individual plant species, polylectic bees, such as bumblebees, collect pollen from many plant families [5, 15].

2.4 Nitrogen-Containing Compounds in Floral Pollen

2.4.1 Protein Content of Floral Pollen

Comparing nectar and pollen as the main food sources for bees, the chemical composition of the latter is much more complex. Pollen is a rich source of nutrients that are vital for the life processes of bees like growth, reproduction, immunocompetence and longevity [39–43]. The protein content of pollen is considered one of the best indicators of nutritional quality [12]. The term "protein" covers many different groups of compounds, which have diverse functions in the life of insects. For example, high molecular weight polypeptides (molecular weight > 10,000 Da) have important role in the immunity of bees, or enzymes can affect the nutritional value of diets by degrading lipids or proteins [5]. The protein content of the pollen may influence the frequency or number of foraging trips to ensure that pollinators provide sufficient nutrients to their offspring [14, 44, 45].

The total nitrogen content and amino acid composition can vary greatly depending on the botanical origin, as well as on the climatic and nutrient conditions of the source plant [46]. The **crude protein** content of floral pollen ranges widely (Table 2.2) between different plant sources. In a comprehensive study of Roulston and Cane (2000), the protein content of hand-collected pollen ranged between 2.5% and 61.0%. The highest protein concentrations were present in vibratile pollinated herbs like *Dodecatheon* (Primulaceae), *Rhexia* (Melastomataceae), and *Solarium* (Solanaceae), as well as bat-pollinated Bombacaceae and bird-pollinated Campanulaceae. Conversely, anemophilous gymnosperms like Cupressaceae and Pinaceae exhibited the lowest protein content [14]. Scientific data suggest that the protein concentration remains relatively consistent within plant genera, families, and divisions [15]. The floral (hand-collected) pollen from different maize genotypes with white, yellow, red, dark red, blue, brown and sweet kernels, for example, showed protein contents ranging from 22.7% to 24.8% [12].

In pollen, several kinds of protein fraction are present like albumins-globulins and other non-protein nitrogen. The former fraction involves mostly enzymes present in certain layers of the pollen grain walls, while non-protein nitrogen usually refers to the free amino acid content of pollen [12]. Twelve enzymes were separated from mature pollen grains of maize (*Zea mays;* inbred S26) by Kalinowski et al. [57], while Li et al. [58] have identified glucanase, xylanase and protease in another maize pollen (*Zea mays;* inbred B73). Regarding the non-protein nitrogen content of floral pollen from maize with different colored kernels, it was 3.8–28.8-fold higher than the albumin-globulin protein fraction (Table 2.3).

Most pollen proteins are probably enzymes needed for pollen tube growth and subsequent fertilisation. Since most flower visitors consume pollen mainly for its protein content, this characteristic may also influence host plant choice. Pollen from

Table 2.2 Crude protein content of floral pollen

Pollen source plant	Crude protein content (%)	References
Zea mays	13.13–20.14	[47]
Medicago sativa	19.45	[48]
Zea mays (red maize)	23.43	[12]
Zea mays (white maize)	24.60	
Zea mays (yellow maize)	23.24	
Zea mays (blue maize)	22.69	
Zea mays (dark red maize)	23.29	
Zea mays (brown maize)	24.84	
Zea mays (sweet maize)	24.83	
Crocus sativus	23.60	[49]
Olea europaea	40.01	[50]
Phoenix dactylifera	39.80	
Phoenix dactylifera	31.11	[46]
Aloe greatheadii	50.8	[51]
Field corn hybrids	22.73–26.88	[52]
Alnus sp.	30.1	[15]
Carengiea gigantea	36.8	
Carya illinoensis	20.0	
Helianthus annuus	30.6	
Ochroma pyramidale	41.7	
Oenocarpus panamensis	23.8	
Opuntia phaecantha	22.1	
Pinus taeda	18.1	
Pseudobombax septanatum	48.1	
Quercus nigra	41.5	
Quercus michauxii	38.4	
Agave deserti	45.8	
Ephedra trifurca	34.7	
Juglans regia	24.4	
Juniperus deppeana	5.7	
Rosa laxa	6.62–8.25	[53]
Typha latifolia L.	13.1	[54]
Virgilia divaricata	25.0	[55]
Opuntia sp.	7.0	[7]
Haplopappus tenuisectus	9.6	
Acacia greggi	10.6	
Parkinsonia aculeata	11.2	
Cereus giganteus	11.3	
Opuntia spp. (green)	12.5	
Opuntia spp. (purple)	12.7	
Prosopis velutina	14.0	
Larrea tridetilata	15.6	
Zea mays	16.47	[56]

Table 2.3 The protein content of floral pollen from maize with kernels of different colors [12]

Pollen samples	Albumins-globulins (% of total proteins)	Non-protein nitrogen (% of total proteins)
Red maize	2.14	53.52
White maize	5.66	47.03
Yellow maize	4.11	49.57
Blue maize	2.00	57.63
Dark red maize	10.22	37.78
Brown maize	2.09	55.20
Sweet maize	5.18	47.17

plant species that need to be visited by bees to pollinate are usually very rich in protein, this may be due to pollinator reward or very small pollen grains. Several research suggest that some polylectic species select pollen based on their protein content [39, 59, 60]. In a study conducted by Regali and co-workers (1995), it was observed that bumblebees (*Bombus terrestris* L.) show a preference for collecting pollen with a high protein content, specifically from oilseed rape (*Brassica napus* L.) compared to pollen with a lower protein concentration found in sunflower (*Helianthus annuus* L.) [60]. Cook and co-workers (2003) demonstrated that honey bees also exhibit a strong preference for the pollen of oilseed rape. In their study, following prior experience, bees clearly favoured oilseed rape pollen over field bean (*Vicia faba* L.) pollen [38]. Despite these resuls, Roulston and co-workers (2000) did not find pollen from animal-pollinated plants to be statistically richer in protein than pollen from wind-pollinated species. Based on their study, pollen with a very high protein content is not necessarily collected by bees. The reason behind this observation may be that besides the protein content of pollen, other factors such as the amount and availability of pollen also affect plant-animal interactions [15]. Besides, the colour, shape, morphology, display area, and odour of flowers [8], the distance between the apiary and pollen sources [61], as well as the amino acid composition of pollens [38] may also play a role in the choice of host plant.

2.4.2 Amino Acid Composition of Floral Pollen

Several studies suggest that the **amino acid composition** is a more accurate indicator of the nutritional value of pollen than protein content, since the nutritional value is lower if essential amino acids are not present in sufficient quantities [37, 43, 46, 51, 53]. Diets with the same protein content but different amino acid composition may have different nutritional values [62]. As mentioned in Sect. 2.3.2., arginine, histidine, isoleucine, leucine, lysine, methionine, phenylalanine, threonine, tryptophan, and valine have been identified as essential for honey bees by de Groot [37]. Therefore, floral pollens rich in these amino acids are proper sources of nutrients for them. Based on the measurements of Jeannerod et al. (2022), the amino acid contents and profiles of floral pollen is quite different and is mainly influenced by the botanical origin. Presumably, due to the similar metabolic processes and flower morphology, the pollens of related plants have a similar nutrient composition [4].

Analysing the pollen of desert plants, McCaughey et al. (1980) found that their protein content is generally low (7.0–15.6%), but their amino acid composition is mostly in agreement with the requirements of honey bees. Authors found glutamic acid, aspartic acid and proline as dominant constituents, and these components showed the greatest variation among species. Lysine and leucine were also found in high amounts in these pollens, however, they were sometimes deficient in tryptophane, phenylalanine, hydroxyproline, tyrosine and aminobutyric acid [7]. Similarly, Jeannerod et al. (2022) found aspartic acid, glutamic acid, glycine, leucine, lysine and proline as the most abundant amino acids in their floral pollen samples. The analysis of pollen from various plant species revealed the presence of all essential amino acids and nearly all other amino acids, except for cysteine [4]. Hassan (2011) identified 8 essential amino acids and 9 nonessential amino acids in Egyptian date palm pollen grains. The main essential amino acids of palm pollen were leucine and lysine (3.3 and 3.0 g/100 g dry weight, respectively), while among nonessential amino acids aspartic acid, alanine and glycine were the most significant constituents (3.6, 2.6 and 2.2 g/100 g dry weight, respectively). Methionine was found to be the limiting essential amino acid in this pollen (0.1 g/100 g dry weight) [46]. In the pollen of nine field corn hybrids, proline was the most abundant amino acid, and the total amino acid content of different hybrids varied between 12.5–68.1% [52]. Results of Jacquemart and co-workers (2018) show that pollen from four different *Tilia* species have a total amino acid content of 209.3–322.1 mg/g and an essential amino acid content of 100.3–136.4 mg/g. The highest values were measured in *T. platyphyllos* and the lowest in *T. tomentosa* [63]. In the research of Qingdian et al. (1997), the crude protein content of *Rosa laxa* Retz pollen collected from different provenances ranged from 6.6% to 8.3% and their amino acid composition was also similar [53].

Available data indicate that the amino acid composition is highly variable between different plant species (Table 2.4), although taxonomically related species usually have similar composition.

Differences in amino acid composition are most evident between families and orders. Besides the botanical origin, the measured quantity of amino acids in floral pollen might be influenced by the analytical method employed. In the research of Jeannerod et al. (2022), ion exchange chromatography was more efficient regarding total amino acid content compared with HPLC. Of the floral pollen samples, the highest total amino acid content was measured in pollen from Boraginaceae species, while pollen from Malvaceae families contained the lowest total amino acid content (Fig. 2.1) [4].

Amino acids in pollen can be present in free and bound forms; those present in large amounts in free form are also abundant in protein-bound form [15]. Of the free amino acids, proline is often the most significant, making up 1–2% of the protein content. This amino acid is particularly important for bees, as this is the main

Table 2.4 Amino acid composition (g/100 g dry weight) of floral pollen

Pollen source	Amino acids (mg g dry weight)																	References
	PHE	HIS	ILE	LEU	LYS	MET	THR	VAL	ALA	ARG	ASP	CYS	GLY	GLU	PRO	SER	TYR	
Phoenix dactylifera	18.90	18.40	25.70	37.20	28.70	6.90	17.20	22.00	82.70	15.90	47.00	n.d.	81.90	47.10	22.80	19.20	15.40	[64]
Phoenix dactylifera	13.30	12.40	13.90	27.70	31.90	4.90	9.00	26.80	69.20	11.80	37.30	n.d.	44.00	39.10	17.30	13.20	9.10	
Phoenix dactylifera	11.50	14.00	17.20	19.90	37.70	4.60	19.10	28.00	81.30	12.50	42.20	n.d.	47.00	44.50	20.70	17.70	13.00	
Phoenix dactylifera	16.30	8.20	13.90	21.50	29.20	8.00	6.20	12.40	24.30	17.40	32.30	n.d.	61.50	39.30	18.10	7.30	3.70	
Tilia cordata	13.30	9.10	12.60	18.70	18.30	6.80	11.70	13.70	11.70	13.40	26.60	2.50	10.70	28.50	22.10	13.40	9.90	[63]
Tilia platyphyllos	9.40	10.80	17.20	25.20	24.10	0.70	15.90	17.50	18.60	15.60	44.00	0.00	16.60	43.60	24.60	20.40	18.00	
Tilia tomentosa	10.80	8.20	10.80	15.70	15.70	5.60	10.20	12.40	10.80	10.90	25.30	1.60	10.00	24.50	16.60	12.20	8.10	
Tilia × *europaea*	14.30	11.10	15.60	19.70	18.80	7.40	12.40	14.90	13.30	13.80	28.30	2.20	12.50	29.60	21.70	14.80	10.40	
Helianthus annuus	10.40	14.40	10.40	16.80	16.90	1.40	11.40	11.50	14.30	11.60	22.40	2.40	14.00	27.30	15.30	12.10	8.70	[65]
Phoenix dactylifera	16.30	16.10	14.90	33.40	29.50	1.10	17.20	18.10	26.10	16.10	35.50	4.20	22.40	17.40	2.80	18.90	15.50	[46]
Aloe greatheadii	10.70	4.90	10.50	18.00	15.60	4.00	11.30	12.40	13.00	13.20	22.10	n.d.	10.30	25.10	16.00	13.60	7.10	[51]
Field corn hybrids	10.50	5.40	11.50	17.70	18.40	n.d.	12.10	14.80	16.90	11.40	24.80	n.d.	13.00	25.40	30.10	12.90	6.90	[52]
Rosa laxa	3.50	1.40	3.20	5.00	4.60	1.80	3.30	4.10	3.60	3.40	9.40	1.00	4.30	10.20	4.40	3.90	2.20	[53]
Rosa laxa	3.50	1.50	3.40	5.00	4.70	2.20	3.50	4.70	3.20	3.50	9.30	1.00	4.10	9.30	4.30	3.70	2.50	
Rosa laxa	3.40	1.30	3.30	4.80	4.50	2.80	3.20	4.00	3.90	3.10	9.50	1.00	4.00	9.00	4.60	4.00	2.40	

Opuntia sp.	35.00	18.90	32.20	52.50	58.10	16.10	28.70	39.90	43.40	29.40	89.60	2.80	35.00	87.50	56.00	33.60	21.70	[7]
Haplopappus tenuisectus	41.30	22.10	41.30	70.10	64.30	13.40	30.70	48.00	42.20	33.60	105.60	3.80	49.00	96.00	89.30	28.80	25.00	
Acacia greggi	31.80	18.00	30.70	51.90	46.60	10.60	26.50	40.30	37.10	31.80	103.90	1.10	40.30	93.30	50.90	36.00	24.40	
Parkinsonia aculeata	40.30	22.40	42.60	71.70	61.60	28.00	43.70	50.40	75.00	61.60	142.20	10.10	50.40	123.20	219.50	42.60	29.10	
Cereus giganteus	52.00	26.00	49.70	83.60	78.00	20.30	21.50	63.30	65.50	48.60	113.00	10.20	53.10	133.30	73.50	44.10	33.90	
Opuntia spp. (green)	51.30	27.50	51.30	83.80	77.50	21.30	42.50	63.80	61.30	50.00	23.80	1.30	66.30	158.80	111.30	53.80	35.00	
Opuntia spp. (purple)	54.60	27.90	54.60	91.40	81.30	22.90	36.80	69.90	67.30	48.30	148.60	6.40	72.40	170.20	127.00	36.80	33.00	
Prosopis velutina	50.40	30.80	49.00	84.00	68.60	15.40	33.60	63.00	56.00	51.80	182.00	5.60	61.60	124.60	74.20	39.20	29.40	
Larrea tridetilata	56.20	25.00	43.70	84.20	73.30	20.30	34.30	62.40	60.80	79.60	120.10	4.70	51.50	265.20	185.60	45.20	31.20	

n.d. no data

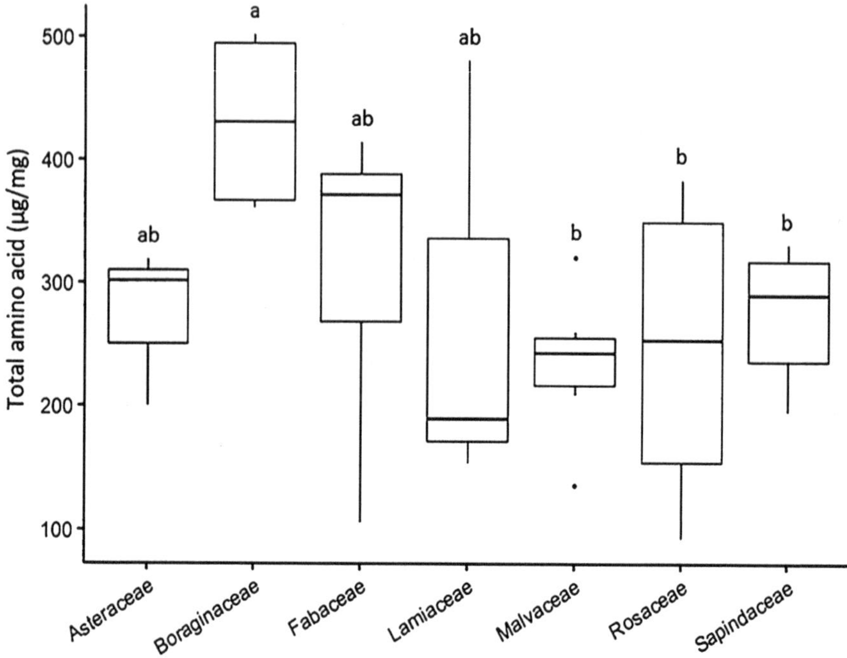

Fig. 2.1 Boxplot of total amino acid content per plant family [4]

constituent of the insect's body tissue proteins [66]. Proline is necessary for egg-laying of the queen and, because it is rapidly metabolised, is used as fuel for bee flight [67, 68]. Howell (1974) assumed that this amino acid is a growth stimulant for other pollen-feeding animals like young bats, being an important component of collagen [69].

In a study by Weiner et al. (2010), the pollen of 142 plant species examined showed significant variation in amino acid composition, both for essential and non-essential amino acids. The amount of water-soluble amino acids in pollen samples ranged from 0.7 to 142.9 mg/g, of which the ratio of essential amino acids were between 2.9% and 60.1%. The same amounts and ratios for protein-bound amino acids were 29.4–248.9 mg/g and 29.9–40.6%, respectively. The highest amount of protein-bound amino acids was measured in *Solanum dulcamara*, while the lowest in *Malva moschata*. Figure 2.2 shows the amino acid profile of two different plant species [62].

The amino acid content of pollen can influence the choice of pollinators of host plant: some sources [59, 70, 71] suggest that bumblebees prefer pollen with a high amino acid content, but other research [72] does not support this observation. Based on the study of Cook and co-workers (2003), bees generally prefer pollen that is rich in essential amino acids especially isoleucine, leucine, and valine [38]. Examining hand-collected pollen samples from 142 plant species Weiner et al. (2010) identified

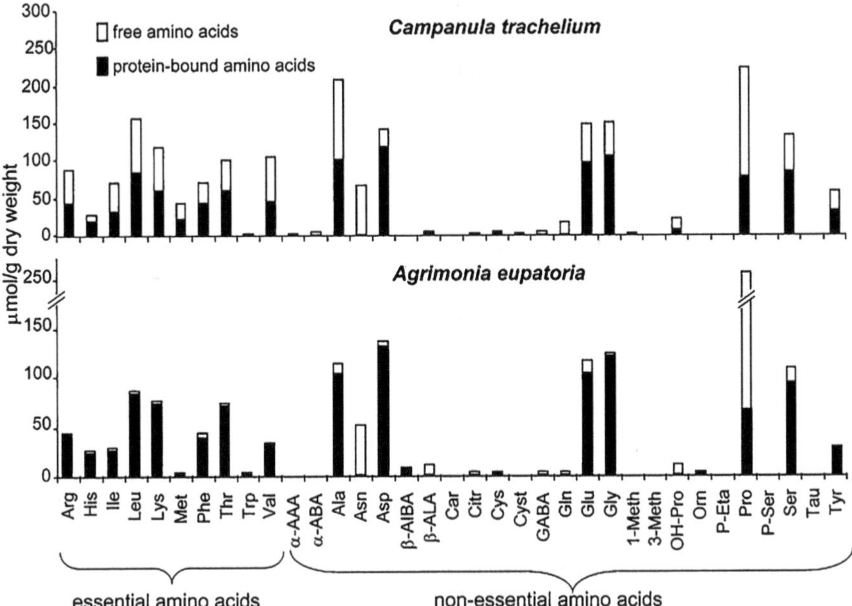

Fig. 2.2 The amino acid composition of two diverse plant species [62]. Arg: arginine, His: histidine, Ile: isoleucine, Leu: leucine, Lys: lysine, Met: methionine, Phe: phenylalanine, Thr: threonine, Trp: tryptophan, Val: valine, α-AAA: α-aminoadipic acid, α-ABA: α-aminobutyric acid, Ala: alanine, Asn: asparagine, Asp: aspartic acid, β-AIBA: β-aminoisobutyric acid, β-Ala: β-alanine, Car: carnosine, Citr: citrulline, Cys: cysteine, Cyst: cystathionine, GABA: γ-aminobutyric acid, Gln: glutamine, Glu: glutamic acid, Gly: glycin, 1-Meth: 1-methylhistidine, 3-Meth: 3-methylhistidine, OH-Pro: hydroxyproline, Orn: ornithine, P-Eta: phosphoethanolamine, Pro: proline, P-Ser: phosphoserine, Ser: serine, Tau: taurine, Tyr: tyrosine

all the essential amino acids in every species, although some were present at very low levels. The total amount of essential amino acids were significantly lower in pollen consumed by oligolectic bees [62]. The amount of essential and nonessential amino acids in pollen is related to pollen germination as reported by Aly [64].

The fact, that floral pollen is an excellent source of proteins and amino acids for honey bees is confirmed by the analytical results showing that many types of pollen have high or maximum **amino acid scores** (AAS = 1). According to the measurements of Jeannerod et al. (2022), several species belonging to the Rosaceae, Lamiaceae and Sapindaceae families had the maximum AAS, while some species were deficient in one or more essential amino acids. For example, pollen of the Boraginaceae and Fabaceae species possessed quite low amino acid scores (<0.1–0.6) [4], despite of the fact that most species of them are highly attractive to bees [73, 74]. These differences in the amino acid composition of plant species may partly explain why bees need to visit several different plant sources during the same foraging trip.

2.5 Nitrogen-Containing Compounds in Bee Pollen

2.5.1 Protein Content of Bee Pollen

Floral pollen and bee-collected pollen from the same plant species do not have the same protein content, as bee-collected pollen contains ingredients other than pure floral pollen (*e.g.* nectar and saliva). Although most of the moisture in the nectar evaporates, its sugar content significantly increases the dry matter content of the pollen. This result in a lower protein content in bee pollen on a dry matter basis compared to flower pollen (Fig. 2.3) [15].

Researchers have quantified the differences between the crude protein content of hand-collected pollen, bee-collected pollen, and pollen stored in honeycombs (bee bread) originated from almond (*Prunus dulcis*) [79], sunflower (*Helianthus annuus*) [65] and aloe (*Aloe greatheadii*) [51]. Results of each study agree that hand-collected pollen is the most abundant in crude proteins with concentrations of 22.6, 26.5 and 50.8 g/100 g for the pollens of almond, sunflower and aloe. After the formation into pollen pellets, these values decreased by 1.8, 46.4 and 38.2% respectively. The protein contents of stored pollens were slightly lower compared to bee-collected pollens, probably due to their honey content.

The concentration of crude protein in bee pollens of the majority of plant species vary between 12 and 30% [80–88]. Protein-rich pollens are regarded as higher quality both in terms of honey bee and human nutrition. The bee pollen standard has

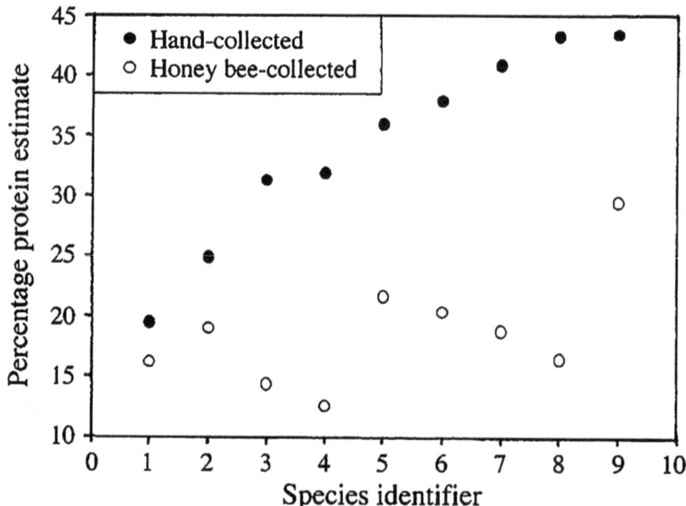

Fig. 2.3 Comparison of protein content of hand-collected pollen and bee pollen from the same plant species [15]. Species represented are (1) *Fagus sylvatica*, (2) *Zea mays*, (3) *Salsola kali*, (4) *Brassica napus*, (5) *Brassica campestris*, (6) *Carnegiea gigantea*, (7) *Trifolium pratense*, (8) *Populus fremontii*, and (9) *Phoenix dactylifera*. Sources of bee pollen data: Standifer [75], Ibrahim [76], McLellan [77], Rabie et al. [78]

recently been published (ISO/FDIS 24382 Bee pollen – Specifications), which specifies the minimum protein content of bee pollens as well as the test method for determination. In the literature, bee pollens are classified as excellent quality (crude protein >25 g/100 g), as average quality (20–25 g/100 g crude protein) or as poor quality (crude protein <20 g/100 g) [82, 89]. The quantity of protein in the pollens of monoculture crops is a very important issue because insufficient protein intake will have a detrimental effect on the development and survival of the colony if no other source of pollen is available [90]. Based on literature data, oilseed rape (*Brassica napus* L.) pollen can be considered as excellent quality because its protein content generally exceeds 25% [87, 89, 91–93]. On the other hand, corn (*Zea mays* L.) and sunflower (*Helianthus annuus* L.) pollens are considered as low quality. The crude protein content of corn pollen was reported to be approximately 17% [20, 82, 87–89, 92], while sunflower pollen usually contains even lower amounts of proteins [65, 87–89, 93]. It is important to note that besides botanical source, environmental and weather conditions may also affect the chemical composition of bee pollens [19]. Nevertheless, Radev (2019) observed low geographical variability of the protein content in pollens of several bee plants of a great importance in Bulgaria [86]. In addition to the above-mentioned factors, the protein content of pollen samples can be influenced by the composition of accompanying plant species, the year of pollen harvesting, and the processing methods employed.

Given their relatively high protein content, bee pollens are potentially valuable and efficient sources of bioactive peptides. To date, several studies have demonstrated that controlled enzymatic hydrolysis of pollen protein is applicable to produce bioactive peptides that may be applicable as functional food ingredients or pharmaceutical drugs [94–99]. Researchers have successfully produced for example pollen peptides with antioxidant [94], antihypertensive [94, 95], anti-inflammatory [96], antibacterial [97], binding [98], enzyme [98], transporter [98], and inhibitor [98] activities. Yan and co-workers (2019) have observed that fermentation also resulted in the formation of small molecular weight peptides (<1000 Da) with potential biological functions, however, further investigations are required in this field [100].

Bee pollens of various plants can potentially be used as functional food ingredients [13], thus, their techno-functional properties, such as protein solubility, carbohydrate solubility, emulsifying activity, emulsion stability, foaming properties, water/oil absorption capacity, wettability and dispersability are very important subjects of pollen research, although only a few studies have assigned these parameters [81, 97, 101–103]. The molecular weight of proteins plays a crucial role regarding several techno-functional parameters of foods [97, 101]. One common technique used to determine the molecular weight of proteins in foods is polyacrylamide gel electrophoresis (PAGE). This method (see Sect. 2.7.5) was already used in the last century for the protein analysis of pollen [104, 105] and has recently started to gain attention again [81, 99, 101, 106, 107].

2.5.2 Amino Acid Composition of Bee Pollen

In Table 2.5, scientific data reported on the total amino acid profiles and crude protein contents of bee-collected pollens of different plants are presented. The dominant plant species is commonly defined as the plant species with at least 45% of its pollen present in the sample. If the proportion of the predominant pollen reaches 80%, it can be referred to as monofloral [11]. As bees usually forage on several plant species at a given day, bee pollens harvested by beekeepers generally contain pollen pellets of a wide range of plants. Separating pollen pellets of different sources is difficult and requires a long time [108], thus, researchers often use multifloral bee pollens or samples the botanical origin of which is not identified.

Generally, bee pollens contain all the essential amino acids, but their amounts vary widely between plant sources [20, 46, 87, 111, 112] and seasons [108, 114]. The percentage of total essential amino acids (EAAs) to total amino acids (TAAs) was found between 36% and 51%. There is a rather low variability between samples regarding this ratio. For example, in the cases of rapeseed, maize and sunflower pollens, the EAA/TAA ratios vary between 39–45%, 38–41% and 36–43%, respectively. The highest essential amino acid ratio was found in the pollen of purple viper's-bugloss (*Echium plantagineum* L.), which is attributed to its exceptionally high (44.9 mg/g dw) histidine content [117].

To identify the predominant amino acid of bee pollen, concentrations were averaged. The mean amino acid concentration values of 62 plant species published by Sommerville and Nicol [89] were weighted for the calculation. Among essential amino acids, leucine and lysine dominated, the mean concentrations of which are 15.4 and 14.6 mg/g dw, respectively. The most abundant non-essential amino acids are glutamic acid, aspartic acid, and proline with mean concentrations of 23.8, 23.5 and 21.1 mg/g dw, respectively. These results are in accordance with previously published data [65, 111, 113]. The most common limiting amino acids for honey bees are methionine and tryptophan [4, 65, 90–92, 111]. Based on the amino acid composition of the reference protein established for humans [31], methionine was found to be the first and phenylalanine (+tyrosine) was found to be the second limiting amino acid in bee pollens harvested in Saudi-Arabia [108, 111].

Approximately 20% of the amino acids are present in free form in bee pollens, and 80% are bounded in proteins [117]. The free amino acid profile of pollens depends on their plant sources, but it is also affected by the conditions of drying and storage [119]. Free amino acids (FAAs) are important as part of nectar taste as an attractant to pollinators (especially bees) [120]. The predominant free amino acid in fresh pollen pellets is glutamic acid, which transforms into proline during the drying process, as a result of the activity of the glutamate dehydrogenase [121]. When the drying temperature is excessively high or the drying process is too long, the free amino acid content decreases (<2 g/100 g), and the free proline content against the total free amino acids increases (>80%). In contrast, if the drying process is correctly done to achieve a final moisture content of 4–8 g/100 g, the free amino acid content remains high (>2.5 g/100 g), and free proline content against the total free amino acids do not exceed 80%. Consequently, the ratio of free proline/total free amino acid can be used as an indicator of the maturity of bee pollen [119, 122].

Table 2.5 Amino acid composition (g/100 g dry weight) of bee pollen

Predominant pollen	Concentration of amino acids (mg/g dry weight)																				TAA	Crude protein (g/100 g)	Reference
	PHE	HIS	ILE	LEU	LYS	MET	THR	TRP	VAL	ALA	ARG	ASP	ASN	CYS	GLY	GLU	GLN	PRO	SER	TYR			
Eucalyptus sp.	9.50	4.90	8.30	14.80	11.60	3.90	9.80	n.d.	10.20	11.40	11.50	20.10	n.d.	4.90	9.40	23.50	n.d.	9.90	9.20	7.60	180.50	17.20	[20]
Trifolium alexandrinum	12.00	5.30	9.80	16.90	11.70	4.60	11.50	n.d.	11.70	13.00	10.80	23.80	n.d.	3.60	10.90	25.60	n.d.	39.40	12.40	9.20	232.20	23.20	
Zea mays	7.80	3.40	7.60	11.70	9.20	3.60	8.80	n.d.	9.70	11.00	8.20	18.10	n.d.	3.50	8.70	16.80	n.d.	20.00	8.70	5.30	162.10	16.36	
Brassica napus	11.10	4.60	12.10	18.20	19.10	3.50	10.60	2.80	13.10	14.20	13.00	25.60	n.d.	3.70	12.10	26.40	n.d.	12.40	11.00	5.70	219.20	27.20	[92]
Bidens pilosa	8.00	7.30	6.90	10.60	10.00	2.10	5.70	1.80	8.40	8.80	5.70	14.50	n.d.	6.90	7.30	14.60	n.d.	6.30	6.30	3.50	134.70	16.40	
Camellia sinensis	13.20	7.40	13.60	20.70	17.60	4.00	10.70	3.00	15.60	16.00	13.60	27.00	n.d.	2.40	13.10	30.40	n.d.	23.10	12.20	6.50	250.10	29.90	
Fraxinus griffithii	8.90	6.70	9.50	14.20	12.10	2.90	7.60	1.30	11.00	11.10	9.10	26.40	n.d.	1.20	9.50	20.80	n.d.	10.90	8.80	5.10	177.10	20.80	
Prunus mume	11.40	5.80	11.50	17.70	16.10	2.80	9.90	2.20	14.40	14.50	12.30	31.40	n.d.	2.60	11.90	28.80	n.d.	24.70	10.30	6.40	234.70	18.30	
Rhus chinensis	11.30	5.40	11.20	16.90	16.60	3.50	9.60	5.10	13.40	13.00	10.50	24.30	n.d.	1.80	11.40	27.40	n.d.	33.50	10.20	6.40	231.50	27.00	
Bombax ceiba	12.90	8.00	15.00	20.50	18.10	4.80	11.90	2.00	17.90	15.60	14.50	32.40	n.d.	1.40	14.20	33.30	n.d.	17.30	11.90	7.30	259.00	32.20	
Hylocereus costaricensis	13.80	6.10	15.00	21.20	20.50	4.60	11.20	3.50	18.50	17.30	14.00	31.90	n.d.	2.90	13.10	34.80	n.d.	9.80	10.70	6.50	255.40	27.80	
Liquidambar formosana	5.80	1.60	6.90	12.40	12.50	2.60	5.10	0.90	2.90	6.60	10.20	16.80	n.d.	2.10	5.80	17.30	n.d.	7.50	5.80	1.80	124.60	15.90	
Nelumbo nucifera	7.90	5.00	9.60	13.40	12.80	2.80	7.30	1.40	10.80	10.00	9.40	23.10	n.d.	0.30	8.20	21.60	n.d.	5.50	8.00	5.60	162.70	18.40	
Zea mays	5.80	3.00	6.70	10.30	10.00	1.20	6.30	1.00	9.90	9.90	7.40	15.70	n.d.	1.50	7.80	16.40	n.d.	19.90	7.00	3.70	143.50	17.20	
Multifloral	7.30	4.00	6.60	11.20	10.00	3.70	5.40	2.60	8.40	7.60	11.10	n.a.	n.d.	2.90	6.70	n.a.	n.d.	20.20	7.50	5.90	121.10	n.a	[109]
Eucalyptus	9.90	6.80	10.80	17.30	8.30	4.60	11.10	4.00	14.30	14.20	14.50	n.a.	n.d.	3.50	11.70	n.a.	n.d.	32.50	14.20	9.40	187.10	n.a	
Brassica napus	5.30	2.40	4.40	6.70	7.20	2.40	5.20	1.20	5.10	5.90	5.60	10.30	n.d.	2.20	5.60	11.40	n.d.	6.10	5.60	3.40	96.00	15.26	[21]
Nelumbo nucifera	4.70	2.50	4.30	6.30	6.40	2.20	4.70	1.40	4.20	5.40	1.80	10.50	n.d.	1.70	5.00	9.60	n.d.	4.30	4.90	4.00	83.90	11.89	
Camellia japonica	7.30	3.50	6.30	10.80	8.10	3.80	6.40	2.00	7.60	8.30	7.20	13.50	n.d.	2.40	6.70	16.30	n.d.	11.90	7.30	5.50	134.90	18.12	
Schisandra chinensis	5.50	2.40	4.70	8.00	8.00	2.50	4.60	1.40	5.50	6.00	5.50	11.90	n.d.	1.80	5.50	12.10	n.d.	20.00	5.20	4.40	115.00	15.51	
Armeniaca vulgaris	6.10	3.40	6.00	10.90	11.30	3.20	6.20	1.70	7.70	8.00	2.90	13.60	n.d.	2.00	6.60	14.10	n.d.	16.60	6.10	5.70	132.10	17.45	

(continued)

Table 2.5 (continued)

Predominant pollen	Concentration of amino acids (mg/g dry weight)																				TAA	Crude protein (g/100 g)	Reference
	PHE	HIS	ILE	LEU	LYS	MET	THR	TRP	VAL	ALA	ARG	ASP	ASN	CYS	GLY	GLU	GLN	PRO	SER	TYR			
Trifolium repens	5.60	8.80	9.60	16.40	15.80	2.40	6.50	n.d.	11.80	14.30	12.00	16.10	n.d.	2.70	11.40	26.50	n.d.	16.10	16.40	7.30	199.70	n.a	[8]
Coreopsis drummondii	3.40	7.70	5.40	9.60	10.80	1.40	4.00	n.d.	5.90	8.40	5.50	8.00	n.d.	2.30	7.30	12.90	n.d.	9.30	8.90	3.60	114.40	n.a	
Erigeron annus	2.40	4.90	4.20	6.60	11.70	0.70	5.20	n.d.	4.70	7.00	4.80	6.60	n.d.	1.80	6.70	11.40	n.d.	6.70	8.10	3.20	96.70	n.a	
Oenothera biennis	2.30	4.20	3.80	5.80	11.20	1.50	3.60	n.d.	4.20	6.10	3.70	6.00	n.d.	2.00	5.90	10.60	n.d.	6.90	8.20	2.90	88.90	n.a	
Brassica napus	10.70	10.30	10.60	17.00	13.40	5.00	10.90	n.d.	12.40	13.40	16.40	16.80	n.d.	1.20	11.30	26.30	n.d.	9.60	13.60	7.60	206.50	26.80	[110]
Multifloral	8.20	8.40	8.30	13.70	10.00	1.40	7.10	n.d.	10.00	11.40	10.70	15.90	n.d.	1.40	9.50	20.50	n.d.	8.10	9.40	5.10	159.10	20.50	
Multifloral	6.10	7.40	5.90	10.40	7.40	2.60	4.50	n.d.	6.90	7.90	10.00	10.00	n.d.	0.40	5.90	14.00	n.d.	16.90	6.50	4.40	127.20	15.00	
Multifloral (spring)	3.19	4.52	6.30	12.85	8.53	0.50	4.58	1.06	10.26	12.56	4.39	16.21	n.d.	1.57	16.74	16.96	n.d.	0.58	7.65	1.95	130.40	20.16	[108]
Multifloral (summer)	2.90	5.17	6.03	11.94	8.26	0.52	4.57	0.96	9.82	12.65	4.23	16.40	n.d.	1.43	16.60	15.83	n.d.	0.63	7.93	1.81	127.68	19.27	
Multifloral (autumn)	2.51	6.19	5.33	10.88	8.38	0.57	4.47	0.82	8.86	12.61	4.28	16.25	n.d.	1.21	16.52	15.45	n.d.	0.59	8.12	1.57	124.61	18.02	
Multifloral (winter)	3.24	4.35	6.05	12.62	9.66	0.47	4.58	1.09	9.98	12.23	4.83	15.63	n.d.	1.68	16.62	16.93	n.d.	0.48	7.34	2.05	129.83	20.14	
Medicago sativa	3.06	3.46	7.20	12.90	6.37	0.44	4.74	1.19	9.95	12.72	3.14	16.44	n.d.	1.79	15.02	16.18	n.d.	0.68	7.63	2.21	125.12	n.a	[111]
Phoenix dactylifera	3.02	3.20	6.53	12.14	8.91	0.36	4.84	1.21	9.11	11.92	4.20	14.16	n.d.	2.42	14.70	17.81	n.d.	0.31	7.01	2.93	124.78	n.a	
Cucurbita pepo	2.57	6.28	5.81	11.70	7.67	0.71	4.00	1.03	8.58	13.49	2.87	16.08	n.d.	1.20	17.64	13.24	n.d.	0.56	7.62	1.55	122.60	n.a	
Helianthus annuus	1.16	6.07	4.62	9.65	7.57	0.36	5.18	0.50	8.29	11.30	3.89	16.00	n.d.	1.22	16.60	18.40	n.d.	0.65	8.86	1.61	121.93	n.a	
Brassica napus	3.02	3.98	6.00	10.87	7.67	0.50	4.38	1.18	9.64	12.12	3.91	16.46	n.d.	1.15	15.84	16.73	n.d.	0.49	7.10	1.34	122.38	n.a	
Papaver	8.47	8.84	8.72	15.31	23.63	n.d.	7.84	n.d.	10.12	11.32	1.67	26.33	n.d.	n.d.	8.47	25.99	0.85	16.97	10.37	4.20	189.10	n.a	[112]
Geranium	4.58	8.14	3.93	10.15	11.17	4.01	6.64	9.22	7.28	10.34	7.49	22.08	n.d.	n.d.	7.68	17.72	0.61	20.59	8.29	4.89	164.81	n.a	
Robinia	8.87	8.06	7.87	17.83	24.87	4.88	10.10	28.93	13.08	14.04	15.72	29.81	n.d.	n.d.	11.7	32.76	0.70	38.47	13.33	7.19	288.21	n.a	
Fraxinus	6.54	8.22	5.33	12.40	20.37	n.d.	6.81	n.d.	8.47	9.05	10.56	24.71	n.d.	n.d.	7.71	22.83	0.59	5.99	8.68	4.88	163.14	n.a	
Verbascum	10.25	8.70	12.30	22.72	31.54	n.d.	11.13	12.10	15.45	16.73	25.89	52.94	n.d.	n.d.	10.28	43.13	1.31	9.79	10.34	4.54	299.14	n.a	
Prunus mume	4.49	8.09	4.88	10.31	10.95	n.d.	5.58	9.60	6.67	7.69	9.89	21.56	n.d.	n.d.	6.51	18.48	0.51	10.26	5.72	4.58	145.77	n.a	
Castanea	7.26	7.93	8.42	16.04	25.13	n.d.	8.01	13.48	10.68	11.45	12.90	23.00	n.d.	n.d.	9.01	30.18	1.00	18.95	8.80	5.21	217.45	n.a	
Rubus + Castanea	9.22	8.52	9.88	19.37	28.36	n.d.	9.46	16.64	12.40	14.03	13.74	37.20	n.d.	n.d.	7.11	33.88	1.05	17.11	9.85	3.97	251.79	n.a	
Multifloral	6.55	8.91	7.01	12.35	20.42	n.d.	6.30	9.67	8.02	8.68	9.65	18.47	n.d.	n.d.	7.81	20.24	0.41	23.42	6.85	4.29	179.05	n.a	

Cocos nucifera	6.54	12.31	7.27	24.66	20.14	4.68	10.42	3.71	9.89	23.25	9.76	28.13	n.d.	2.54	18.36	30.27	n.d.	21.42	9.08	4.43	246.86	25.39	[113]
Coriandrum sativum	9.17	5.84	10.09	19.95	18.68	1.23	8.41	2.57	10.26	17.15	10.32	26.49	n.d.	2.31	12.39	27.82	n.d.	23.74	8.26	3.87	218.55	22.06	
Brassica napus	8.26	7.65	3.86	17.61	13.48	3.45	7.37	2.91	9.47	18.24	10.84	23.17	n.d.	2.09	2.16	23.55	n.d.	25.19	5.24	4.03	188.57	19.63	
Multifloral	8.80	3.46	9.55	22.59	17.67	2.71	5.26	4.59	7.14	15.87	11.98	24.32	n.d.	1.86	15.45	25.55	n.d.	20.81	8.46	3.63	209.70	21.89	
Multifloral (spring)	6.50	4.21	6.27	11.15	11.12	4.32	6.16	1.98	6.70	9.49	7.16	15.54	n.d.	1.92	7.72	17.66	n.d.	16.54	8.57	3.86	146.87	21.00	[114]
Multifloral (summer)	7.15	4.58	7.17	12.59	11.69	5.18	6.73	2.13	8.42	11.54	8.75	16.07	n.d.	2.84	8.33	18.15	n.d.	21.24	9.89	4.63	167.08	20.92	
Multifloral (autumn)	6.00	4.02	5.67	10.67	9.78	5.12	6.00	1.78	6.76	9.78	7.15	14.01	n.d.	2.76	7.12	15.63	n.d.	21.52	8.97	4.00	146.74	19.59	
Multifloral (winter)	6.96	3.87	6.71	12.17	10.65	5.55	6.62	2.12	7.52	9.22	8.84	16.60	n.d.	2.19	7.99	18.85	n.d.	23.64	8.76	4.28	162.54	22.80	
Quercus sp.	2.00	4.00	9.00	15.00	7.00	n.d.	9.00	n.d.	9.00	11.00	11.00	16.00	n.d.	3.00	9.00	21.00	n.d.	7.00	9.00	6.00	148.00	23.20	[91]
Actinidia arguta	12.00	6.00	13.00	20.00	12.00	n.d.	4.00	n.d.	14.00	14.00	15.00	27.00	n.d.	1.00	12.00	22.00	n.d.	7.00	11.00	6.00	196.00	26.50	
No data	2.80	5.36	6.70	12.81	8.62	0.54	5.22	n.d.	10.21	13.88	4.05	17.88	n.d.	n.d.	18.12	18.57	n.d.	0.61	8.70	2.13	136.20	19.20	[115]
Helianthus annuus	5.54	8.40	5.51	8.97	8.93	0.44	6.22	0.24	6.16	7.63	5.94	12.35	n.d.	0.60	7.54	13.77	n.d.	8.79	6.69	5.35	119.07	14.20	[65]
Brassica napus	10.80	8.10	11.10	17.00	15.30	4.20	9.70	12.80	12.90	12.50	14.00	22.10	n.d.	2.50	11.70	25.00	n.d.	15.70	12.30	6.80	224.50	27.27	[87]
Citrullus lanatus	9.70	7.00	10.10	15.50	11.20	3.20	7.50	18.20	11.90	11.20	13.20	25.70	n.d.	2.70	9.50	22.00	n.d.	11.20	10.50	6.30	206.60	20.68	
Camellia japonica	13.00	9.30	13.30	20.60	15.50	6.20	9.90	1.40	15.70	14.50	25.80	25.80	n.d.	3.00	12.50	28.70	n.d.	30.90	12.50	8.50	267.10	28.95	
Dendranthema indicum	5.30	5.50	6.00	8.90	8.60	2.90	4.50	11.70	7.30	7.50	8.30	12.40	n.d.	2.50	6.30	12.90	n.d.	20.10	6.30	3.50	140.50	14.86	
Fagopyrum esculentum	5.30	4.20	5.60	8.80	7.20	2.00	4.60	7.70	7.00	7.30	7.20	12.40	n.d.	1.50	5.50	18.10	n.d.	14.80	5.80	3.60	128.60	14.26	
Helianthus annuus	4.60	6.90	5.10	7.90	7.00	1.20	4.30	7.00	6.40	7.00	10.50	11.30	n.d.	2.40	6.40	11.10	n.d.	9.50	5.10	2.80	116.50	15.34	
Nelumbo nucifera	7.90	6.50	8.60	12.50	9.50	2.50	6.60	9.40	10.20	9.60	10.20	20.20	n.d.	2.30	8.00	17.70	n.d.	7.40	8.10	5.10	162.30	16.96	
Papaver rhoeas	11.20	9.10	11.80	18.40	13.50	4.30	9.10	13.80	14.10	12.90	13.40	23.60	n.d.	2.60	11.50	26.80	n.d.	27.20	11.60	7.40	242.30	24.99	
Rosa rugosa	6.70	5.30	6.80	11.00	8.00	2.50	5.50	10.30	8.40	8.70	9.80	19.90	n.d.	2.00	7.20	16.40	n.d.	11.80	7.30	4.60	152.20	22.94	
Schisandra chinensis	11.10	7.00	10.40	16.20	14.20	2.50	8.20	10.50	12.60	11.80	12.70	22.00	n.d.	2.90	10.50	25.00	n.d.	49.50	10.50	6.90	244.50	27.40	
Vicia faba	11.20	7.20	11.90	17.90	15.90	4.00	9.20	12.50	13.70	12.90	13.80	23.00	n.d.	2.20	11.00	25.50	n.d.	59.50	11.80	7.20	270.40	27.85	
Zea mays	6.00	4.40	6.50	10.10	9.20	2.20	5.70	8.60	8.40	9.10	9.00	14.20	n.d.	2.20	7.20	14.10	n.d.	22.10	7.60	4.20	150.80	17.89	

(continued)

Table 2.5 (continued)

Predominant pollen	Concentration of amino acids (mg/g dry weight)																				TAA	Crude protein (g/100 g)	Reference
	PHE	HIS	ILE	LEU	LYS	MET	THR	TRP	VAL	ALA	ARG	ASP	ASN	CYS	GLY	GLU	GLN	PRO	SER	TYR			
Brassica napus	4.89	2.90	4.96	9.28	9.85	<LOD	5.83	n.d.	6.40	6.50	7.27	10.60	n.d.	<LOD	6.28	9.50	n.d.	11.40	8.26	3.86	107.78	n.a.	[116]
Brassica napus	5.25	3.89	5.02	9.86	10.36	0.45	6.82	n.d.	9.07	7.13	6.45	14.10	n.d.	0.56	6.46	11.85	n.d.	12.17	7.97	4.27	121.68	n.a.	
Various[a]	10.36	6.84	11.09	17.51	16.47	6.06	10.88	n.d.	13.21	13.55	13.86	27.97	n.d.	4.71	12.07	27.71	n.d.	27.71	13.18	7.64	240.82	25.90	[89]
Aloe greatheadii	12.97	6.66	12.72	22.11	19.94	6.81	14.79	0.50	15.35	18.87	16.11	28.42	n.d.	n.d.	13.38	29.42	n.d.	21.67	16.96	8.64	265.29	31.40	[51]
Multifloral	9.65	6.84	9.22	10.81	10.97	4.10	4.17	n.d.	7.26	10.68	5.03	15.10	3.87	n.d.	6.4	17.88	5.91	22.88	2.74	7.43	160.94	n.a.	[117]
Cistus ladanifer	8.80	3.78	8.57	8.51	9.88	3.69	4.21	n.d.	5.54	9.66	4.26	13.52	3.43	n.d.	5.69	16.09	4.98	23.91	3.05	5.71	143.28	n.a.	
Echium plantagineum	27.16	44.90	15.98	16.87	37.11	7.23	5.50	n.d.	15.76	13.31	8.27	32.34	5.68	n.d.	11.2	16.36	14.23	19.39	2.41	37.56	331.26	n.a.	
Multifloral	9.10	5.00	9.60	14.70	13.00	3.50	9.60	n.d.	10.90	11.80	9.90	20.10	0.20	n.d.	10.60	22.10	n.d.	n.d.	11.00	6.10	167.20	n.a.	[118]
Prunus dulcis	11.60	5.30	12.10	19.30	18.40	5.30	9.60	n.d.	15.60	15.10	12.30	29.80	n.d.	0.60	12.60	30.20	n.d.	29.00	10.00	7.50	244.30	22.20	[79]

n.d. no data

[a]Representing the mean amino acid profile of 62 plant species

2.6 Digestibility and Bioavailability/Bioaccessability of Pollen Proteins

The digestibility of proteins, as well as the bioavailability or bioaccessability of amino acids are important factors of the nutritional quality of foods [27, 123]. These terms are related to the utilization of dietary protein by the human body. Digestibility is a measure of how efficiently the body can digest and absorb the proteins from a particular food source [124]. Bioavailability refers to the proportion of amino acids that becomes available for use and storage in the body [125], while bioaccessability represents the proportion of amino acids that is released from the food matrix and is potentially available for absorption in the small intestine [123]. These properties are influenced by certain food matrix parameters including the concentration and composition of fatty acids, carbohydrates, as well as the presence of antinutritional factors such as glucosinolates, trypsin inhibitors, phytates and tannins [124, 126]. A large variety of processing procedures are applicable to improve the digestibility as well as the bioavailability and bioaccessability of foods which include cooking, autoclaving, germination, microwave, irradiation, spray-drying, freeze-drying and fermentation [27].

Available *in vitro* and *in vivo* studies suggest that the digestibility of pollen proteins is limited, because the cell wall of pollen grains (Fig. 2.4) is very complex and resistant which cannot be broken by human and animal gastrointestinal digestion [14, 21, 127]. The outermost layer of the cell wall of pollen is the pollenkitt, a semi-solid coating mainly composed of glycerides, free fatty acids, sterols, sterol esters, hydrocarbons, non-sterol terpenoids, and carotenoid pigments. Owing to its high concentration of volatiles and pigments, pollenkitt plays a crucial role in attracting pollinators [128]. Inside the pollenkitt, there is a highly resistant and firm membrane known as the exine. This layer is primarily composed of a biopolymer called sporopollenin [129]. The exine is resistant to decay and digestion, but it typically

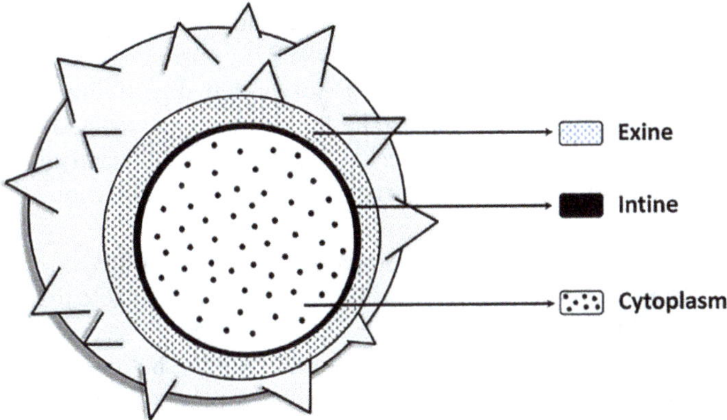

Fig. 2.4 Illustration of the multi-layered structure of pollen grains [127]

contains one or more pores that lead to the inner cell wall layer known as the intine [14]. The intine layer, composed of cellulose and pectin, surrounds the nutrient-rich cytoplasm, is less resistant than the exine layer [127, 129].

Based on the results of previous studies, the protein digestibility of raw bee pollens of different origins is approximately 60% [15, 21, 129–132]. Differences in digestibility among pollens of different plant species may be attributed to the variations between pollen wall porosity, thickness, and composition [14]. Pollen grains that have a higher number and larger size of pores could potentially be digested more easily and quickly. This is due to the fact that digestive enzymes can penetrate more effectively into the cytoplasmic content, leading to a more efficient release of the inner contents of the pollen grains [127].

Recently, several successful experiments have been conducted on processes aimed to break the exine of bee pollen through mechanical, physical, and biotechnological methods. These include fermentation [132], as well as alkaline, thermal [129], ultrasonic [21, 103], freeze thawing [103], mechanic [131], and enzymatic [103, 129, 133] pretreatments. Benavides-Guevara and co-workers (2017) have compared the effect of four different pretreatment methods (enzymatic, alkaline, dry thermal, and wet thermal) on the *in vitro* digestibility of different nutrients and found that the digestibility increased from 62% to 85–98% as a result of the treatments [129]. Xue and Li (2023) investigated the effect of different cell wall disruption methods on the physiochemical properties and techno-functional parameters of camellia bee pollen. In comparation to physical methods (ultrasonication, freeze-thawing and their combination), enzymatic hydrolysis resulted in significantly higher yield of proteins and better solubility, emulsifying properties, digestibility and gelation property owing to the partial hydrolysis of proteins induced by protease [103]. Zuluaga and co-workers (2015) also demonstrated the remarkable effectiveness of enzymatic hydrolysis. They found that subjecting the sample to thermal treatment at 121 °C for 10 min, combined with an enzymatic hydrolysis using a protease, resulted in a significant 10% enhancement in overall digestibility [133]. Fermentation of bee pollen with lactic acid bacteria provides another effective biotechnological solution for this issue. In the study of Di Cagno and co-workers (2019), the utilization of a mixed inoculum of selected *L. kunkeei* PF12, PL13, and PF15 strains, and *Hanseniaspora uvarum* AN8Y27B, under controlled conditions, enabled a significant increasement of the *in vitro* protein digestibility of bee pollen [132]. Results of another study, obtained by *in vitro* and *in vivo* digestions, suggest that wall disruption by ultrasonication is effective to significantly increase the digestibility of bee pollens originating from rape, lotus, camellia, wuweizi and apricot [21].

As mentioned above, controlled fermentation of bee pollen is a promising method to enhance its digestibility. However, it is important to note that bee pollen undergoes natural fermentation in the hive if not harvested by beekeepers. The pollen collected by foragers is not directly consumed by bees, but it is stored in comb cells. Under anaerobic conditions, fermentation takes place as a metabolic process driven by specific microorganisms, including bacteria (e.g. *Pseudomonas* spp., *Lactobacillus* spp.) and yeasts (e.g. *Saccharomyces* spp.). Bees mix the stored pollen with their secretion containing digestive enzymes. This process promotes the transformation of pollen into a more digestible product with an extended shelf life, known as bee bread [100, 127]. Bee bread is not only consumed by bees but is also recognised as a valuable food supplement and apitherapeutic agent [134].

2.7 Pollen Sampling and Analytical Techniques

2.7.1 Sampling of Pollen

Investigating the chemical composition of floral pollen is a complex task, as it requires a relative large amount of sample (at least 100–200 mg), which is difficult to collect directly from the flowers. Presumably, this is the reason of more information available about the composition of bee pollen, *i.e.* pollen pellets collected and brought back to the hive by bees [4, 14]. Bee pollen is usually obtained from pollen traps. In case of insect-pollinated (entomophilous) plants, this is an easier way to collect pollen than gathering fresh floral pollen [43]. Floral pollen samples are usually collected directly from the plant in the field [7, 12, 15, 46, 49, 51–53, 55, 57, 58, 62, 135, 136], during anthesis. Floral pollen is most often investigated from plants from which pollen is easy to collect, such as conifers, angiosperm trees (oak, birch, hickory), grasses or other economically important plant species, *e.g.* maize [12, 15, 47, 58, 136]. Immediately after collection, the pollen is sometimes dried [15, 53] at low temperature (36–40 °C), because the chemical composition of pollen is rather sensitive for heating. After drying, pollen is usually cleaned, sieved and stored in an airtight bag in the freezer until analysis to avoid nutrient degradation [4, 7, 12, 15, 47, 49, 51, 52, 57, 62–64, 136]. On occasion, it is not possible to collect pollen outdoors, in which case samples are collected using various instruments (paintbrush, tweezers) after transport to the laboratory. Vanderplanck and co-workers (2020) for example, have applied tuning fork to collect pollen from the flowers and the samples were analysed after cleaning and lyophilization [6].

2.7.2 Analysis of Protein Content

For the investigation of crude protein content of pollen samples, the determination on the basis of nitrogen content is the general method [137]. These measurements can be carried out using automatic equipment like nitrogen analyzer or micro-Kjeldahl apparatus [7, 15]. Using these methods to calculate the protein content of the sample, the total nitrogen content is usually multiplied by a conversion factor. The results are given as percentage. In addition to determining the total protein content, Žilić and co-workers [12] also measured the protein fraction of maize pollen. The pollen was consecutively extracted by the Osborne procedure applied by

Lookhart and Bean [138] with modifications. Different solvents were applied for the extraction of various protein fractions: deionized water for albumins, 0.5 M NaCl for globulins, 70% aqueous ethanol solution for the gliadin fraction and 50% 1-propanol +1% dithiothreitol (DTT) mixture for the extraction of glutenins. The extraction steps were followed by the precipitation of albumins-globulins with 10% trichloroacetic acid.

2.7.3 *Nitrogen-to-Protein Conversion Factor*

After Kjeldahl determination, the nitrogen content is converted to protein content by a factor considering the estimated concentration of N in the investigated protein. The nitrogen-to-protein (NP) conversion factor is usually set at 6.25 since proteins contain approximately 16% N on average. However, it has long been acknowledged that there are considerable variations between the amino acid compositions of diverse proteins, thus their N content is also different [139]. Plant-derived foods typically contain higher amounts of nitrogen-containing non-protein compounds (e.g. free amino acids, acid amides, peptides, nucleic acids, nitrogenous lipids, ammonium salts, and secondary compounds) compared to foods of animal origin. Consequently, using a NP conversion factor of 6.25 for the determination of protein content of plant-derived foods often lead to the over-estimation of crude protein content [140].

Although apiculture products can be considered as animal-derived foods, pollen undergo minor transformations by honey bees during collection [141], thus, the concentration of nitrogen-containing non-protein compounds may be high in bee pollen. This may raise the question of how accurate crude protein determination can be achieved when applying the universal (6.25) NP conversion factor. To the best of our knowledge, two research has been published on this issue. Based on the analysis of samples originating from different (n = 21) plant species, Rabie and co-workers (1983) reported that the mean nitrogen content of bee pollen protein is approximately 18%. In this regard, a NP conversion factor of 5.6 (calculated as 100/18) was suggested to be applied for calculating the crude protein content of pollen samples following the Kjeldahl determinations [78]. According to Buchmann (1986) the value of the conversion factor for hand-collected pollen may be even lower (4.5–5.25) [142]. Based on the above information, it is likely to overestimate the protein content of bee pollens when calculating with a NP conversion factor of 6.25. It is recommended to use a conversion factor of 5.6, which may allow a more accurate determination of the protein content of pollens.

2.7.4 *Analysis of Amino Acid Composition*

Several analytical techniques exist for the determination of amino acid profiles of pollen samples, among which ion exchange chromatography (IEX) is the most widespread [4, 5, 46, 52, 62, 63, 143]. Before the analysis of the (total) amino acid profile of pollen by ion exchange chromatography, some sample preparation is required. Initially, the pollen samples have to be hydrolysed. Usually, two diverse steps are required due to the different nature of amino acids. The majority of these components can be hydrolysed with concentrated acids (*e.g.* 6 M HCl) [4, 5, 7, 46, 51, 64, 143], whereas the separation of tryptophan requires an alkaline method applying NaOH, LiOH or $Ba(OH)_2$ as a solvent [144–149]. This latter amino acid is very sensitive to acidic environments and its oxidative degradation makes it impossible to quantify [144]. The hydrolysis step is usually followed by drying, pH adjustment and filtration [4]. Sample preparation also requires acid hydrolysis before HPLC measurement. The extraction of free amino acids requires additional analytical steps like precipitation, centrifugation and filtration [5]. In the studies of Jeannerod et al. (2022), IEX chromatography proved to be more effective than HPLC for measuring the amino acid content in floral pollen [4].

2.7.5 Electrophoretic Analysis of Pollen Proteins

Gel electrophoresis is an analytical technique which can be effectively used for the separation of pollen proteins based on their molecular weight [101, 104–107, 150]. This technique exploits the differences in charge, molecular size, and shape of proteins to separate them in an electric field within a gel matrix. The proteins migrate through the gel at different rates, enabling to their distinct separation and identification based on their unique characteristics [151]. Based on previous studies, different running buffers, such as tris-glycine [81, 101] and phosphate buffered saline (PBS) [105], have been shown to be applicable for providing an electrically conductive environment and maintaining a stable pH during the electrophoretic analysis of pollen proteins. The next step involves preparing a polyacrylamide gel, which allows for the separation of proteins dissolved in the extraction solution. Sodium dodecyl sulphate-polyacrylamide gel electrophoresis (SDS-PAGE) has been applied by more researchers for the separation of pollen proteins [99, 105–107]. Before analysis, the protein extract is subjected to denaturation, which can be conducted for example by keeping samples in boiling water for 5 min [81, 105]. During electrophoresis, the electrodes create an electric field within the gel, resulting in the migration of charged proteins [151]. The identification of proteins can be conducted by several techniques, including staining with Coomassie Brilliant Blue (CBB) [81, 99, 101, 105], immunoblotting [106] and mass spectrometric analysis [107].

2.8 Conclusions and Suggestions

This book chapter aims to provide a comprehensive overview of the amino acids, peptides, and proteins found in floral pollen and bee-collected pollen, presenting available scientific information. The manuscript covers various aspects, such as protein content, bound and free amino acids, non-essential and essential amino acids, as well as the digestibility and bioaccessability of pollen proteins. To accommodate readers less acquainted with the subject, concise general descriptions of the chemical properties of these components and commonly used indexes for evaluating the nutritional value and quality of protein sources are also included. Additionally, this chapter offers insights into analytical determination techniques commonly applied in this field.

The protein content of pollens varies usually between 12% and 30%. However, it's important to note that the Kjeldahl method, often used with the universal nitrogen-to-protein conversion factor of 6.25, might lead to an overestimation of crude protein content. This discrepancy arises due to the presence of relatively high levels of non-protein nitrogen in pollen. Based on literature research, it is recommended to use a conversion factor of 5.6 to estimate the protein content of a pollen sample more reliable. While the protein content of bee pollens primarily depends on their botanical origin, it's interesting to observe that the ratio of essential amino acids to total amino acids remains constant across different plant species. The first limiting amino acid in pollens is usually methionine, but the low concentration of tryptophan is also a frequent limiting factor for pollen proteins both for humans and honey bees. Free amino acid analysis of bee pollen is a useful indicator of the freshness or the adequacy of processing methods. Pollen proteins have limited digestibility due to the complex and resistant cell wall composition. The digestibility of pollen protein is influenced by the morphological structure of pollen grains, so pollens of different plants may differ regarding their digestibility. Numerous studies have investigated various procedures to enhance pollen digestibility, including fermentation, alkaline, thermal, ultrasonic, mechanic, and enzymatic pretreatments. However, in addition to assessing their effectiveness, it is important for further research to encompass evaluating the safety, sustainability, and economic implications associated with these procedures.

References

1. Dar SA, Hassan GI, Padder BA, Wani AR, Parey SH (2017) Pollination and evolution of plant and insect interaction. J Pharmacogn Phytochem 6:304–311
2. Chauhan T, Sarkar S, Singh A (2021) Review on honeybee: miracle agent of pollination. Plant Arch 21:2205–2209
3. Kumar S, Joshi PC, Nath P, Singh VK (2018) Impacts of insecticides on pollinators of different food plants. Entomol Ornithol Herpetol 7:2161–0983
4. Jeannerod L, Carlier A, Schatz B, Daise C, Richel A, Agnan Y et al (2022) Some bee-pollinated plants provide nutritionally incomplete pollen amino acid resources to their pollinators. PLoS One 17:e0269992
5. Vanderplanck M, Moerman R, Rasmont P, Lognay G, Wathelet B, Wattiez R, Michez D (2014) How does pollen chemistry impact development and feeding behaviour of polylectic bees? PLoS One 9:e86209
6. Vanderplanck M, Zerck P-L, Lognay G, Michez D (2020) Sterol addition during pollen collection by bees: another possible strategy to balance nutrient deficiencies? Apidologie 5:826–843
7. McCaughey WF, Gilliam M, Standifer LN (1980) Amino acids and protein adequacy for honey bees of pollens from desert plants and other floral source. Apidologie 11:75–86
8. Ghosh S, Jeon H, Jung C (2020) Foraging behaviour and preference of pollen sources by honey bee (Apis mellifera) relative to protein contents. J Ecol Environ 44:1–7

9. Rusnáková M, Hrouzek J, Hrouzková S (2023) Present state and perspectives in analytical methods for pesticide residues analysis in bee pollen: an overview. J Apic Res 62:76–96
10. Mohamed E, Ali MA, Ghazala N (2022) Evaluation of the efficiency different types of bee pollen-collection traps in honey bee colonies during summer season. Arab Univ J Agri Sci 30:141–146
11. Campos MGR, Bogdanov S, Almeida-Muradian LB, Szczesna T, Mancebo Y, Frigerio C, Ferreira F (2008) Pollen composition and standardisation of analytical methods. J Apic Res 47:156–163
12. Žilić S, Vančetović J, Janković M, Maksimović V (2014) Chemical composition, bioactive compounds, antioxidant capacity and stability of floral maize (Zea mays L.) pollen. J Funct Foods 10:65–74
13. Kostić AŽ, Milinčić DD, Barać MB, Ali Shariati M, Tešić ŽL, Pešić MB (2020) The application of pollen as a functional food and feed ingredient – the present and perspectives. Biomol Ther 10:84
14. Roulston TH, Cane JH (2000) Pollen nutritional content and digestibility for animals. Plant Syst Evol 222:187–209
15. Roulston TH, Cane JH, Buchmann SL (2000) What governs the protein content of pollen grains: pollinator preferences, pollen-pistil interactions, or phylogeny? Ecol Monogr 70:617–643
16. Aličić D, Flanjak I, Ačkar Đ, Jašić M, Babić J, Šubarić D (2020) Physicochemical properties and antioxidant capacity of bee pollen collected in Tuzla Canton (B&H). J Cent Eur Agric 21:42–50
17. Barbieri D, Gabriele M, Summa M, Colosimo R, Leonardi D, Domenici V, Pucci L (2020) Antioxidant, nutraceutical properties, and fluorescence spectral profiles of bee pollen samples from different botanical origins. Antioxidants 9:1001
18. Kostić AŽ, Pešić MB, Mosić MD, Dojčinović BP, Natić MM, Trifković JĐ (2015) Mineral content of bee pollen from Serbia. Arh Hig Rada Toksikol 66:251–258
19. Saraiva LC, Cunha FV, Léllis DR, Nunes LC (2018) Composition, biological activity and toxicity of bee pollen: state of the art. Boletin Latinoamericano y del Caribe de Plantas Medicinales y Aromaticas 17:426–440
20. Alshallash KS, Abolaban G, Elhamamsy SM, Zaghlool A, Nasr A, Nagib A et al (2023) Bee pollen as a functional product – chemical constituents and nutritional properties. J Ecol Eng 24
21. Wu W, Qiao J, Xiao X, Kong L, Dong J, Zhang H (2021) In vitro and in vivo digestion comparison of bee pollen with or without wall-disruption. J Sci Food Agric 101:2744–2755
22. Whitford D (2005) Proteins – structure and functions. Wiley, Chichester, pp 1–83
23. Amaya-Farfan J, Bertoldo Pacheco MT (2003) Amino acid – properties and occurrence. In: Caballero B (ed) Encyclopedia of food sciences and nutrition. Academic, Oxford, pp 181–191
24. Gubicskóné Kisbenedek A, Szabó Z (2015) Élelmiszer-tudományi Ismeretek. Medicina Könyvkiadó Zrt, Budapest, pp 17–39
25. López-García G, Dublan-García O, Arizmendi-Cotero D, Gómez Oliván LM (2022) Antioxidant and antimicrobial peptides derived from food proteins. Molecules 27:1343
26. Zaloga GP, Siddiqui RA (2004) Biologically active dietary peptides. Mini Rev Med Chem 4:815–821
27. Sá AGA, Moreno YMF, Carciofi BAM (2020) Food processing for the improvement of plant proteins digestibility. Crit Rev Food Sci Nutr 60:3367–3386
28. Phillips SM, Paddon-Jones D, Layman DK (2020) Optimizing adult protein intake during catabolic health conditions. Adv Nutr 11:S1058–S1069
29. Prandi B, Faccini A, Lambertini F, Bencivenni M, Jorba M, Van Droogenbroek B et al (2019) Food wastes from agrifood industry as possible sources of proteins: a detailed molecular view on the composition of the nitrogen fraction, amino acid profile and racemisation degree of 39 food waste streams. Food Chem 286:567–575

30. WHO/FAO/UNU (2007) Protein and amino acid requirements in human nutrition, World Health Organization technical report series (935). World Health Organization, Geneva
31. FAO (2013) Food and agriculture Organization of the United Nations: dietary protein quality evaluation in human nutrition, report of an FAO expert consultation. Food and Agriculture Organization, Rome
32. Forester SM, Jennings-Dobbs EM, Sathar SA, Layman DK (2023) Perspective: developing a nutrient-based framework for protein quality. J Nutr 153:2137
33. Wolfe RR, Rutherfurd SM, Kim IY, Moughan PJ (2016) Protein quality as determined by the digestible indispensable amino acid score: evaluation of factors underlying the calculation. Nutr Rev 74:584–599
34. Hrassnigg N, Crailsheim K (2005) Differences in drone and worker physiology in honeybees (Apis mellifera). Apidologie 36:255–277
35. Paoli PP, Donley D, Stabler D, Saseendranath A, Nicolson SW, Simpson SJ, Wright GA (2014) Nutritional balance of essential amino acids and carbohydrates of the adult worker honeybee depends on age. Amino Acids 46:1449–1458
36. Barraud A, Barascou L, Lefebvre V, Sene D, Le Conte Y, Alaux C et al (2022) Variations in nutritional requirements across bee species. Front Sustain Food Syst 6
37. De Groot AP (1953) Protein and amino acid requirements of the honey bee (Apis mellifera). Laboratory of Comparative Physiology, University of Utrecht, p 268
38. Cook SM, Awmack CS, Murray DA, Williams IH (2003) Are honey bees' foraging preferences affected by pollen amino acid composition? Ecol Entomol 28:622–627
39. Hanley ME, Franco M, Pichon S, Darvill B, Goulson D (2008) Breeding system, pollinator choice and variation in pollen quality in British herbaceous plants. Funct Ecol 22:592–598
40. Roulston TH, Cane JH (2002) The effect of pollen protein concentration on body size in the sweat bee Lasioglossum zephyrum (Hymenoptera: Apiformes). Evol Ecol 16:49–65
41. Alaux C, Ducloz F, Crauser D, Le Conte Y (2010) Diet effects on honeybee immunocompetence. Biol Lett 6:562–565
42. Lee KP, Simpson SJ, Wilson K (2008) Dietary protein-quality inuences melanization and immune function in an insect. Funct Ecol 22:1052–1061
43. Nicolson SW (2011) Bee food: the chemistry and nutritional value of nectar, pollen and mixtures of the two. Afr Zool 46:197–204
44. Danforth BN (1990) Provisioning behavior and the estimation of investment ratios in a solitary bee, calliopsis (Hypomacrotera) persimilis (Cockerell) (Hymenoptera: Andrenidae). Behav Ecol Sociobiol 27:159–168
45. Neff JL, Simpson BB (1997) Nesting and foraging behavior of Andrena (Callandrena) rudbeckiae Robertson (Hymenoptera: Apoidea: Andrenidae) in Texas. J Kansas Entomol Soc 70:100–113
46. Hassan HMM (2011) Chemical composition and nutritional value of palm pollen grains. Glob J Biotechnol Biochem 6:1–7
47. Bujang JS, Zakaria MH, Ramaiya SD (2021) Chemical constituents and phytochemical properties of floral maize pollen. PLoS One 16:e0247327
48. Taha E-KA (2015) The impact of feeding certain pollen substitutes on maintaining the strength and productivity of honeybee colonies (Apis mellifera L.). Bull Entomol Soc Egypt 41:63–74
49. Sani AM, Kakhki AH, Moradi E (2013) Chemical composition and nutritional value of saffron's pollen (Crocussativus L.). Nutr Food Sci 43:490–495
50. Basuny AM, Arafat SM, Soliman HM (2013) Chemical analysis of olive and palm pollen: antioxidant and antimicrobial activation properties. J Food Technol 1:14–21
51. Human H, Nicolson SW (2006) Nutritional content of fresh, bee-collected and stored pollen of aloe greatheadii var. davyana (Asphodelaceae). Phytochemistry 67:1486–1492
52. Lundgren JG, Wiedenmann RN (2004) Nutritional suitability of corn pollen for the predator Coleomegilla maculata (Coleoptera: Coccinellidae). J Insect Physiol 50:567–575

53. Qingdian L, Ying L, Jianping L (1997) Yield and nutritional value of Rosa laxa Retz pollen. Sci Hortic 71:43–48
54. Schmidt JO, Thoenes SC, Levin MD (1987) Survival of honey bees, Apis mellifera (Hymenoptera: Apidae), fed various pollen sources. Ann Entomol Soc Am 80:176–183
55. Louw GN, Nicolson SW (1983) Thermal, energetic and nutritional considerations in foraging and reproduction of the carpenter bee Xylocopa capitata. J Entomol Soc South Afr 46:227–240
56. Andronescu DI (1915) The physiology of the pollen of *Zea mays* with special regard to vitality. Dissertation, University of Illinois
57. Kalinowski A, Radłowski M, Bartkowiak S (2002) Maize pollen enzymes after two-dimensional polyacrylamide gel electrophoresis in the presence or absence of sodium dodecyl sulfate. Electrophoresis 23:138–143
58. Li Y, Suen DF, Huang C-Y, Kung S-Y, Huang AHC (2012) The maize Tapetum employs diverse mechanisms to synthesize and store proteins and flavonoids and transfer them to the pollen surface. Plant Physiol 158:1548–1561
59. Leonhardt SD, Blüthgen N (2012) The same, but different: pollen foraging in honeybee and bumblebee colonies. Apidologie 43:449–464
60. Regali A, Rasmont P (1995) New bioassays to evaluate diet in orphan colonies of Bombus terrestris [French]. Apidologie 26:273–281
61. Ghramh HA, Khan KA (2023) Honey bees prefer pollen substitutes rich in protein content located at short distance from the apiary. Animals 13:885
62. Weiner CN, Hilpert A, Werner M, Linsenmair KE, Blüthgen N (2010) Pollen amino acids and flower specialisation in solitary bees. Apidologie 41:476–487
63. Jacquemart A-L, Moquet L, Ouvrard P, Quetin-Leclercq J, Hérent M-F, Quinet M (2018) Tilia trees: toxic or valuable resources for pollinators? Apidologie 49:538–550
64. Aly HSH (2018) Evaluation of pollen grains germination, viability and chemical composition of some date palm males. Middle East J Agri Res 7:235–247
65. Nicolson SW, Human H (2013) Chemical composition of the 'low quality' pollen of sunflower (Helianthus annuus, Asteraceae). Apidologie 44:144–152
66. Van Der Moezel PG, Dells JC, Pate JS, Loneragan WA, Bell DT (1987) Pollen selection by honeybees in shrublands of the Northern Sandplains of Western Australia. J Apic Res 26:224–232
67. Bertazzini M, Medrzycki P, Bortolotti L, Maistrello L, Forlani G (2010) Amino acid content and nectar choice by forager honeybees (Apis mellifera L.). Amino Acids 39:315–318
68. Teulier L, Weber J-M, Crevier J, Darveau C-A (2016) Proline as a fuel for insect flight: enhancing carbohydrate oxidation in hymenopterans. Proc R Soc B Biol Sci 283:1–8
69. Howell DJ (1974) Bats and pollen: physiological aspects of the syndrome of chiropterophily. Comp Biochem Physiol A Physiol 48:263–276
70. Somme L, Vanderplanck M, Michez D, Lombaerde I, Moerman R, Wathelet B, Wattiez R, Lognay G, Jacqemart A-L (2014) Pollen and nectar quality drive major and minor floral choices of bumble bees. Apidologie 46:92–106
71. Kriesell L, Hilpert A, Leonhardt SD (2016) Different but the same: bumblebee species collect pollen of different plant sources but similar amino acid profiles. Apidologie 48:102–116
72. Moerman R, Vanderplanck M, Fournier D, Jacquemart A-L, Michez D (2017) Pollen nutrients better explain bumblebee colony development than pollen diversity. Insect Conserv Divers 10:171–179
73. Baude M, Kunin WE, Boatman ND, Conyers S, Davies N, Gillespie MAK et al (2016) Historical nectar assessment reveals the fall and rise of floral resources in Britain. Nature 530:85–88
74. Goodwin R, Cox H, Taylor M, Evans L, McBrydie H (2011) Number of honey bee visits required to fully pollinate white clover (Trifolium repens) seed crops in Canterbury, New Zealand. N Z J Crop Hortic Sci 39:7–19

75. Standifer LN (1966) Some lipid constituents of pollens collected by honeybees. J Apic Res 5:93–98
76. Ibrahim SH (1974) Composition of pollen gathered by honeybees from some major sources. Agric Res Rev (Cairo) 52:121–123
77. McLellan AR (1977) Minerals, carbohydrates and amino acids of pollens from some woody and herbaceous plants. Ann Bot 41:1225–1232
78. Rabie AL, Wells JD, Dent LK (1983) The nitrogen content of pollen protein. J Apic Res 22:119–123
79. Standifer LN, McCaughey WF, Dixon SE, Gilliam M, Loper GM (1980) Biochemistry and microbiology of pollen collected by honey bees (Apis mellifera L.) from almond, Prunus dulcis. II. Protein, amino acids and enzymes. Apidologie 11:163–171
80. Forcone A, Aloisi PV, Ruppel S, Muñoz M (2011) Botanical composition and protein content of pollen collected by Apis mellifera L. in the north-west of Santa Cruz (Argentinean Patagonia). Grana 50:30–39
81. Kostić AŽ, Barać MB, Stanojević SP, Milojković-Opsenica DM, Tešić ŽL, Šikoparija B, Radišić P, Prentović M, Pešić MB (2015) Physicochemical composition and techno-functional properties of bee pollen collected in Serbia. LWT Food Sci Technol 62:301–309
82. Liolios V, Tananaki C, Dimou M, Kanelis D, Goras G, Karazafiris E, Thrasyvoulou A (2015) Ranking pollen from bee plants according to their protein contribution to honey bees. J Apic Res 54:582–592
83. Martins MC, Morgano MA, Vicente E, Baggio SR, Rodriguez-Amaya DB (2011) Physicochemical composition of bee pollen from eleven Brazilian states. J Apic Sci 55:107–116
84. Nogueira C, Iglesias A, Feás X, Estevinho LM (2012) Commercial bee pollen with different geographical origins: a comprehensive approach. Int J Mol Sci 13:11173–11187
85. Prđun S, Svečnjak L, Valentić M, Marijanović Z, Jerkovic I (2021) Characterization of bee pollen: physico-chemical properties, headspace composition and FTIR spectral profiles. Foods 10:2103
86. Radev Z (2019) Pollen protein content from different regions in Bulgaria suggests low variability. Bee World 96:108–110
87. Yang K, Wu D, Ye X, Liu D, Chen J, Sun P (2013) Characterization of chemical composition of bee pollen in China. J Agric Food Chem 61:708–718
88. Conti I, Medrzycki P, Argenti C, Meloni M, Vecchione V, Boi M, Mariotti MG (2016) Sugar and protein content in different monofloral pollens-building a database. Bull Insectol 69:318–320
89. Somerville DC, Nicol HI (2006) Crude protein and amino acid composition of honey bee-collected pollen pellets from south-east Australia and a note on laboratory disparity. Aust J Exp Agric 46:141–149
90. Dufour C, Fournier V, Gioveenazzo P (2020) Diversity and nutritional value of pollen harvested by honey bee (Hymenoptera: Apidae) colonies during lowbush blueberry and cranberry (Ericaceae) pollination. Can Entomol 152:622–645
91. Ghosh S, Jung C (2017) Nutritional value of bee-collected pollens of hardy kiwi. Actinidia arguta (Actinidiaceae) and oak, Quercus sp.(Fagaceae). J Asia Pac Entomol 20:245–251
92. Hsu PS, Wu TH, Huang MY, Wang DY, Wu MC (2021) Nutritive value of 11 bee pollen samples from major floral sources in Taiwan. Foods 0:22291
93. Végh R, Puter D, Vaskó Á, Csóka M, Mednyánszky Z (2022) Examination of the nutrient content and color characteristics of honey and pollen samples. J Food Investig 68:3793–3806
94. Maqsoudlou A, Sadeghi Mahoonak A, Mora L, Mohebodini H, Ghorbani M, Toldrá F (2019) Controlled enzymatic hydrolysis of pollen protein as promising tool for production of potential bioactive peptides. J Food Biochem 43:e12819
95. Zhu S, Wang S, Wang L, Huang D, Chen S (2021) Identification and characterization of an angiotensin-I converting enzyme inhibitory peptide from enzymatic hydrolysate of rape (Brassica napus L.) bee pollen. LWT 147:111502

96. Saisavoey T, Sangtanoo P, Chanchao C, Reamtong O, Karnchanatat A (2021) Identification of novel anti-inflammatory peptides from bee pollen (Apis mellifera) hydrolysate in lipopolysaccharide-stimulated RAW264. 7 macrophages. J Apic Res 60:280–289
97. Sarabandi K, Akbarbaglu Z, Peighambardoust SH, Ayaseh A, Jafari SM (2023) Physicochemical, antibacterial and bio-functional properties of persian poppy-pollen (Papaver bracteatum) protein and peptides. J Food Measurement Charact:1–12
98. Guo J, Yan J, Guo M, Jin Y (2014) Application of reversed-phase liquid chromatography-tandem mass spectrometry in the identification of protein and bioactivity peptides from rape bee pollen. Chin J Chromatogr 32:284–289
99. Pasarin D, Rovinaru C (2023) Chemical analysis and protein enzymatic hydrolysis of poly-floral bee pollen. Emir J Food Agric
100. Yan S, Li Q, Xue X, Wang K, Zhao L, Wu L (2019) Analysis of improved nutritional composition of bee pollen (Brassica campestris L.) after different fermentation treatments. Int J Food Sci Technol 54:2169–2181
101. Kostić AŽ, Milinčić DD, Trifunović BDŠ, Stanojević SP, Lević S, Nedić N, Nedović V, Tešić ZL, Pešić MB (2020) Nutritional and techno-functional properties of monofloral bee-collected sunflower (Helianthus annuus L.) pollen. Emir J Food Agric 32:768–777
102. Thakur M, Nanda V (2020) Exploring the physical, functional, thermal, and textural properties of bee pollen from different botanical origins of India. J Food Process Eng 43:e12935
103. Xue F, Li C (2023) Effects of ultrasound assisted cell wall disruption on physicochemical properties of camellia bee pollen protein isolates. Ultrason Sonochem 92:106249
104. Payne RC, Fairbrothers DE (1973) Disc electrophoretic study of pollen proteins from natural populations of Betula populifolia in New Jersey. Am J Bot 60:182–189
105. Greenhouse VY, Xolocotzin EH, de Cuadra CR, Larralde C (1982) Electrophoretic and immunological characterization of pollen protein of Zea mays races. Econ Bot 36:113–123
106. Castro AJ, Suárez C, Zienkiewicz K, Alché JDD, Zienkiewicz A, Rodríguez-García MI (2013) Electrophoretic profiling and immunocytochemical detection of pectins and arabinogalactan proteins in olive pollen during germination and pollen tube growth. Ann Bot 112:503–513
107. Borutinskaitė V, Treigytė G, Matuzevičius D, Čeksterytė V, Kurtinaitienė B, Serackis A, Navakauska D, Navakauskienė R (2019) Proteomic studies of honeybee-and manually-collected pollen. Zemdirbyste-Agriculture 106:183
108. Al-Kahtani SN, Taha EK, Khan KA, Ansari MJ, Farag SA, Shawer DM, Elnabawy ESM (2020) Effect of harvest season on the nutritional value of bee pollen protein. PLoS One 15:e0241393
109. Hendriksma HP, Pachow CD, Nieh JC (2019) Effects of essential amino acid supplementation to promote honey bee gland and muscle development in cages and colonies. J Insect Physiol 117:103906
110. Ghosh S, Jung C (2020) Changes in nutritional composition from bee pollen to pollen patty used in bumblebee rearing. J Asia Pac Entomol 23:701–708
111. Taha EKA, Al-Kahtani S, Taha R (2019) Protein content and amino acids composition of bee-pollens from major floral sources in Al-Ahsa, eastern Saudi Arabia. Saudi J Biol Sci 26:232–237
112. Themelis T, Gotti R, Orlandini S, Gatti R (2019) Quantitative amino acids profile of monofloral bee pollens by microwave hydrolysis and fluorimetric high performance liquid chromatography. J Pharm Biomed Anal 173:144–153
113. Thakur M, Nanda V (2018) Assessment of physico-chemical properties, fatty acid, amino acid and mineral profile of bee pollen from India with a multivariate perspective. J Food Nutr Res 57:328–340
114. Negrao AF, Orsi RO (2018) Harvesting season and botanical origin interferes in production and nutritional composition of bee pollen. An Acad Bras Cienc 90:325–332
115. Farag SA, El-Rayes T (2016) Research article effect of bee-pollen supplementation on performance, carcass traits and blood parameters of broiler chickens. Asian J Anim Vet Adv 11:168–177

116. You J, Liu L, Zhao W, Zhao X, Suo Y, Wang H, Li Y (2007) Study of a new derivatizing reagent that improves the analysis of amino acids by HPLC with fluorescence detection: application to hydrolyzed rape bee pollen. Anal Bioanal Chem 387:2705–2718
117. Paramás AMG, Bárez JAG, Marcos CC, García-Villanova RJ, Sánchez JS (2006) HPLC-fluorimetric method for analysis of amino acids in products of the hive (honey and bee-pollen). Food Chem 95:148–156
118. Yook HS, Lim SI, Byun MW (1998) Changes in microbiological and physicochemical properties of bee pollen by application of gamma irradiation and ozone treatment. J Food Prot 61:217–220
119. Castagna A, Benelli G, Conte G, Sgherri C, Signorini F, Nicolella C, Ranieri A, Canale A (2020) Drying techniques and storage: do the affect the nutritional value of bee-collected pollen? Molecules 25:4925
120. Bayram NE, Gercek YC, Çelik S, Mayda N, Kostić AŽ, Dramićanin AM, Özkök A (2021) Phenolic and free amino acid profiles of bee bread and bee pollen with the same botanical origin–similarities and differences. Arab J Chem 14:103004
121. Gardana C, Del Bo C, Quicazán MC, Corrrea AR, Simonetti P (2018) Nutrients, phytochemicals and botanical origin of commercial bee pollen from different geographical areas. J Food Compos Anal 73:29–38
122. Serra Bonvehí J, Escolà Jordà R (1997) Nutrient composition and microbiological quality of honeybee-collected pollen in Spain. J Agric Food Chem 45:725–732
123. Grundy MML, Edwards CH, Mackie AR, Gidley MJ, Butterworth PJ, Ellis PR (2016) Re-evaluation of the mechanisms of dietary fibre and implications for macronutrient bioaccessibility, digestion and postprandial metabolism. Br J Nutr 116:816–833
124. Boye J, Wijesinha-Bettoni R, Burlingame B (2012) Protein quality evaluation twenty years after the introduction of the protein digestibility corrected amino acid score method. Br J Nutr 108:S183–S211
125. Ojha S, Bekhit AED, Grune T, Schlüter OK (2021) Bioavailability of nutrients from edible insects. Curr Opin Food Sci 41:240–248
126. Gilani GS, Xiao CW, Cockell KA (2012) Impact of antinutritional factors in food proteins on the digestibility of protein and the bioavailability of amino acids and on protein quality. Br J Nutr 108:S315–S332
127. Aylanc V, Falcão SI, Ertosun S, Vilas-Boas M (2021) From the hive to the table: nutrition value, digestibility and bioavailability of the dietary phytochemicals present in the bee pollen and bee bread. Trends Food Sci Technol 109:464–481
128. Dobson HE (1988) Survey of pollen and pollenkitt lipids–chemical cues to flower visitors? Am J Bot 75:170–182
129. Benavides-Guevara RM, Quicazan MC, Ramírez-Toro C (2017) Digestibility and availability of nutrients in bee pollen applying different pretreatments. Ingeniería y competitividad 19:119–128
130. Aylanc V, Falcão SI, Vilas-Boas M (2023) Bee pollen and bee bread nutritional potential: chemical composition and macronutrient digestibility under in vitro gastrointestinal system. Food Chem 413:135597
131. Omar EM, Darwish HY, Othman AA, El-Seedi HR, Al Naggar Y (2022) Crushing corn pollen grains increased diet digestibility and hemolymph protein content while decreasing honey bee consumption. Apidologie 53:52
132. Di Cagno R, Filannino P, Cantatore V, Gobbetti M (2019) Novel solid-state fermentation of bee-collected pollen emulating the natural fermentation process of bee bread. Food Microbiol 82:218–230
133. Zuluaga C, Serrato JC, Quicazan M (2015) Bee-pollen structure modification by physical and biotechnological processing: influence on the availability of nutrients and bioactive compounds. Chem Eng Trans 43:79–84

134. Kieliszek M, Piwowarek K, Kot AM, Błażejak S, Chlebowska-Śmigiel A, Wolska I (2018) Pollen and bee bread as new health-oriented products: a review. Trends Food Sci Technol 71:170–180
135. Zu P, Koch H, Schwery O, Pironon S, Phillips C, Ondo I et al (2021) Pollen sterols are associated with phylogeny and environment but not with pollinator guilds. New Phytol 230:1169–1184
136. Wu SS, Suen DF, Chang HC, Huang AHC (2002) Maize tapetum xylanase is synthesized as a precursor, processed and activated by a serine protease, and deposited on the pollen. J Biol Chem 277:49055–49064
137. AOAC (2019) Official methods of analysis of the Association of Official Analytical Chemists. Latimer GW Jr (ed), Rockville, USA
138. Lookhart G, Bean S (1995) Separation and characterization of wheat protein fractions by high-performance capillary electrophoresis. Cereal Chem 72:527–532
139. Mariotti F, Tomé D, Mirand PP (2008) Converting nitrogen into protein – beyond 6.25 and Jones' factors. Crit Rev Food Sci Nutr 48:177–184
140. Levey DJ, Bissell HA, O'keefe SF (2000) Conversion of nitrogen to protein and amino acids in wild fruits. J Chem Ecol 26:1749–1763
141. Formicki G, Greń A, Stawarz R, Zyśk B, Gał A (2013) Metal content in honey, propolis, wax, and bee pollen and implications for metal pollution monitoring. Pol J Environ Stud 22:99
142. Buchmann SL (1986) Vibratile pollination in solarium and Lycopersicon: a look at pollen chemistry. In: D'Arcy WG (ed) Solanaceae: biology and systematics. Columbia University Press, New York, p 23
143. Spitz HD (1973) A new approach for sample preparation of protein hydrolyzates for amino acid analysis. Anal Biochem 56:66–73
144. Bech-Andersen S (1991) Determination of tryptophan with HPLC after alkaline hydrolysis in autoclave using α-methyl-tryptophan as internal standard. Acta Agric Scand 41:305–309
145. Hugli TE, Moore S (1972) Determination of the tryptophan content of proteins by ion exchange chromatography of alkaline hydrolysates. J Biol Chem 247:2828–2834
146. La Cour R, Jørgensen H, Schjoerring JK (2019) Improvement of tryptophan analysis by liquid chromatography-single quadrupole mass spectrometry through the evaluation of multiple parameters. Front Chem 7:1–7
147. Lucas B, Sotelo A (1980) Effect of different alkalies, temperature, and hydrolysis times on tryptophan determination of pure proteins and of foods. Anal Biochem 109:192–197
148. Yust MM, Pedroche J, Girón-Calle J, Vioque J, Millán F, Alaiz M (2004) Determination of tryptophan by high-performance liquid chromatography of alkaline hydrolysates with spectrophotometric detection. Food Chem 85:317–320
149. Zhang JZ, Xue XF, Zhou JH, Chen F, Wu LM, Li Y et al (2009) Determination of tryptophan in bee pollen and royal jelly by high-performance liquid chromatography with fluorescence detection. Biomed Chromatogr 23:994–998
150. Kostić AŽ, Milinčić DD, Stanisavljević NS, Gašić UM, Lević S, Kojić MO, Pešić ŽL, Nedović V, Barać MB, Pešić MB (2021) Polyphenol bioaccessibility and antioxidant properties of in vitro digested spray-dried thermally-treated skimmed goat milk enriched with pollen. Food Chem 351:129310
151. Smith DM (2017) Protein separation and characterization procedures. Food Anal 15:431–453

Chapter 3
Bee Pollen Carbohydrates Composition and Functionality

Jasna Bertoncelj, Nataša Lilek, and Mojca Korošec

3.1 Introduction

Bee pollen is a bee product that has a beneficial nutritional value. It represents the male gametes of flowers that honey bees collect in special combs and stick together with their saliva and nectar from the honey stomach. To ensure adequate nutrition of the bee larvae in their early stage of development, it is then transferred into the hive [1–5]. Besides honey and water, pollen is the bees' main source of nutrients, especially proteins, which enable the physiological development and survival of the bee colony, and carbohydrates [6].

Bee pollen is rich in nutrients as it contains simple carbohydrates (sugars), dietary fibre, proteins, lipids, minerals and various bioactive compounds such as vitamins, enzymes, carotenoids and phenolic compounds. It is a balanced food that can be used as a food on its own or as a food supplement or, due to its bioactive properties, as a medicine. Research into its functional properties and its effects on human well-being and even on curing some health problems suggests that bee pollen is an important contributor to functional foods [5, 7–11]. Carbohydrates are the main nutritional component of bee pollen, with a very variable content according to published studies. The type of plant, together with climatic conditions and geographical origin, plays a fundamental role in the carbohydrate content of bee pollen [1, 10, 12–14]. Carbohydrates fulfil various important functions in living organisms. Monosaccharides are the main source of energy for metabolism, while

J. Bertoncelj (✉) · M. Korošec
Department of Food Science and Technology, Biotechnical Faculty, University of Ljubljana, Ljubljana, Slovenia
e-mail: jasna.bertoncelj@bf.uni-lj.si; mojca.korosec@bf.uni-lj.si

N. Lilek
R&D Department, Medex d.o.o., Ljubljana, Slovenia
e-mail: natasa.lilek@medex.si

N. Ecem Bayram et al. (eds.), *Pollen Chemistry & Biotechnology*,
https://doi.org/10.1007/978-3-031-47563-4_3

polysaccharides store energy and are structural components [15]. Simple carbohydrates, mainly the sugars fructose, glucose and sucrose, usually account for more than 90% of all sugars in bee pollen. Bee pollen is also a good source of dietary fibre, which is found in the inner and outer layers of the pollen grain. It is well known that dietary fibre has a beneficial effect on certain physiological functions in the digestive tract and the prevention of non-communicable diseases therefore including this bee product in the diet can help to achieve the recommended daily intake [1, 10, 16–18]. The contents of sugars and dietary fibre in bee pollen can vary within wide limits depending on botanical and geographical origin and also on the harvesting methods [19–21]. The use of different analytical methods also contributes to the variability of the data, making it difficult to compare the results of different studies [22].

Based on numerous studies over the past decade, bee pollen has proven to be an excellent food supplement, that has therapeutic effects (antioxidant, antimicrobial, anti-inflammatory) in addition to its high nutritional value. The nutritional composition of bee pollen ensures the functioning of numerous metabolic processes and naturally stimulates the functioning of the body, which is true for a healthy person and even for people with special nutritional needs. Therefore, bee pollen is also gaining importance in apitherapy [3, 8, 9, 23–25]. The efficacy of various bioactive compounds in bee pollen that show positive effects on human health and well-being should be further confirmed in properly designed clinical trials [26, 27].

This chapter reviews previous research on the composition and functionality of carbohydrates in bee pollen. This includes general information on sugars and dietary fibre and their content in bee pollen of different botanical and geographical origins. The importance of bee pollen carbohydrates for human nutrition is also discussed.

3.2 Bee Pollen Characteristics

Pollen is a microscopic structure that is characteristic and specific to each botanical species [28]. Pollen is the basis for sexual reproduction in plants, as it represents the male fertilising cells [4]. Pollen grains measure between 2.5 and 250 μm, depending on the plant species, and vary in size, shape, colour and weight [8, 12, 29].

The pollen grain is surrounded by an outer wall called the exine, which consists largely of the complex carbohydrate sporopollenin [4, 30, 31]. The exine strongly protects the pollen grain from physicochemical factors; it is surrounded by a layer of lipids, carbohydrates, terpenoids and carotenoid pigments. The exine is often perforated, and these so-called germination pores lead to the inner pollen wall, the intine. The intine, composed mainly of cellulose and pectin, is the final barrier to the nutrient-rich cytoplasm of the pollen grain [8, 31].

3.2.1 Pollen Collection

The raw material for the production of bee pollen is found by bees on plants at their stamens, which contain numerous tiny pollen grains. When the bees visit the flowers, tiny pollen grains stick to the hairs on their bodies. The bees clean themselves during flight and moisten the tiny pollen grains with the saliva and nectar from their stomachs, creating a pollen pellet which they deposit in a special structure (corbiculae) on their hind legs. The bees bring the bee pollen into the hive, where it is stored in honeycomb cells and used for their nutrition [4, 28, 32, 33]. The flower of a plant species influences the colour of the pollen grain, which ranges from white or creamy white and yellow to orange, red, green, grey and dark brown [12].

To collect bee pollen for human nutrition, bee pollen traps are used, which are placed at the entrance or at the bottoms of beehives. When passing a bee pollen trap, which contains small openings, the pollen pellets are stripped from the bee legs and fall into the collection tray [34, 35]. About 10% of the bee pollen is taken from the bees with bee pollen traps. The bees deposit the rest in combs, add the lactic acid bacteria from their stomach and seal them with a thin layer of honey. The anaerobic conditions lead to lactic acid fermentation, which converts bee pollen into bee bread, a food reserve for developing larvae and adult bees [36]. Collecting bee bread from honeycombs or pollen directly from plants is difficult. Therefore, most researchers use pollen collected from bees to determine the chemical composition of different pollens [31, 37].

Fresh bee pollen is a microbiologically susceptible food that is highly hygroscopic and contains between 15% and 30% water [1, 28, 38–40]. Stability of bee pollen is usually achieved by direct freezing of the product or by different drying techniques. Dried bee pollen is most often used in human nutrition. The drying process requires a suitable drying temperature at controlled conditions in order to maintain the nutritional quality and safety of the bee pollen and to minimize the loss of bioactive compounds and negative impacts on the sensory quality [4, 16, 40–42].

Multifloral bee pollen (Fig. 3.1, left) is the most represented product available on the market for human consumption. Bees almost never collect pollen from only one

Fig. 3.1 Multifloral (left) and monofloral (C. sativa) (right) bee pollen. (Photo credit: N. Lilek)

plant species; therefore, the daily harvest is usually a mixture of bee pollen pellets from different botanical origins. Therefore, it differs in its chemical composition and functional properties. Obtaining the monofloral bee pollen for human consumption is not easy. It requires the transport of bees to large areas planted with a particular monoculture. In addition, pellets of multifloral bee pollen need to be manually separated or mechanically sorted by colour. Each botanical type of bee pollen has its own characteristic features. Monofloral bee pollen (Fig. 3.1, right: *Castanea sativa* bee pollen) has sensory and biochemical properties similar to those of the plant from which it is derived, while multifloral bee pollen has different properties derived from different botanical species [2].

3.2.2 Bee Pollen Composition

Bee pollen is a natural source of sugars, proteins, lipids and dietary fibre, and also contains small amounts of amino acids, saturated and unsaturated fatty acids, minerals, vitamins, carotenoids, and phenolic compounds. The content of these ingredients is influenced by various factors, such as the botanical and geographical origin of the bee pollen, the type of soil, the weather conditions, the climate, the season, the processing of the fresh bee pollen and the storage conditions. As a result, there are significant differences in the nutritional composition of bee pollen, both in multifloral and monofloral varieties [1, 2, 4, 9, 10, 12, 14, 24, 25, 43, 44]. Differences also arise from the use of different analytical methods [45].

According to [46] bee pollen is used in human nutrition because it contains nutrients and various bioactive compounds in a relatively high concentration and in a balanced ratio. By reviewing more than 100 studies, Thakur and Nanda [12] summarised the average composition of bee pollen based on data from both fresh and dried products. According to their findings, bee pollen contains on average 54.22% (18.50–84.25%) carbohydrates, 21.30% (4.50–40.70%) proteins, 5.31% (0.41–13.50%) lipids, 8.75% (0.15–31.26%) dietary fibre and 2.91% (0.50–7.75%) ash.

In much smaller quantities, bee pollen also contains all essential amino acids, saturated and unsaturated fatty acids, elements (K, P, S, Ca, Mg, Zn, Cu, Fe, Mn), vitamins (B vitamins, β-carotene, vitamin E, vitamin C), pigments (lycopene and zeaxanthin) and secondary plant metabolites, such as flavonoids (quercetin, kaempferol, catechin, isorhamnetin, galangin, myricetin) and phytosterols, which can have antioxidant, antimicrobial, anticancer and anti-inflammatory effects [1, 3, 4, 8, 24, 47].

3.2.3 Carbohydrates

Carbohydrates are the most widespread and abundant organic compounds on the Earth. They are a major source of energy in the diet and comprise a range of compounds, all containing carbon, hydrogen and oxygen. Carbohydrates are probably best known for their role in energy metabolism, where simple sugars (glucose and fructose) can be used directly for energy or serve as reserve compounds (starch, glycogen) in plants and animals. As macronutrients, carbohydrates provide fuel for the central nervous system and energy for working muscles. Some other compounds may be involved in the formation of plant or bacterial cell walls, and are classified as structural carbohydrates (cellulose, pectin) [48, 49]. In a healthy diet, carbohydrates should make up 45–60% of daily energy intake [50].

Simple and complex carbohydrates contribute to different functional properties of foods. Carbohydrates have a wide range of rheological and other properties, including solubility, sweetening effect, hygroscopicity, crystallisation inhibition and aroma enhancement. These properties are based on the chemical structure of carbohydrates and interactions with other molecules in foods, such as lipids and proteins [51].

According to chemical classification, carbohydrates are divided into three groups depending on the degree of polymerisation (DP): (i) sugars (DP 1–2; mono- and disaccharides), (ii) oligosaccharides (DP 3–9; malto-oligosaccharides, non-digestible oligosaccharides) and (iii) polysaccharides (DP >9; starch and non-starch polysaccharides) [48, 52]. From a nutritional perspective, carbohydrates can be classified according to their digestion and absorption in the body. Digestible carbohydrates are absorbed and digested in the human small intestine and provide sugars to the body's cells, while non-digestible carbohydrates are resistant to hydrolysis in the small intestine but can be partially fermented by the gut microbiota in the large intestine. The term dietary fibre is used for non-digestible carbohydrates [50, 52].

Dietary fibre has a positive effect on certain physiological functions in the digestive tract and the prevention of non-communicable diseases. Dietary fibre includes heterogeneous components such as non-starch polysaccharides (cellulose, hemicelluloses, pectins, hydrocolloids, β-glucans), resistant oligosaccharides, resistant starch and lignin, which occur naturally in foods (mostly of plant origin) but can also be extracted from foods by chemical, physical or enzymatic methods, or synthesized, and have been shown to have physiological effect or a beneficial effect on health [50, 53]. Fermentation of dietary fibre in the colon under the influence of the intestinal microbiota produces short-chain fatty acids, which have a direct positive effect on the health of the epithelial cells of the colon. The most commonly documented health benefits include colon function with a reduction in transit time, an increase in stool mass and colon fermentation (production of short-chain fatty acids), a reduction in blood cholesterol levels, a reduction in blood glucose levels, an improvement in the gut microbiota and a reduction in blood pressure [53, 54]. Natural sources of dietary fibre are whole or minimally processed plant foods, including whole grains, fruits, vegetables, legumes, nuts and seeds [55].

Due to the chemical diversity of carbohydrates, specific methods for the analysis of different carbohydrates in foods have only recently become routinely available. Therefore, carbohydrate contents are often still determined "by difference" [50]. Various studies on bee pollen composition use this calculation for carbohydrate content. These values can represent total carbohydrates: 100 – (g fat + g protein + g ash + g water) with dietary fibre included [1, 2, 4, 32, 39, 56, 57] or available/

digestible carbohydrates: 100 – (g fat + g protein + g ash + g dietary fibre + g water) if dietary fibre is analysed and considered in the calculation [35].

The carbohydrates in bee pollen are derived from the flower pollen and nectar, with which the bee mixes the flower pollen to form bee-collected pollen [56]. They are the main component of bee pollen and account about two-thirds of its dry weight (dw) [5, 25]. Due to the added nectar, carbohydrates constitute a large part of the nutrient content. The addition of nectar to fresh flower pollen results in increased carbohydrate content in collected bee pollen from 35.0% to 61.0% dw [56]. Besides low molecular weight sugars (fructose, glucose and sucrose), polysaccharides (starch, callose, pectin, cellulose and sporopollenin) are also the carbohydrate components of pollen grains [58]. There is not much data on the sugar composition of flower pollen. In 'Oblačinska' sour cherry (*Prunus cerasus* L.) pollen, glucose was the most abundant sugar (8.58–13.67 g/100 g), determined by high-performance anion-exchange chromatography with pulsed amperometric detection, followed by fructose (5.67–9.27 g/100 g) and sucrose (3.19–7.51 g/100 g). Other sugars detected in much lower amounts were trehalose, arabinose, isomaltose, turanose, maltose and maltotriose. The total carbohydrate content (various sugars and sorbitol) ranged from 22.68 to 32.49 g/100 g [58].

Due to the different botanical and geographical origins of bee pollen, as well as climatic conditions and different processing and storage techniques, the reported values for total carbohydrates in bee pollen vary widely. Some literature data on total carbohydrate content are shown in Table 3.1. The reported values range from 42.33 to 84.25 g/100 g dw. The calculated carbohydrate values are higher than the results obtained with high-performance liquid chromatography (HPLC) and gas chromatography (GC), as these methods generally do not determine components such as dietary fibre and starch [42]. To determine the content of total soluble carbohydrates, the Anthrone method was also used. The values for total carbohydrates for 15 commercial bee pollen samples from Spain ranged from 62.80 to 75.32 g/100 g [59].

3.2.3.1 Sugars in bee pollen

Sugars in bee pollen have sometimes been ignored in previous studies, or have been included in the total carbohydrate, which also includes dietary fibre and starch [2, 21, 43, 57, 60, 62]. Among the sugars in bee pollen, fructose, glucose and sucrose account for more than 90% of all low molecular weight sugars, the proportion also depends on the plant species from which the pollen was collected [63, 64].

The most commonly used methods for the determination of sugars in bee pollen are chromatographic methods with different detections – the use of liquid chromatography [13, 19, 22, 64–67] predominates, and to a lesser extent gas chromatography is also used [63, 68]. Less frequently capillary electrophoresis for single sugars [69] and spectrophotometric methods with phenol-sulfuric acid for total sugars/total soluble carbohydrate content are used for analysis [62, 70, 71].

Table 3.1 The carbohydrate content in bee pollen of different botanical and geographical origin

Carbohydrate content[a] (g/100 g dw)	Number of samples	Botanical origin	Geographical origin	References
47.4–73.1	35	multifloral	Slovenia	[35]
62.5–71.3	10	sweet chestnut (*C. sativa*)	Slovenia	[35]
54.85–73.98	32	multifloral	Slovenia	[39]
59.43–77.82	12	monofloral, different types	China	[2]
64.42–81.84	26	multifloral	Serbia	[57]
69.68–84.25	8	multifloral	Portugal, Spain	[32]
61.2–70.6	22	multifloral	Portugal	[43]
60.82–70.76	5	multifloral	Portugal	[60]
55.81–66.49	3	monofloral (*Zea mays, Trifolium alexandrinum, Eucalyptus* sp.)	Egypt	[11]
42.33–46.16	25	monofloral (*Cocos nucifera, Coriandrum sativum, Brassica napus*)	India	[61]
46.07 (average)	10	multifloral	India	[61]

[a]Expressed as 100 – (g fat + g protein + g ash + g water), *dw* dry weight

Several studies reporting the sugar profile of bee pollen generally show higher levels of fructose and glucose, followed by other sugars such as sucrose and maltose [63, 67, 68, 70, 72]. The sucrose content in some samples of Croatian monofloral and multifloral bee pollen [13] was relatively high compared to other studies, with maximum values around 20 g/100 g dw as shown in Table 3.2. Most bee pollen contains more fructose than glucose. The ratio of fructose to glucose in bee pollen ranges widely, between 1.0 and 2.5 [42, 68]. In contrast, Bertoncelj et al. [67] found more glucose than fructose in different types of Slovenian bee pollen in 68% of the samples. The results of some published studies from different countries on the sugar content of bee pollen are presented in Table 3.2. They show the variability of fructose, glucose, sucrose, maltose, melezitose and total sugar contents.

Liolios et al. [22] investigated the sugar profile of Greek monofloral bee pollen using high performance liquid chromatography with refractive index detector. Analysis of 117 samples (30 different bee pollen species) revealed high levels of glucose (13.59 to 27.69 g/100 g dw) and fructose (15.53 to 33.48 g/100 g dw), accounting for 94% of total sugars. Sucrose, turanose, maltose, trehalose, melibiose and melezitose were found in much lower contents. The average content of total sugars was 42.10 g/100 g dw, ranging from 34.71 g/100 g dw for ivy (*Hedera helix*) bee pollen to 63.53 g/100 g dw for kiwi (*Actinidia chinensis*) bee pollen. A similar result for total sugar content (32.62 g/100 g dw) was reported for Slovenian monofloral ivy bee pollen [67].

Sattler et al. [20] investigated the nutritional composition of Brazilian bee pollen. The contents of fructose and glucose determined by the HPLC method were much lower compared to other studies, ranging from 3.4 g/100 g dw to 7.0 g/100 g dw and

Table 3.2 The sugar content in bee pollen of different botanical and geographical origin

Sugar content (g/100 g dw)	Number of samples	Method	Botanical origin	Geographical origin	References
Fructose					
13.36–27.84	18	LC-MS	multifloral	Slovenia	[67]
18.15–26.48	7	LC-MS	sweet chestnut (*C. sativa*)	Slovenia	[67]
10.79–23.88	16	HPLC-RID	multifloral	Croatia	[13]
10.87–25.65	48	HPLC-RID	monofloral, different types	Croatia	[13]
12.59–23.62	154	HPLC-RID	multifloral	Brazil	[66]
16.42 (average)	29	HPLC-RID	multifloral	Morocco	[70]
18.46–21.44	24	HPLC-RID	multifloral	Romania	[72]
18.10–21.30	196	HPLC-RID	multifloral	Colombia	[18]
17.1–23.0	3	UHPLC-MS	multifloral	Italy, Spain, Colombia	[73]
15.20–22.40	20	GC-FID	multifloral	Spain	[63]
13.38–21.80	95	GC-FID	multifloral	Poland	[68]
15.9–19.9	5	HPLEC-PAD	multifloral	Israel, China, Romania, Spain	[19]
Glucose					
10.60–26.59	18	LC-MS	multifloral	Slovenia	[67]
11.94–28.49	7	LC-MS	sweet chestnut (*C. sativa*)	Slovenia	[67]
8.92–18.70	16	HPLC-RID	multifloral	Croatia	[13]
8.96–19.86	48	HPLC-RID	monofloral, different types	Croatia	[13]
6.99–21.85	154	HPLC-RID	multifloral	Brazil	[66]
11.44 (average)	29	HPLC-RID	multifloral	Morocco	[70]
9.49–17.40	24	HPLC-RID	multifloral	Romania	[72]
11.60–20.30	196	HPLC-RID	multifloral	Colombia	[18]
14.1–15.9	3	UHPLC-MS	multifloral	Italy, Spain, Colombia	[73]
10.86–17.90	20	GC-FID	multifloral	Spain	[63]
5.99–18.17	95	GC-FID	multifloral	Poland	[68]
8.2–13.1	5	HPLEC-PAD	multifloral	Israel, China, Romania, Spain	[19]
Sucrose					
0.05–0.17	18	LC-MS	multifloral	Slovenia	[67]
0.07–0.28	7	LC-MS	sweet chestnut (*C. sativa*)	Slovenia	[67]
2.62–22.04	16	HPLC-RID	multifloral	Croatia	[13]
1.62–20.33	48	HPLC-RID	monofloral, different types	Croatia	[13]
1.38 (average)	29	HPLC-RID	multifloral	Morocco	[70]

(continued)

Table 3.2 (continued)

Sugar content (g/100 g dw)	Number of samples	Method	Botanical origin	Geographical origin	References
0.13–1.45	24	HPLC-RID	multifloral	Romania	[72]
4.50–9.00	196	HPLC-RID	multifloral	Colombia	[18]
5.10–6.20	3	UHPLC-MS	multifloral	Italy, Spain, Colombia	[73]
4.20–9.40	20	GC-FID	multifloral	Spain	[63]
3.74–7.59	95	GC-FID	multifloral	Poland	[68]
14.8–18.4	5	HPLEC-PAD	multifloral	Israel, China, Romania, Spain	[19]
Maltose					
0.16–6.03	18	LC-MS	multifloral	Slovenia	[67]
ND–5.85	7	LC-MS	sweet chestnut (*C. sativa*)	Slovenia	[67]
0.65–4.98	16	HPLC-RID	multifloral	Croatia	[13]
0.44–5.48	48	HPLC-RID	monofloral, different types	Croatia	[13]
0.21–1.06	24	HPLC-RID	multifloral	Romania	[72]
0.79–3.20	20	GC-FID	multifloral	Spain	[63]
0.96–3.32	95	GC-FID	multifloral	Poland	[68]
Melezitose					
ND–0.98	18	LC-MS	multifloral	Slovenia	[67]
ND–0.45	7	LC-MS	sweet chestnut (*C. sativa*)	Slovenia	[67]
0.01–1.54	16	HPLC-RID	multifloral	Croatia	[13]
0.05–2.16	48	HPLC-RID	monofloral, different types	Croatia	[13]
0.90–2.99	24	HPLC-RID	multifloral	Romania	[72]
0.10–0.29	20	GC-FID	multifloral	Spain	[63]
Total sugars					
24.71–55.19	18	LC-MS	multifloral	Slovenia	[67]
36.01–55.35	7	LC-MS	sweet chestnut (*C. sativa*)	Slovenia	[67]
35.79–51.29	16	HPLC-RID	multifloral	Croatia	[13]
26.20–52.58	48	HPLC-RID	monofloral, different types	Croatia	[13]
34.6–48.8	20	GC-FID	multifloral	Spain	[63]
41.1–60.4	36	spectrophotometry	multifloral	Brazil	[62]
18.52–46.44[a]	6	spectrophotometry	multifloral	Morocco	[71]
31.69 (average)[a]	29	spectrophotometry	multifloral	Morocco	[70]

LC-MS liquid chromatography-mass spectrometry, *HPLC-RID* high-performance liquid chromatography-refractive index detector, *UHPLC-MS* ultra high-performance liquid chromatography-mass spectrometry, *GC-FID* gas chromatography-flame ionisation detector, *HPLEC-PAD* high-performance ligand exchange chromatography with pulsed amperometric detection, *dw* dry weight, *ND* not detected

[a]total sugar content/soluble carbohydrates content expressed as a gram of glucose equivalents/100 g dw

4.3 g/100 g to 8.9 g/100 g dw, respectively, in twenty bee pollens. Kalaycıoglu et al. [69] also reported particularly low levels of fructose and glucose in bee pollen from Turkey, with values ranging from 5.50 to 6.22 g fructose/100 g and 2.42 to 4.93 g glucose/100 g bee pollen.

As the main reason for the variability in the results for sugar composition, the use of different analytical methods is indicated [22]. The differences between studies may also be due to the fact that the nectar and honey added by the bees during the formation of bee pollen pellets influences the sugar profile, and the botanical origin of the bee pollen is also an important factor in the sugar content of the collected bee pollen [21, 33, 56, 68].

Some studies also report the content of other sugars, namely turanose, isomaltose, maltulose, trehalose, erlose, maltotriose and raffinose, which are present in bee pollen in very small amounts [63, 64, 72]. The contents of these minor sugars are also very different among studies.

3.2.3.2 Dietary Fibre in Bee Pollen

There are relatively few studies related to the dietary fibre content in bee pollen, some of them only provide information on crude fibre content [61], which is only part of the insoluble fibre fraction, and this requires attention when comparing data.

Bee pollen has been shown to be a relatively good source of dietary fibre, most of present dietary fibre is insoluble [2, 4, 67]. Dietary fibre content is usually lower than that of sugars, but it still plays an important role. The content of dietary fibre in bee pollen varies. There are differences between bee pollen of different botanical origin and also due to the use of different analytical methods [4]. Pollen grains are surrounded by an inner protective layer of pectin and cellulose, which are part of the cell walls, and an outer protective layer consisting of persistent carbohydrate polymer – sporopollenin (protects against desiccation). This polymer contains phenolic compounds and fatty acids that are covalently linked by ether and ester bonds [74]. The components mentioned are counted among the dietary fibre [3, 4].

Several methods have been developed and officially prescribed for the determination of dietary fibre content, both for the determination of total dietary fibre, soluble dietary fibre and insoluble dietary fibre, and for the determination of the individual fibre components. The methods have been modified and updated as the definition of dietary fibre has evolved. The original enzyme-gravimetric methods are relatively simple, affordable, quite rapid and robust enough for routine analysis, but do not provide a detailed profile of the different fibre components and do not include the low molecular weight components of dietary fibre (resistant oligosaccharides). Quantitative determination of dietary fibre content by HPLC, included in the newer fibre methods, allows the determination of all components of dietary fibre, including low molecular weight carbohydrate polymers (with 3–9 monomer units), and is ideal, when the food in question contains an unknown amount and type of dietary fibre [75]. However, these new methods are time-consuming and very expensive and, to our knowledge, have not yet been used to determine dietary fibre in bee pollen. For the determination of dietary fibre in bee pollen, most authors use the enzyme-gravimetric method (AOAC 991.43), which determines soluble and insoluble dietary fibre separately [2, 35, 62, 67, 76], but does not include low molecular weight components. On average, bee pollen contains between 7.0 and 20.0 g/100 g dw of total dietary fibre [2, 12, 17, 67].

In published studies where soluble and insoluble dietary fibre content is reported separately, insoluble fibre accounts for a greater proportion, usually more than 70% of total dietary fibre. According to the data presented in Table 3.3, the range for soluble dietary fibre was between 0.5 and 5.92 g/100 g dw, while the insoluble fibre content ranged from 6.22 to 30.12 g/100 g dw. As reported by [76] very low total dietary fibre contents were found in Saudi bee pollen. Values between 0.15 and 1.70 g/100 g dw were determined for five monofloral bee pollens using the AOAC 991.43 method.

Table 3.3 The dietary fibre content in bee pollen of different botanical and geographical origin

Dietary fibre (g/100 g dw)	Number of samples	Method	Botanical origin	Geographical origin	References
Total dietary fibre					
8.41–25.9	35	AOAC 991.43	multifloral	Slovenia	[35]
16.7–20.5	10	AOAC 991.43	sweet chestnut (*C. sativa*)	Slovenia	[35]
17.60–31.26	12	AOAC 991.43	monofloral, different types	China	[2]
7.8–18.1	196	AOAC 985.29	multifloral	Colombia	[18]
10.6–15.9	20	Englyst	multifloral	Spain	[63]
12.84 (average)	50	AOAC 985.29	multifloral	Colombia	[17]
14.50–14.65	5	enzymatic-gravimetric	multifloral and monofloral	Spain	[16]
Insoluble dietary fibre					
6.22–18.7	35	AOAC 991.43	multifloral	Slovenia	[35]
14.4–17.1	10	AOAC 991.43	sweet chestnut (*C. sativa*)	Slovenia	[35]
13.39–30.12	12	AOAC 991.43	monofloral, different types	China	[2]
6.5–17.6	196	AOAC 985.29	multifloral	Colombia	[18]
9.0–13.1	20	Englyst	multifloral	Spain	[63]
10.63 (average)	50	AOAC 985.29	multifloral	Colombia	[17]
Soluble dietary fibre					
0.55–7.33	35	AOAC 991.43	multifloral	Slovenia	[35]
1.95–3.80	10	AOAC 991.43	sweet chestnut (*C. sativa*)	Slovenia	[35]
0.86–5.92	12	AOAC 985.29	monofloral, different types	China	[2]
0.5–4.6	196	AOAC 985.29	multifloral	Colombia	[18]
1.59–3.66	20	Englyst	multifloral	Spain	[63]
2.21 (average)	50	AOAC 991.43	multifloral	Colombia	[17]

3.2.4 Importance of Bee Pollen Carbohydrates for Human Nutrition

The use of bee pollen in human nutrition increased after the World War II, when bee pollen traps were developed and became readily available [45]. Knowledge of the functional properties and nutritional value of bee pollen and its effects on certain medical conditions have increased consumer awareness [11, 24, 27, 43–45, 59, 60, 63, 64, 77, 78].

According to [79], a nutritious food provides beneficial nutrients (e.g., proteins, vitamins, minerals, essential amino acids, essential fatty acids, dietary fibre) and minimises potentially harmful effects of certain compounds (e.g. anti-nutrients, amounts of sodium, saturated fat, added sugars). Bee pollen is an important source of nutrients and energy and can be used as a supplement to the daily diet.

The dietary guidelines state that complex carbohydrates should make up about half of the calories in a balanced diet, while sugars (i.e. simple carbohydrates) should be limited to no more than 5–10% of total energy intake [80]. The limit for sugars refers to free/added sugars and not to sugars that are naturally present in food [81]. Bee pollen with only naturally occurring sugars therefore provides a rapidly utilisable source of energy. It can also resist oxidation, regulate blood sugar and lower blood lipids [82].

According to the European Food Safety Authority [50], a daily intake of 25 g of dietary fibre is adequate for normal bowel function in adults. Other European recommendations state that dietary fibre intake should be at least 30 g per day [53, 75]. Due to the increased consumption of processed foods, which are often low in dietary fibre, and the associated decline in the intake of unprocessed foods of plant origin, the intake of dietary fibre in the population of industrialised countries is significantly below the recommendations. A healthy diet should include a variety of fibre-rich whole foods such as whole grains, fruits, vegetables, legumes, nuts and seeds, which contribute not only to the recommended daily intake of fibre but also of other important nutrients. As bee pollen has been shown to be a relatively good source of dietary fibre its consumption can contribute to the recommended daily intake [2, 4, 12, 25, 35].

Bee pollen improves metabolism and general physical performance, boosts energy and is very suitable for recovering from diseases and for people who are underweight [46, 83]. For prevention and improvement of health, 10–20 g of bee pollen is recommended daily, usually twice for 3 months a year [84]. In apitherapy, the daily dose of bee pollen is higher, from 20 to 40 g per day for an adult to benefit from the desired therapeutic effects [24, 84]. Such amounts of bee pollen can provide up to 25% of the recommended dietary fibre intake.

Long-term consumption of bee pollen can improve health, promote blood circulation, delay the ageing process, strengthen the immune system and increase physical and mental performance. Further studies on metabolic pathways and biomedical interactions are needed to prove the bioactivity of bee pollen in controlling body functions and preventing diseases [10]. Due to the large variability in bee pollen composition, it is difficult to realistically assess the dietary intake of individual macro- and micronutrients by consuming the daily recommended amount of multifloral bee pollen. Nevertheless, the characterisation of monofloral bee pollen with a more consistent composition in terms of chemical and functional properties can lead to a more accurate intake of nutrients for the specific nutritional needs of the individual and better valorisation of certain monofloral bee pollens.

For better sensory acceptability of bee pollen, it is recommended to mix it with honey, juice, smoothies or dairy products. A strong outer layer of the pollen grain is also very effective against human digestion. Nevertheless, bee pollen is digestible; the degree of digestibility also depends on the botanical origin of the bee pollen [5, 12]. To increase the digestibility of bee pollen and thus the availability of individual nutrients and functionality, it is recommended to grind or soak bee pollen grains in warm water or other liquid before consumption, which makes the pollen grain layer more permeable [3, 8, 24, 44, 84].

By adding bee pollen to various foods such as dairy, meat and bakery products as well as beverages, we can increase the intake of this bee product and at the same time also improve the nutritional and functional properties of the foods enriched with bee pollen. The influence of pollen components, including carbohydrates, on physical, thermal, textural and techno-functional properties is also of great interest to food technologists [9].

Despite all the benefits of bee pollen, it should be mentioned that pollen may contain anti-nutritional compounds such as allergens, pyrrolizidine alkaloids, toxic and potentially toxic elements, and mycotoxins [9]. Allergic reactions to the consumption of bee pollen are rare but cannot be excluded, especially in people allergic to anemone pollen. The allergy is usually triggered by inhaling such pollen, whereas it rarely occurs when bee pollen is ingested. It is recommended to start including bee pollen in the diet in small amounts, increasing daily to the recommended doses [10, 45, 84]. According to [85] fermentation is an efficient approach for reducing or eliminating food allergenicity. Study based on proteomics indicated that allergenicity of bee pollen could be decreased with fermentation process [86].

3.3 Future Perspectives

Demand for healthy diets, prevention of diet-related diseases and concerns about synthetic food additives have led to increased consumption of bee pollen by consumers. Bee pollen not only has the role of a dietary supplement and functional food, but also that of a potential medicine. Bee pollen is used to prevent and treat many chronic diseases, especially metabolic disorders. It could play a preventive role in various diseases such as diabetes, obesity, hyper-dyslipidaemia, cardiovascular diseases, prostatitis and immune disorders [8, 10, 23, 87]. The exact extent of biological activity is difficult to determine due to the large variability in the composition of this bee product, which depends mainly on botanical origin. Quality standards for bee pollen need to be established for clinical use, which would facilitate the use of bee pollen for medicinal purposes [3, 14]. In the future, further appropriate clinical studies are needed to confirm the beneficial effects of bee pollen or its ingredients on human health and well-being [10, 87]. Further studies are also needed to assess bio-accessibility, bioavailability and metabolism, as well as the influence of the gut microbiota, and to measure safe and potentially toxic dosages [14].

Bee pollen contains all the essential nutrients and various bioactive compounds, but previous studies indicate that it has limited digestibility due to a complex outer protective layer called exine. For this reason, many methods have been tested to better utilise the nutrients and increase consumption. Among them are biotechnological methods such as fermentation, which give remarkable results and are more affordable than other techniques [10, 82].

Oligosaccharides and polysaccharides are very important components of bee pollen that contribute to the regulation of various biological functions. These types of compounds are commonly referred to as insoluble and soluble dietary fibres, which are considered to have a prebiotic effect by selectively promoting the growth of beneficial bacteria within the human microbiota. Therefore, future research should also focus on the development of rapid, reliable, easy-to-use and cost-effective analytical methods for the routine analysis of carbohydrates from bee pollen, especially different types of dietary fibre with proven beneficial effects on health. In the absence of information on the direct effects of complex carbohydrates from bee pollen on the human microbiota, the bioactivity of these compounds should be investigated in future scientific and clinical studies.

To realise the potential of bee pollen in human nutrition and its beneficial effects on health, it is necessary to promote the consumption of bee pollen among consumers. Based on the total dietary fibre content, European legislation [88] allows the use of the nutrition claim 'high fibre' (at least 6 g of dietary fibre per 100 g of a food). The use of such labelling for bee pollen may lead to higher acceptance and consumption of this bee product [67, 83]. The development of functional foods and food supplements with the addition of bee pollen may also increase the consumption of this bee product. However, due to their complexity and variability, there are limitations to the use of bee pollen-based products, highlighting the need for standardisation before safe use in human nutrition and apitherapy.

3.4 Conclusions

Interest in bee pollen is growing, mainly due to health-oriented consumers who want to use natural foods for a healthy diet or for their therapeutic effects. Bee pollen is a rich natural source of nutrients and biologically active compounds and in many cases can be considered a functional food or functional ingredient that enhances the nutritional value of other foods. An important component of bee pollen is carbohydrates, including sugars and dietary fibre, which have a positive effect on health. Bee pollen is a natural food that, due to its composition helps to achieve the recommended daily intake of basic nutrients, including dietary fibre, and is also a good source of important bioactive compounds and therefore undoubtedly part of a balanced diet. As bee pollen is a complex biological system in which carbohydrates, especially various dietary fibre components, very likely contribute to its bioactivity, further analysis and also clinical studies are needed to confirm its nutritional and biological value.

References

1. Campos MGR, Bogdanov S, Almeida-Muradian LB, Szczesna T, Mancebo Y, Frigerio C, Ferreira F (2008) Pollen composition and standardization of analytical methods. J Apicult Res Bee World 47:154–161
2. Yang K, Wu D, Ye X, Liu D, Chen J, Sun P (2013) Characterization of chemical composition of bee pollen in China. J Agric Food Chem 61:708–718
3. Denisow B, Denisow-Pietrzyk M (2016) Biological and therapeutic properties of bee pollen: a review. J Sci Food Agric 96:4303–4309
4. Bogdanov S (2017) The pollen book: collection, harvest, composition and quality http://www.bee-hexagon.net. Accessed 12 Dec 2012
5. Aylanc V, Falcão SI, Ertosun S, Vilas-Boas M (2021) From the hive to the table: nutrition value, digestibility and bioavailability of the dietary phytochemicals present in the bee pollen and bee bread. Trends Food Sci Technol 109:464–481
6. Brodschneider R, Crailsheim K (2010) Nutrition and health in honey bees. Apidologie 41:278–294
7. Guiné RPF (2015) Bee pollen: chemical composition and potential beneficial effects on health. Curr Nutr Food Sci 11:301–308
8. Komosinska-Vassev K, Olczyk P, Kafmierczak J, Mencner L, Olczyk K (2015) Bee Pollen: chemical composition and therapeutic application. Evid Based Complement Alternat Med 297425:1–6
9. Kostić A, Milinčić D, Barać M, Shariati MA, Tešić Ž, Pešić M (2020) The application of pollen as a functional food and feed ingredient—the present and perspectives. Biomol Ther 10:84
10. Khalifa SAM, Elashal MH, Yosri N, Du M, Musharraf SG, Nahar L, Sarker SD, Guo Z, Cao W, Zou X et al (2021) Bee pollen: current status and therapeutic potential. Nutrients 13:1876
11. Alshallash KS, Abolaban G, Elhamamsy SM, Zaghlool A, Nasr A, Nagib A (2023) Bee pollen as a functional product – chemical constituents and nutritional properties. J Ecol Eng 24:173–183
12. Thakur M, Nanda V (2020) Composition and functionality of bee pollen: a review. Trends Food Sci Technol 98:82–106
13. Prđun S, Svečnjak L, Valentić M, Marijanović Z, Jerković I (2021) Characterization of bee pollen: physico-chemical properties, headspace composition and FTIR spectral profiles. Foods 10:2103
14. Giampieri F, Quiles JL, Cianciosi D, Forbes-Hernández TY, Orantes-Bermejo FJ, Alvarez-Suarez JM, Battino M (2022) Bee products: An emblematic example of underutilized sources of bioactive compounds. J Agric Food Chem 70:6833–6848
15. Ares AM, Valverde S, Bernal JL, Nozal MJ, Bernal J (2018) Extraction and determination of bioactive compounds from bee pollen. J Pharm Biomed Anal 147:110–124
16. Domínguez-Valhondo D, Gil DB, Hernández MT, González-Gómez D (2011) Influence of the commercial processing and floral origin on bioactive and nutritional properties of honeybee-collected pollen. Int J Food Sci Tech 46:2204–2211
17. Díaz MC, Zuluaga C, Morales C, Quicazán M (2012) Determinación de fibra dietaría en polen apícola Colombiano. Vitae 19:454–456
18. Fuenmayor CA, Zuluaga-Domínguez CM, Diaz-Moreno AC, Quicazán MC (2014) Evaluation of the physicochemical and functional properties of Colombian bee pollen. Rev MVZ Cordoba 19:4003–4014

19. Qian WL, Khan Z, Watson DG, Fearnley J (2008) Analysis of sugars in bee pollen and propolis by ligand exchange chromatography in combination with pulsed amperometric detection and mass spectrometry. J Food Compos Anal 21:78–83
20. Sattler JAG, Melo ILP, Granato D, Araújo E, Silva-Freitas A, Barth OM (2015) Impact of origin on bioactive compounds and nutritional composition of bee pollen from southern Brazil. Food Res Int 77:82–91
21. Conti I, Medrzycki P, Argenti C, Meloni M, Vecchione V, Boi M, Mariotti MG (2016) Sugar and protein content in different monofloral pollens – building a database. Bull Insectology 69:318–320
22. Liolios V, Tananaki C, Dimou M, Kanelis D, Rodopoulou M-A, Thrasyvoulou A (2018) Exploring the sugar profile of unifloral bee pollen using high performance liquid chromatography. J Food Nutr Res 57:341–350
23. Cornara L, Biagi M, Xiao J, Burlando B (2017) Therapeutic properties of bioactive compounds from different honeybee products. Front Pharmacol 8:1–20
24. Kieliszek M, Piwowarek K, Kot AM, Błażejak S, Chlebowska-Śmigiel A, Wolska I (2018) Pollen and bee bread as new health-oriented products: a review. Trends Food Sci Technol 71:170–180
25. Li QQ, Wang K, Marcucci MC, Sawaya ACHF, Hu L, Xue XF, Wu LM, Hu FL (2018) Nutrient-rich bee pollen: a treasure trove of active natural metabolites. J Funct Foods 49:472–484
26. Weis WA, Ripari N, Lopes Conte F, da Silva Honorio M, Sartori AA, Matucci RH, Sforcin JM (2022) An overview about apitherapy and its clinical applications. Phytomedicine Plus 2:100239
27. Baky MH, Abouelela MB, Wang K, Farag MA (2023) Bee pollen and bread as a super-food: a comparative review of their metabolome composition and quality assessment in the context of best recovery conditions. Molecules 28:715
28. Almeida-Muradian LB, Pamplona LC, Coimbra S, Barth OM (2005) Chemical composition and botanical evaluation of dried bee pollen pellets. J Food Compos Anal 18:105–111
29. Shubharani R, Roopa P, Sivaram V (2013) Pollen morphology of selected bee forage plants. Global J Biosci Biotechnol 2:82–90
30. Kesseler R, Harley M (2006) Pollen the hidden sexuality of flowers. Papadakis Publisher, London
31. Roulston TH, Cane JH (2000) Pollen nutritional content and digestibility for animals. Plant Syst Evol 222:187–209
32. Nogueira C, Iglesias A, Feás X, Estevinho LM (2012) Commercial bee pollen with different geographical origins: a comprehensive approach. Int J Mol Sci 13:11173–11187
33. Dimou M, Liolios V, Rodopoulou M-A, Tananaki C (2020) The effect of pollen grain morphology on sugars content of bee-collected pollen: the significance of size and exine ornamentation. Grana 59:389–395
34. Krell R (1996) Value-added products from beekeeping. Food and Agricultural Organisation of the United Nations, Rome
35. Lilek N (2020) Influence of the botanical origin on nutritional value of bee pollen. Dissertation, University of Ljubljana
36. Di Cagno R, Filannino P, Cantatore V, Gobbetti M (2019) Novel solid-state fermentation of bee-collected pollen emulating the natural fermentation process of bee bread. Food Microbiol 82:218–230
37. Weiner CN, Hilpert A, Werner M, Linsenmair KE, Blüthgen N (2010) Pollen amino acids and flower specialisation in solitary bees. Apidologie 41:476–487
38. Morgano MA, Milani RF, Martins MCT, Rodriguez-Amaya D (2011) Determination of water content in Brazilian honeybee-collected pollen by Karl Fischer titration. Food Control 22:1604–1608
39. Lilek N, Pereyra Gonzales A, Kandolf Borovšak A, Božič J, Bertoncelj J (2015) Chemical composition and content of free tryptophan in Slovenian bee pollen. J Food Nutr Res 54:323–333
40. Castagna A, Benelli G, Conte G, Sgherri C, Signorini F, Nicolella C, Ranieri A, Canale A (2020) Drying techniques and storage: do they affect the nutritional value of bee-collected pollen? Molecules 25:4925

41. Isik A, Ozdemir M, Doymaz I (2019) Effect of hot air drying on quality characteristics and physicochemical properties of bee pollen. Food Sci Technol 39:224–231
42. Campos MG, Anjos O, Chica M, Campoy P, Nozkova J, Almaraz-Abarca N, Barreto LMRC, Nordi JC, Estevinho LM, Pascoal A et al (2021) Standard methods for pollen research. J Apicult Res 60:1–109
43. Feás X, Pilar Vazquez-Tato M, Estevinho L, Seijas JA, Iglesias A (2012) Organic bee pollen: botanical origin, nutritional value, bioactive compounds, antioxidant activity and microbiological quality. Molecules 17:8359–8377
44. Yücel B, Topal E, Kösoğlu M (2017) Bee products as functional food. In: Waisundara VY, Shiomi N (eds) Superfood and functional food – an overview of their processing and utilization. InTechOpen, pp 15–33
45. Campos MGR, Frigerio C, Lopes J, Bogdanov S (2010) What is the future of beepollen? J ApiProd ApiMed Sci 2:131–144
46. Ulbricht C, Conquer J, Giese N, Khalsa KPS, Sklar J, Weissner W, Woods J (2009) An evidence-based systematic review of bee pollen by the Natural Standard Research Collaboration. J Diet Suppl 6:290–312
47. Lilek N, Kandolf Borovšak A, Bertoncelj J, Vogel-Mikuš K, Nečemer M (2022) Use of EDXRF elemental fingerprinting for discrimination of botanical and geographical origin of Slovenian bee pollen. X-Ray Spectrom 51:186–197
48. Boyer R (2002) Concepts in Biochemistry. Wiley, New York
49. Eggleston G, Finley JW, deMan JM (2018) Carbohydrates. In: Principles of food chemistry, Food science text series. Springer, Cham
50. EFSA (2010) EFSA Panel on Dietetic Products, Nutrition, and Allergies (NDA); Scientific Opinion on Dietary Reference Values for carbohydrates and dietary fibre. EFSA J 8(3):1462, 77 p
51. Chinachoti (1995) Carbohydrates: functionality in foods. Am J Clin Nutr 61:922S–929S
52. SACN – Scientific advisory committee on nutrition (2015) Carbohydrates and health report. The Stationery Office, London
53. Barber TM, Kabisch S, Pfeier AFH, Weickert MO (2020) The health benefits of dietary fibre. Nutrients 12:3209
54. Evans CEL (2020) Dietary fibre and cardiovascular health: a review of current evidence and policy. Proc Nutr Soc 79:61–67
55. Dreher ML (2018) Introduction to dietary fiber. In: Dietary fiber in health and disease. Humana Press, Basel
56. Human H, Nicolson SW (2006) Nutritional content of fresh, bee-collected and stored pollen of *Aloe greatheadii* var. davyana (Asphodelaceae). Phytochemistry 67:1486–1492
57. Kostić AŽ, Barać MB, Stanojević SP, Milojković-Opsenica DM, Tešić LLJ, Šikoparija B, Radišić P, Prentović M, Peršić MB (2015) Physicochemical composition and techno-functional properties of bee pollen collected in Serbia. LWT – Food Sci Technol 62:301–309
58. Fotirić Akšić M, Gašić U, Dabić Zagorac D, Sredojević M, Tosti T, Natić M, Meland M (2019) Chemical fingerprint of 'Oblačinska' sour cherry (*Prunus cerasus* L.) pollen. Biomolecules 9:391
59. Villanueva MTO, Marquina AD, Serrano RB, Abellan GB (2002) The importance of bee collected pollen in the diet: study of its composition. Int J Food Sci Nutr 53:217–224
60. Estevinho ML, Rodrigues S, Pereira AP, Feás X (2012) Portuguese bee pollen: palynological study, nutritional and microbiological evaluation. Int J Food Sci Technol 47:429–435
61. Thakur M, Nanda V (2018) Assessment of physico-chemical properties, fatty acid, amino acid and mineral profile of bee pollen from India with a multivariate perspective. J Food Nutr Res 57:328–340
62. Carpes ST, Mourão GB, Alencar SM, Masson ML (2009) Chemical composition and free radical scavenging activity of *Apis mellifera* bee pollen from Southern Brazil. Braz J Food Technol 12:220–229

63. Serra-Bonvehí J, Escolà-Jordà R (1997) Nutrient composition and microbiological quality of honeybee-collected pollen in Spain. J Agric Food Chem 45:725–732
64. Bobiş O, Marghitas LA, Dezmirean D, Morar O, Bonta V, Chirila F (2010) Quality parameters and nutritional value of different commercial bee products. Anim Sci Biotechnol 67:91–96
65. Szczêsna T (2007) Study on the sugar composition of honeybee-collected pollen. J Apic Sci 51:15–21
66. Martins MCT, Morgano AM, Vicente E, Baggio SR, Rodriguez-Amaya DB (2011) Physicochemical composition of bee pollen from eleven Brazilian states. J Apic Sci 55:107–116
67. Bertoncelj J, Polak T, Pucihar T, Lilek N, Kandolf Borovšak A, Korošec M (2018) Carbohydrate composition of Slovenian bee pollens. Int J Food Sci Technol 53:1880–1888
68. Szczêsna T, Rybak-Chmielewska H, Chmielewski W (2002) Sugar composition of pollen loads harvested at different periods of the beekeeping season. J Apic Sci 46:107–114
69. Kalaycıoglu Z, Kaygusuz H, Doker S, Kolaylı S, Bedia Erim F (2017) Characterization of Turkish honeybee pollens by principal component analysis based on their individual organic acids, sugars, minerals, and antioxidant activities. LWT – Food Sci Technol 84:402–408
70. Bakour M, Laaroussi H, Ferreira-Santos P, Genisheva Z, Ousaaid D, Teixeira JA, Lyoussi B (2022) Exploring the palynological, chemical, and bioactive properties of non-studied bee pollen and honey from Morocco. Molecules 27:5777
71. Laaroussi H, Ferreira-Santos P, Genisheva Z, Bakour M, Ousaaid D, El Ghouizi A, Teixeira JA, Lyoussi B (2023) Unveiling the techno-functional and bioactive properties of bee pollen as an added-value food ingredient. Food Chem 405:134958
72. Oroian M, Dranca F, Ursachi F (2022) Characterization of Romanian bee pollen – an important nutritional source. Foods 11:2633
73. Gardana C, Del Bo C, Quicazan MC, Correa AR, Simonetti P (2018) Nutrients, phytochemicals and botanical origin of commercial bee pollen from different geographical areas. J Food Compos Anal 73:29–38
74. Liu L, Fan X (2013) Tapetum: regulation and role in sporopollenin biosynthesis in Arabidopsis. Plant Mol Biol 83:165–175
75. Stephen AM, Champ MMJ, Cloran SJ, Fleith M, van Lieshout L, Mejborn H, Burley VJ (2017) Dietary fibre in Europe: current state of knowledge on definitions, sources, recommendations, intakes and relationships to health. Nutr Res Rev 30:149–190
76. Taha EA (2015) Chemical composition and amounts of mineral elements in honeybee-collected pollen in relation to botanical origin. J Apic Sci 59:75–81
77. Soares de Arruda VA, Santos Pereira AA, Silva de Freitas A, Marth MO, Almeida-Muradian LB (2013) Dried bee pollen: B complex vitamins, physicochemical and botanical composition. J Food Compos Anal 29:100–105
78. Salazar-González C, Díaz-Moreno C (2016) The nutritional and bioactive aptitude of bee pollen for a solid-state fermentation process. J Apic Res 55:161–175
79. Neufeld L, Hendriks SL, Hugas M (2021) Healthy diet: a definition for the United Nations Food systems summit 2021. A paper from the scientific group of the UN food systems summit. UN, New York
80. Lamothe LM, Lê K-A, Abou Samra R, Roger O, Green H, Ke M (2019) The scientific basis for healthful carbohydrate profile. Crit Rev Food Sci Nutr 59:1058–1070
81. WHO World Health Organization (2015) Guideline: sugars intake for adults and children. World Health Organization, Geneva
82. Wu W, Qiao J, Xiao X, Kong L, Dong J, Zhang H (2021) *In vitro* and *In vivo* digestion comparison of bee pollen with or without wall-disruption. J Sci Food Agric 101:2744–2755
83. Korošec M, Bertoncelj J (2020) The importance of bee products in human nutrition. Acta Agric Slov 115:223–235
84. Bogdanov S (2017) The pollen book: nutrition, functional properties and health. http://www.bee-hexagon.net. Accessed 12 Dec 2012
85. Pi X, Yang Y, Cui Q, Wan Y, Fu G, Chen H, Cheng J (2021) Recent advances in alleviating food allergenicity through fermentation. Crit Rev Food Sci Nutr 62:7255–7268

86. Yin S, Tao Y, Jiang Y, Meng L, Zhao L, Xue X, Li Q, Wu L (2022) A combined proteomic and metabolomic strategy for allergens characterization in natural and fermented brassica napus bee pollen. Front Nutr 9:822033
87. Algethami JS, El-Wahed AAA, Elashal MH, Ahmed HR, Elshafiey EH, Omar EM, Naggar YA, Algethami AF, Shou Q, Alsharif SM et al (2022) Bee pollen: clinical trials and patent applications. Nutrients 14:2858
88. European Commission (2006) Regulation (EC) No 1924/2006 of the European Parliament and of the Council of 20 December 2006 on nutrition and health claims made on foods. Off J Eur Union L404:9–25

Chapter 4
Lipids in Pollen

Aleksandar Ž. Kostić and Sofija Kilibarda

4.1 Introduction

Apart from carbohydrates and proteins, lipids are the main pollen components [1] with significant energetic and nutritional importance. The average lipid content in bee-collected pollen is estimated to be between 1 and 13% [2], depending on the botanical/geographical origin. Lipids in pollen originate from both diploid tapetal cells and haploid vegetative cells [3]. They accumulate not only in the outer layer of pollen grain but also in the inner zones, exhibiting differences in chemical composition and functionality [3]. This chapter provides a comprehensive overview of the different lipid zones within the pollen grain. Specifically, it presents a detailed analysis of the fatty acid composition and volatile profile of various pollen samples, including both floral and bee-collected types. Other lipophilic substances, such as carotenoids and lipophilic vitamins, will be discussed in subsequent chapters of this book.

4.2 Pollen Wall Structure with Lipid Zones

The pollen wall, known as the sporoderm, consists of three separate zones/membranes: exine, intine, and pollen coat [4]. It represents one of the most intricate outer layers among cells of higher plants. One of the most important components of the

A. Ž. Kostić (✉)
University of Belgrade, Faculty of Agriculture, Chair of Chemistry and Biochemistry, Belgrade, Serbia
e-mail: akostic@agrif.bg.ac.rs

S. Kilibarda
University of Belgrade, Faculty of Agriculture, Department of Crop and Vegetable Science, Belgrade, Serbia

N. Ecem Bayram et al. (eds.), *Pollen Chemistry & Biotechnology*,
https://doi.org/10.1007/978-3-031-47563-4_4

exine membrane is sporopollenin, a complex natural biopolymer, that continues to intrigue scientists. From a chemical standpoint, sporopollenin is a polyalcohol with a possible stoichiometry as follows: $C_{90}H_{144}O_{27}$ [5]. However, there is also an opinion that this material is a much more complex cross-linked polymer representing a mixture of different carotenoids, their esters, aliphatic chains (both linear and branched), oxygenated phenolic rings, and phenylpropanoid units [5, 6]. Regardless of how it is defined, sporopollenin contains microcapsules with a hydrophobic nature capable of associating aliphatic moieties and forming conjugated systems with phenolic compounds, carboxylic acids, alcohols or some lactones [5]. Due to its high stability and resistance, this natural biopolymer is easily applicable as a carrier for removing pollutants from environments. This is based on the following characteristics: (a) sporopollenin has high bonding capacity; (b) it possesses appropriate functional groups for further modifications and preparation of different derivatives; (c) it exhibits relatively low swelling capacity and straining properties [5]. Most recently, sporopollenin has been recognized as a possible candidate for developing natural microcapsules as carriers of drugs in pharmaceutical industries [7]. Sporopollenin exine capsules (SECs) are washed out of proteins (to prevent allergic reactions) [7, 8] and can be filled with some hydrophobic or hydrophilic molecules or with probiotic bacteria, which can be targeted for delivery into intestinal mucus or some other receptors in our body [8]. In addition to this, exine is characterized by the presence of different polysaccharides [4].

Apart from sporopollenin and the exine membrane, the pollen grain of entomophilous plants is usually covered with an oily layer known as pollenkitt, enriched with lipophilic substances [4]. The pollenkitt layer is characteristic of all angiosperms, while a similar layer called tryphine, is found exclusively on Brassicaceae pollens [9]. While pollenkitt is entirely hydrophobic and composed of lipids, tryphine also contains hydrophilic substances [9]. Botanically, pollenkitt originates from tapetal elaiosome and spherosome structures found in all angiosperms pollinated by animals, particularly entomophilous plants, but not in gymnosperms [9]. This layer is extremely important for the pollination process as it makes the entomophilous pollen grain sticky and easy to adhere to animal bodies, ensuring the transfer and participation in the pollination of another plant. This is why it is distributed on the surface and throughout the exine. On the other hand, in the case of anemophilous plants, the pollen grain is lighter, less sticky, and the pollenkitt layer is not located on the exine but between the columellae and within the interior of the exine [9]. A significant comprehensive study on the chemical composition of pollenkitt can be found in the research by Dobson [10]. Clear differences in lipid composition were observed between the pollenkitt layer and the internal part of the pollen grain. Specifically, while the pollenkitt contained a complex mixture of various neutral lipids, polar lipids, such as phospholipids, glycolipids, and sulfolipids, were exclusively found in the inner zone, mainly as part of cell membranes [10]. Chromatographic analysis revealed that different hydrocarbons (mostly aliphatic), and non-sterol alcohols were solely found in the pollenkitt layer. However, triglycerides and sterols were predominantly observed in the internal pollen grain fraction. Additionally, different esters of sterols were found in both fractions, but mostly in the pollenkitt layer. Interestingly, free fatty acids and their methyl esters were almost exclusively present in the pollenkitt [10]. It is worth noting that a recent article unveiled a detailed metabolomics profile of bee-collected pollen from rape seed before and after the fermentation process. The profile was determined using UPLC-QToF-MS analysis, and included several lipids, particularly different phospholipids such as choline alfoscerate, LysoPe 16:0, LysoPC 16:0, LysoPE16:1, LysoPE16:3, LysoPC 18:2, LysoPE 18:3, which represented various lysophosphatidyl derivatives [11]. These compounds were most likely liberated from cell membranes. Furthermore, highly volatile substances, such as terpenoids, were almost exclusively found in the pollenkitt [10] contributing to the aroma of pollen. The pollen aroma plays a significant role in communication with pollinators, providing "guidance" for insects to find flowers as a pollen source and aiding in some specific detection by pollen-specialist species [12]. Certain data suggest that pollenkitt lipids are plant species-specific, providing precise information for communication between pollinators and plants [13]. Recent research also indicates that pollenkitt is responsible for the differences observed among plant species in the Malvoideae family with collectable and non-collectable pollen grains [14]. Pollenkitt volatiles also include α-methyl alcohols and ketones, offering protection to the plant's flower from intruders and predators [12]. A detailed volatile profile for some pollens will be discussed in the following subsection.

4.3 Volatile Profile of Pollen

Pollen odor is usually not related to flower aroma and can be significantly different [15]. One of the first reports on the volatile profile of pollen revealed the presence of 31 different compounds in the pollen of *Rosa rugosa* and *R. canina,* including terpenoids, aromatic hydrocarbons and aliphatic hydrocarbons [15]. Interestingly, different profiles were observed despite the species belonging to the same genus. This leads to the conclusion that bees can differentiate between flowers and pollen of similar species based on volatiles. Differences in the chemical volatile profile between pollen and other plant organs were also observed in the case of *Citrus limon*. It was found that pollen did not contain limonene, which was the predominant compound in several plant organs. In total, pollen contained ten different compounds, with sesquiterpenoids, *trans*-nerolidol, and β-caryophyllene being the predominant ones [16]. In an analysis of volatile compounds obtained from stingless bees in Brazil, the compounds were concentrated using SPME and determined via GC/MS. More than 100 components were identified, with hydrocarbons and esters being the most predominant, followed by terpenoids, ketones and alcohols. The primary source of pollen in this study was the plant *Mimosa caesalpiniifolia* Benth., and the most frequently detected compounds were kaur-16-ene, methyl and ethyl hexadecanoate, methyl linoleate, and heneicosane [17]. When comparing different bee products, it was observed that pollen contained the highest number (33) of identified components (HS-SPME/GC/MS) compared to beeswax, bee bread, and honey [18]. Additionally, differences in chemical profiles were also observed. Unlike honey, other bee products contained different hydrocarbons in the highest ratio among the 55 identified compounds [18]. In the case of lotus bee-collected pollen, HS-SPME/GC/MS analysis revealed the presence of 58 volatile compounds, with the following compounds being predominant: acetic M, ethanol M, ethyl acetate and sotolon (lactone) [19]. Based on SPME/GC/MS analysis of Moroccan bee-collected pollen samples, 47 compounds were determined including aldehydes, esters, hydrocarbons, ketones, terpenes, carboxylic acids, and ethers [20]. The most predominant compounds were hexanal and 3,5-octadien-2-one. However, a significant influence of botanical origin was observed. For instance, in the case of a monofloral sample originating from *Coriandrum* plants, 3,5-octadien-2-one and octanal were identified as the most predominant compounds [20].

4.4 Total Lipids and Fatty Acids in Pollen

The determination of total lipid content (TLC) represents one of the basic nutritional parameters for various foodstuffs, including pollen. The lipid content in pollen can vary significantly and is usually in the range of 1 to 13 g/100 g (dry weight) [2]. It is strongly influenced by the botanical and geographical origin of the samples, as well as storage conditions or preparation procedures. For example, the average TLC determined for ten different Brazilian bee-collected pollen samples was 7 g/100 g (dry weight) [21], while for seven samples with different geographical origin, the mean TLC value was found to be 5.39 g/100 g (dry weight) [22]. In a study of 26 bee-collected pollen samples from Serbia with different botanical origins, an average TLC of 2.66 g/100 g (dry weight) was determined, ranging from 1.31 to 6.78 g/100 g (dry weight) with the highest TLC observed in the Brassicaceae monofloral sample and the lowest in the polyfloral sample (*Sophora* and *Helianthus annuus* L.) [23]. In line with this, monofloral bee-collected sunflower pollen from Serbia contained only 1.62 g/100 g of lipids, expressed on dry weight [24]. TLC ranged from 0.91 (*Cistus/Castanea* sp.) to 2.13 (*Hedera* sp.) g/100 g in different monofloral samples, including a *Rubus* sp. (1.31 g/100 g) pollen sample from Italy [25]. A moderate TLC content (5.8 g/100 g) was determined in the case of saffron floral pollen [26]. As mentioned, it has been found that pollen-processing technology can affect TLC. Grinding the pollen sample was found to increase TLC from 10.1 to 11.4 g/100 g, although the obtained results were not statistically significantly different [27].

Apart from determining TLC, one of the most examined qualitative parameters for pollen (both floral and bee-collected) is the analysis of the fatty acids (FAs) profile and its quantitative composition. Besides their importance for the human diet, FAs in pollen are also important for bees. It has been reported that the presence of myristic, linolenic, linoleic, and dodecanoic acids is important in order to prevent the growth of several bee pathogens, while pollens enriched with palmitic and oleic acids are excellent for bee nutrition [28]. A wealth of published data is available from the 1980s and 1990s [29–32] until the present day (Table 4.1). Based on the available and collected data, the number of identified/quantified different FAs in

Table 4.1 Predominant fatty acids determined in different pollen samples

Reference	Pollen type and geographical/botanical origin	Fatty acid		
[29]	Bee-collected (BCP) USA *Prunus dulcis* (Almond)	Linoleic (18:2)	Palmitic (16:0)	Linolenic (18:3)
[27]		Oleic (18:1)	Linolenic (18:3)	Palmitic (16:0)
[31]	BCP Spain No data	Palmitic (16:0)	Linoleic (18:2)	Linolenic (18:3)
[32]	BCP Iberian Peninsula *Castanea sativa* Mill. (Sweet chestnut); *Erica* sp.; *Eucalyptus* sp.; *Helianthemum alyssoides* (Lam.) Vent. (Rockrose); *Quercus robur* L. (Pedunculate oak); *Raphanus raphanistrum* L. (Wild radish)	Palmitic (16:0)	Linoleic (18:2)	Linolenic (18:3)
[33]	BCP Portugal Boraginaceae, Fagaceae, Ericaceae, Cistaceae, Rosaceae, Asteraceae, Mimosaceae, Lamiaceae, Fabaceae	α-Linolenic (18:3)	Linoleic (18:2)	Palmitic (16:0)
[34]	BCP China *Brassica napus* L. (Rapeseed); *Camellia japonica* L. (Japanese camellia); *Dendranthema indicum* L. (Chrysanthemum); *Helianthus annuus* L. (Sunflower); *Rosa rugosa* Thunb. (Beach rose); *Vicia faba* L. (Faba bean); *Zea mays* L. (Maize)	Linolenic (18:3)	Palmitic (16:0)	Oleic (18:1)
[35]	BCP Brazil *Zea mays* L. (Maize)	Oleic (18:1)	Linoleic (18:2)	Arachidic (20:0)
[36]	Floral pollen (FP) Czech Republic *Sambucus nigra* L. (Elder)	Palmitic (16:0)	Oleic (18:1)	Linoleic (18:2)

(continued)

Table 4.1 (continued)

Reference	Pollen type and geographical/botanical origin	Fatty acid		
[37]	BCP South Africa *Aloe greatheadii Schönland var. davyana* (Schönland) Glen & D.S.Hardy (Spotted aloe)	Gadoleic (20:1)	Palmitic (16:0)	Oleic (18:1)
[38]	BCP/FP South Africa *Helianthus annuus* L. (Sunflower)	Lauric (12:0)	Palmitic (16:0)	α-Linolenic (18:3n3)
[39]	BCP Serbia *Cynara scolymus* (Artichoke)	Eicosapentaenoic EPA (20:5n3)	Oleic (18:1)	Pentadecanoic (15:0)
[24]	BCP Serbia *Helianthus annuus* L. (Sunflower)	Stearic (18:0)	α-Linolenic (18:3n3)	Pentadecanoic (15:0)
[40]	BCP Serbia Brassicaceae; Fabaceae; *Fraxinus* sp.; *Helianthus annuus* L. (Sunflower); *Salix* sp.; Rosaceae; *Vitis* sp.; *Plantago* sp.; *Ambrosia* sp.; Apiaceae, *Artemisia* sp.; Asteraceae, *Zea mays* L. (Maize); Rannunuculace; *Tilia* sp.; Moraceae; *Fenestrate* sp.; *Juglans* sp.; Poaceae; *Viola* sp.; *Plantago* sp.; *Sambucus* sp.; Chenopodiaceae; *Rumex* sp.; *Carduus* sp.; *Robinia* sp.; *Sophora* sp.; Cannabaceae	Palmitic (16:0)	Oleic (18:1)	Linoleic (18:2)
[41]	FP Serbia *Zea mays* L. (Maize)	Palmitic (16:0)	Henicosanoic acid (21:0)	Oleic (18:1)
[42]	BCP Italy *Castanea* sp. (chestnut); *Salix* L. (Willow)	Linoleic (18:2)	α-Linolenic (18:3n3)	Palmitic (16:0)

[43]	BCP Turkey *Centaurea* sp.; Brassicaceae; *Castanea sativa* Mill. (Sweet chestnut); *Cistus* sp.	Palmitic (16:0)	Linolenic (18:3)	Oleic (18:1)
[44]	BCP Romania *Crataegus monogyna* Jacq. (Common hawthorn); *Filipendula ulmaria* (L.) Maxim. (Meadowsweet); *Malus domestica* Miller (Apple); *Trifolium repens* L. (White clover); *Prunus* sp.; *Calluna vulgaris* (L.) Hull (Ling); *Calendula officinalis* L. (Pot marigold); *Taraxacum officinale* (L.) Weber (Dandelion); *Anthilis* sp.; *Rosa canina* L. (Dog rose); *Brassica* sp.; *Calluna vulgaris* (L.) Hull (Ling); *Salix* sp.; *Robinia pseudoacacia* L. (Black locust)	Linoleic (18:2)	Palmitic (16:0)	Oleic (18:1)
[45]	BCP Croatia *Taraxacum officinale* (L.) Weber (Dandelion); *Salix* sp.; *Prunus spinosa* L. (Blackthorn); *Aesculus hippocastanum* L. (Horse chestnut); *Prunus mahaleb* L. (Mahaleb cherry); *Quercus pubescens* Willd. (Downy oak); *Filipendula vulgaris* Moench (Dropwort)	α-Linolenic (18:3n3)	Linoleic (18:2)	Palmitic (16:0)
[46]	BCP Brazil Not specified	Palmitic (16:0)	Linoleic (18:2)	α-Linolenic (18:3n3)
[47]	BCP Taiwan *Brassica napus* L. (Rape); *Bidens pilosa* var. *radiata* Scherff (Beggatick); *Camellia sinensis* (L.) Kuntze (Tea tree); *Fraxinus griffithii* C.B.Clarke (Evergreen ash); *Prunus mume* (Siebold) Siebold & Zucc. (Japanese apricot); *Rhus chinensis* var. *roxburghii* (DC.) Rehder. (Nutgall tree); *Bombax ceiba* L. (Red cotton tree); *Hylocereus costaricensis* (F.A.C.Weber) S.Arias & N.Korotkova (Pitaya); *Liquidambar formosana* Hance (Formosan gum); *Nelumbo nucifera* Gaertn. (Lotus); *Zea mays* L. (Maize)	Palmitic (16:0)	Stearic (18:0)	Linoleic (18:2)

(continued)

Table 4.1 (continued)

Reference	Pollen type and geographical/botanical origin	Fatty acid		
[48]	BCP Brazil *Cocos nucifera* L. (Cocos); *Miconia* spp.; *Spondias* spp.; *Eucalyptus* spp.	Linolenic (18:3)	α-Linolenic (18:3n3)	Palmitic (16:0)
[49]	BCP India *Cocos nucifera* L. (Cocos); *Coriandrum sativum* L. (Coriander); *Brassica napus* L. (Rape); *Zea mays* L. (Maize); *Cenchrus americanus* (L.) Morrone (Pearl millet)	α-Linolenic (18:3n3)	Linoleic (18:2)	Arachidic (20:0)
[25]	BCP Italy *Rubus* sp.; *Cistus* sp.; *Castanea* sp.	Oleic (18:1)	Linoleic (18:2)	Margaric (17:0)
[50]	BCP Colombia Not specified	α-Linolenic (18:3n3)	Palmitic (16:0)	Linoleic (18:2)
[51]	FP Russia *Pinus silvestris (Scots pine)*; *P. sibirica (Siberian pine)*; *P. pumila (Siberian dwarf pine)*	Palmitic (16:0)	Stearic (18:0)	Arachidic (20:0)

pollen samples is as follows: 20 saturated (SFA), 12 monounsaturated (MUFA) and 18 polyunsaturated (PUFA) fatty acids. Given the strong differences and the vast quantity of data, only the predominant fatty acids have been extracted and presented in Table 4.1.

It can be seen that among SFAs, the most predominant FAs in the examined pollen samples were palmitic, stearic and arachidic acids. Among MUFAs, oleic acids and gadoleic acids were frequently found in the pollen samples while among PUFAs, the following acids were the most abundant: linoleic and α-linolenic acids. Curiously, a strong predominance of eicosapentaenoic acid (EPA) was observed in the artichoke bee-collected pollen sample from Serbia [39] making it an important source of this rare yet vital nutrient. The influence of both botanical and geographical origin was observed. However, some of the investigated FAs are uncommon in pollen samples. For instance, butyric acid was reported only in the bee-collected pollen samples from Brazil [48] and India [49]. Moreover, long-chain SFAs, such as hexacosanoic (C26:0) and triacontanoic (C30:0) acids were only reported in the floral pollen samples of *Pinus* (Russia) [51] and *Sambucus nigra* L. (Czech Republic) [36] species. Among MUFAs, *cis*-10-pentadecenoic acid was reported only in maize floral pollen samples collected from different hybrids [41] along with elaidic acid (C18:1ω9t) which was also observed exclusively in floral pollen samples obtained from maize [41] and *Pinus* plants [51]. Heneicosenoic acid was only reported for Indian monofloral bee-collected pollen samples originated from coconut, coriander and rapeseed [49] as well as *Sambucus nigra* L. floral pollen (Czech Republic) for octacosenoic long chain fatty acid [36]. Among PUFAs, *trans* fatty acid, linolelaidic acid, was reported only in Serbian floral maize pollen [41] as well as eicosatrienoic acid in bee-collected pollen from Croatia [45], and stearidonic acid for monofloral *Castanea* and *Salix* sp. bee-collected pollen samples from Italy [25]. In addition, one other *trans* FA was reported in rape seed bee-collected pollen from China- palmitelaidic acid, which is a *trans* isomer of palmitoleic acid [11]. The powerful metabolomics analysis applied in this case also revealed the presence of two quite specific FAs in this pollen sample (not included in Fig. 4.1): 2-hydroxylinolenic acid (hydroxyl derivative of linolenic acid) and (10E,12Z)-9-hydroperoxy-10,12-octadecadienoic acid, containing quite specific peroxide O-O bond at C-9 position [11] which is probably a result of peroxidation of the C-9 double bond. In Fig. 4.1, a short overview of all obtained data related to identified/quantified FAs has been made.

There are ideas regarding the application of certain "rare and unusual fatty acids" in the field of chemotaxonomic, as they can be linked with the botanical origin of samples and specific plant families, genera, or species [52–54]. According to the literature data and suggestions, all plants can be classified into two groups, based on the presence of specific FAs: "18:3" and "16:3" plants [55]. In the first case, plants contain linolenic acid as an essential component, while the second group of plants contains at least 2% of *cis*-7,10,13-hexadecatrienoic acid [54, 55]. "18:3" plants are at a higher stage of evolutionary development, whereas "16:3" plants are at a lower stage since this fatty acid is also present in prokaryotic organisms such as cyanobacteria and algae [55]. Based on these suggestions, and with the application of modern

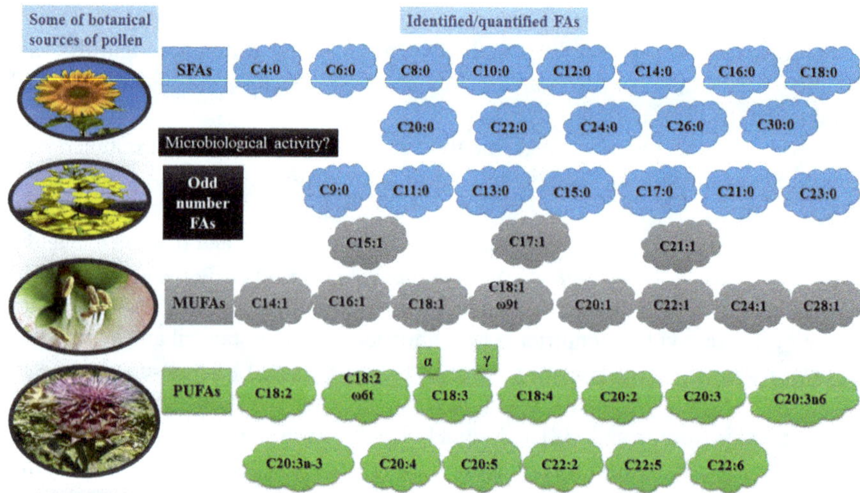

Fig. 4.1 List of identified/quantified fatty acids in floral/bee-collected pollen samples

statistical tools, it may be possible to determine whether some of the "rare fatty acids" present in pollen samples are suitable for chemotaxonomic analysis and the determination of their botanical origin.

The presence of some long-chain highly unsaturated FAs like γ-linolenic acid (C18:3n-6), eiocosatrienoic acid (C20:3n-3) and dihomo-γ-linolenic acid (C20:3n-6) (Fig. 4.1) can be related to microorganism activity [56]. Since plants and mammals do not have the ability to biosynthetize these acids, due to the lack of appropriate enzymes for C- chain elongation, it is possible that the presence of microorganisms from bee's saliva tricks this process [57]. Similarly, some of the observed odd-chain fatty acids (Fig. 4.1) can be associated with microbial activity. Pentadecanoic acid (C15:0) and margaric acid (C17:0) are relatively uncommon in the plant and animal kingdoms [58, 59]. However, for humans, the main sources are dairy products and ruminant microbiome activity during fermentation [60]. Given that fermentation is a common occurrence during pollen collection by bees, as well as during storage and processing, it is possible that these FAs originate from these activities. In the FA biosynthesis process, the acetyl coenzyme A unit is a basic building unit dictating the formation of fatty acids with an even number of C atoms. Conversely, the propionyl unit is quite rare, except in the case of certain microorganisms that enable them to start from this basic unit and biosynthetize odd-chain FAs [61].

4.5 Concluding Remarks

Pollen metabolome research has garnered significant attention in recent decades. Among various aspects, extensive studies have been conducted on pollen lipophilic substances, including fatty acids, non-polar components, which form part of the volatile fraction responsible for communication with the environment. However, these volatiles can also play a crucial role in the sensorial properties of pollen when applied in the food or cosmetic industries. Nonetheless, further research should focus on the application of certain advanced technological tools to enhance bioaccessibility and bioavailability of all metabolites. This is particularly important due to the destruction of both strong pollen membranes, exine and intine, during digestion. Additionally, understanding the fate and properties of these metabolites during the digestive process is essential. Furthermore, the specific and excellent properties of sporopollenin, an important pollen component, are not nearly used enough. Its excellent encapsulation capabilities have begun finding applications in the food sector, enabling targeted delivery of sensitive molecules within our bodies. By leveraging sporopollenin stability, these hydrophobic or hydrophilic compounds can be protected from the effects of enzymes or adverse effects of the environment during digestion. Such advancements will open a new "window" in pollen application in the food industry as functional ingredient.

References

1. Thakur M, Nanda V (2020) Composition and functionality of bee pollen: a review. Trends Food Sci Technol 98:82–106. https://doi.org/10.1016/j.tifs.2020.02.001
2. Campos MGR, Bogdanov S, de Almeida-Muradian LB et al (2008) Pollen composition and standardisation of analytical methods. J Apic Res 47:154–161. https://doi.org/10.1080/00218839.2008.11101443
3. Ischebeck T (2016) Lipids in pollen — they are different. Biochim Biophys Acta (BBA) - Mol Cell Biol Lipid 1861:1315–1328. https://doi.org/10.1016/j.bbalip.2016.03.023
4. Wiermann R, Gubatz S (1992) Pollen Wall and Sporopollenin. Int Rev Citol 140:35–72. https://doi.org/10.1016/S0074-7696(08)61093-1
5. Yaacob SFFS, Jamil RZR, Suah FBM (2022) Sporopollenin based materials as a versatile choice for the detoxification of environmental pollutants — a review. Int J Biol Macromol 207:990–1004. https://doi.org/10.1016/j.ijbiomac.2022.03.206
6. Mackenzie G, Boa AN, Diego-Taboada A et al (2015) Sporopollenin, the least known yet toughest natural biopolymer. Front Mater 2:66. https://doi.org/10.3389/fmats.2015.00066
7. Ageitos JM, Robla S, Valverde-Fraga L et al (2021) Purification of hollow Sporopollenin microcapsules from sunflower and chamomile pollen grains. Polymers (Basel) 13:2094. https://doi.org/10.3390/polym13132094
8. Schouten PJ, Soto-Aguilar D, Aldalbahi A et al (2022) Design of sporopollenin-based functional ingredients for gastrointestinal tract targeted delivery. Curr Opin Food Sci 44:100809. https://doi.org/10.1016/j.cofs.2022.100809
9. Pacini E, Hesse M (2005) Pollenkitt – its composition, forms and functions. Flora – Morphol, Distribut, Funct Ecol Plant 200:399–415. https://doi.org/10.1016/j.flora.2005.02.006
10. Dobson HEM (1988) Survey of pollen and Pollenkitt lipids – chemical cues to flower visitors? Am J Bot 75:170. https://doi.org/10.2307/2443884
11. Zhang H, Zhu X, Huang Q et al (2023) Antioxidant and anti-inflammatory activities of rape bee pollen after fermentation and their correlation with chemical components by ultra-performance liquid chromatography-quadrupole time of flight mass spectrometry-based untargeted metabolomics. Food Chem 409:135342. https://doi.org/10.1016/j.foodchem.2022.135342
12. Dobson HEM, Bergström G (2000) The ecology and evolution of pollen odors. Plant Syst Evol 222:63–87. https://doi.org/10.1007/BF00984096

13. Piskorski R, Kroder S, Dorn S (2011) Can pollen headspace volatiles and pollenkitt lipids serve as reliable chemical cues for bee pollinators? Chem Biodivers 8:577–586. https://doi.org/10.1002/cbdv.201100014
14. Konzmann S, Neunkirchen M, Voigt D et al (2023) Pollenkitt is associated with the collectability of Malvoideae pollen for corbiculate bees. J Pollinat Ecol 32:128–138. https://doi.org/10.26786/1920-7603(2023)754
15. Dobson HEM, Bergström J, Bergström G, Groth I (1987) Pollen and flower volatiles in two *Rosa* species. Phytochemistry 26:3171–3173. https://doi.org/10.1016/S0031-9422(00)82464-4
16. Flamini G, Tebano M, Cioni PL (2007) Volatiles emission patterns of different plant organs and pollen of *Citrus limon*. Anal Chim Acta 589:120–124. https://doi.org/10.1016/j.aca.2007.02.053
17. de Sousa LNJ, Lopes JAD, Moita Neto JM et al (2017) Volatile compounds and palynological analysis from pollen pots of stingless bees from the mid-north region of Brazil. Braz J Pharm Sci 53(2). https://doi.org/10.1590/s2175-97902017000214093
18. Starowicz M, Hanus P, Lamparski G, Sawicki T (2021) Characterizing the volatile and sensory profiles, and sugar content of beeswax, beebread, bee pollen, and honey. Molecules 26:3410. https://doi.org/10.3390/molecules26113410
19. Ni J-B, Bi Y-X, Vidyarthi SK et al (2023) Non-thermal electrohydrodynamic (EHD) drying improved the volatile organic compounds of lotus bee pollen via HS-GC-IMS and HS-SPME-GC-MS. LWT-Food Sci Technol 176:114480. https://doi.org/10.1016/j.lwt.2023.114480
20. Aylanc V, Larbi S, Calhelha R et al (2023) Evaluation of antioxidant and anticancer activity of mono- and polyfloral Moroccan bee pollen by characterizing phenolic and volatile compounds. Molecules 28:835. https://doi.org/10.3390/molecules28020835
21. Almeida-Muradian LB, Pamplona LC, Coimbra S, Barth OM (2005) Chemical composition and botanical evaluation of dried bee pollen pellets. J Food Compos Anal 18:105–111. https://doi.org/10.1016/j.jfca.2003.10.008
22. de Arruda VAS, Pereira AAS, de Freitas AS et al (2013) Dried bee pollen: B complex vitamins, physicochemical and botanical composition. J Food Compos Anal 29:100–105. https://doi.org/10.1016/j.jfca.2012.11.004
23. Kostić AŽ, Barać MB, Stanojević SP et al (2015) Physicochemical composition and techno-functional properties of bee pollen collected in Serbia. LWT – Food Sci Technol 62:301–309. https://doi.org/10.1016/j.lwt.2015.01.031
24. Kostić AŽ, Milinčić DD, Trifunović BDŠ et al (2020) Nutritional and techno-functional properties of monofloral bee-collected sunflower (*Helianthus annuus* L.) pollen. Emir J Food Agric 32(11):768–777. https://doi.org/10.9755/ejfa.2020.v32.i11.2188
25. Sagona S, Pozzo L, Peiretti PG et al (2017) Palynological origin, chemical composition, lipid peroxidation and fatty acid profile of organic Tuscanian bee-pollen. J Apic Res 56:136–143. https://doi.org/10.1080/00218839.2017.1287995
26. Mohamadi Sani A, Hemmati Kakhki A, Moradi E (2013) Chemical composition and nutritional value of saffron's pollen (*Crocus sativus* L.). Nutr Food Sci 43:490–495. https://doi.org/10.1108/NFS-04-2012-0040
27. Chehraghi M, Jafarizadeh-Malmiri H, Javadi A, Anarjan N (2023) Effects of planetary ball milling and ultrasonication on the nutrients and physico–chemical and biological properties of the honey bee pollen. J Food Measur Character 17(4):1–10. https://doi.org/10.1007/s11694-023-01913-9
28. Manning R (2001) Fatty acids in pollen: a review of their importance for honey bees. Bee World 82:60–75. https://doi.org/10.1080/0005772X.2001.11099504
29. Loper GM, Standifer LN, Thompson MJ, Gilliam M (1980) Biochemistry and microbiology of bee-collected almond (*Prunus dulcis*) pollen and bee bread. I- Fatty acids, sterols, vitamins and minerals. Apidologie 11:63–73. https://doi.org/10.1051/apido:19800108
30. Shawer MB, Ali SM, Abdellatif MA, El-Refai AA (1987) Biochemical studies of bee-collected pollen in Egypt 2. Fatty acids and non-saponifiables. J Apic Res 26:133–136. https://doi.org/10.1080/00218839.1987.11100749

31. Serra Bonvehí J, Escolà Jordà R (1997) Nutrient composition and microbiological quality of honeybee-collected pollen in Spain. J Agric Food Chem 45:725–732. https://doi.org/10.1021/jf960265q
32. Saa-Otero MP, Díaz-Losada E, Fernández-Gómez E (2000) Analysis of fatty acids, proteins and ethereal extract in honeybee pollen – considerations of their floral origin. Grana 39:175–181. https://doi.org/10.1080/00173130051084287
33. Estevinho LM, Rodrigues S, Pereira AP, Feás X (2012) Portuguese bee pollen: palynological study, nutritional and microbiological evaluation. Int J Food Sci Technol 47:429–435. https://doi.org/10.1111/j.1365-2621.2011.02859.x
34. Yang K, Wu D, Ye X et al (2013) Characterization of chemical composition of bee pollen in China. J Agric Food Chem 61:708–718. https://doi.org/10.1021/jf304056b
35. Markowicz Bastos DH, Monika Barth O, Isabel Rocha C et al (2004) Fatty acid composition and palynological analysis of bee (*Apis*) pollen loads in the states of São Paulo and Minas Gerais, Brazil. J Apic Res 43:35–39. https://doi.org/10.1080/00218839.2004.11101107
36. Stránský K, Valterová I, Fiedler P (2001) Nonsaponifiable lipid components of the pollen of elder (*Sambucus nigra* L.). J Chromatogr A 936:173–181. https://doi.org/10.1016/S0021-9673(01)01313-9
37. Human H, Nicolson SW (2006) Nutritional content of fresh, bee-collected and stored pollen of *Aloe greatheadii* var. *davyana* (Asphodelaceae). Phytochemistry 67:1486–1492. https://doi.org/10.1016/j.phytochem.2006.05.023
38. Nicolson SW, Human H (2013) Chemical composition of the 'low quality' pollen of sunflower (*Helianthus annuus*, Asteraceae). Apidologie 44:144–152. https://doi.org/10.1007/s13592-012-0166-5
39. Kostić AŽ, Milinčić DD, Nedić N et al (2021) Phytochemical profile and antioxidant properties of bee-collected artichoke (*Cynara scolymus*) pollen. Antioxidants 10:1091. https://doi.org/10.3390/antiox10071091
40. Kostić AŽ, Pešić MB, Trbović D et al (2017) The fatty acid profile of Serbian bee-collected pollen–a chemotaxonomic and nutritional approach. J Apic Res 56:533–542. https://doi.org/10.1080/00218839.2017.1356206
41. Kostić AŽ, Mačukanović-Jocić MP, Špirović Trifunović BD et al (2017) Fatty acids of maize pollen – quantification, nutritional and morphological evaluation. J Cereal Sci 77:180–185. https://doi.org/10.1016/j.jcs.2017.08.004
42. Conte G, Benelli G, Serra A et al (2017) Lipid characterization of chestnut and willow honeybee-collected pollen: impact of freeze-drying and microwave-assisted drying. J Food Compos Anal 55:12–19. https://doi.org/10.1016/j.jfca.2016.11.001
43. Mayda N, Özkök A, Ecem Bayram N et al (2020) Bee bread and bee pollen of different plant sources: determination of phenolic content, antioxidant activity, fatty acid and element profiles. J Food Measur Character 14:1795–1809. https://doi.org/10.1007/s11694-020-00427-y
44. Mărgăoan R, Mărghitaş LAl, Dezmirean DS et al (2014) Predominant and secondary pollen botanical origins influence the carotenoid and fatty acid profile in fresh honeybee-collected pollen. J Agric Food Chem 62:6306–6316. https://doi.org/10.1021/jf5020318
45. Primorac L, Bilić Rajs B, Gal K et al (2023) The specificity of monofloral bee pollen fatty acid composition from Croatia and its nutritional value. J Central Eur Agric 24:104–114. https://doi.org/10.5513/JCEA01/24.1.3784
46. Melo BKC de, Silva JA da, Gomes RD da S, et al (2023) Physicochemical composition and functional properties of bee pollen produced in different locations. Braz J Food Technol 26:e2022006. doi:https://doi.org/10.1590/1981-6723.00622
47. Hsu P-S, Wu T-H, Huang M-Y et al (2021) Nutritive value of 11 bee pollen samples from major floral sources in Taiwan. Foods 10:2229. https://doi.org/10.3390/foods10092229
48. Araújo J, Chambó E, Costa M et al (2017) Chemical composition and biological activities of mono- and heterofloral bee pollen of different geographical origins. Int J Mol Sci 18:921. https://doi.org/10.3390/ijms18050921

49. Thakur M, Nanda V (2018) Assessment of physico-chemical properties, fatty acid, amino acid and mineral profile of bee pollen from India with a multivariate perspective. J Food Nutr Res 57:328–340
50. Fuenmayor BC, Zuluaga DC, Díaz MC et al (2014) Evaluation of the physicochemical and functional properties of Colombian bee pollen. Rev MVZ Cordoba 19:4003–4014. https://doi.org/10.21897/rmvz.120
51. Erdyneeva SA, Shiretorova VG, Tykheev ZA, Radnaeva LD (2021) Fatty acid composition of pollen from *Pinus sylvestris*, *P. sibirica*, and *P. pumila*. Chem Nat Compd 57:741–742. https://doi.org/10.1007/s10600-021-03462-3
52. Aitzetmüller K (1993) Capillary GLC fatty acid fingerprints of seed lipids—a tool in plant chemotaxonomy? J High Resolut Chromatogr 16:488–490. https://doi.org/10.1002/jhrc.1240160809
53. Bağci E, Bruehl L, Aitzetmuller K, Altan Y (2003) Chemotaxonomic approach to the fatty acid and tocochromanol content of *Cannabis sativa* L. (Cannabaceae). Turk J Botan 27:7. https://journals.tubitak.gov.tr/botany/vol27/iss2/7
54. Tsydendambaev VD, Christie WW, Brechany EY, Vereshchagin AG (2004) Identification of unusual fatty acids of four alpine plant species from the Pamirs. Phytochemistry 65:2695–2703. https://doi.org/10.1016/j.phytochem.2004.08.021
55. Mongrand S, Bessoule J-J, Cabantous F, Cassagne C (1998) The C16:3\C18:3 fatty acid balance in photosynthetic tissues from 468 plant species. Phytochemistry 49:1049–1064. https://doi.org/10.1016/S0031-9422(98)00243-X
56. Alonso DL, Maroto FG (2000) Plants as 'chemical factories' for the production of polyunsaturated fatty acids. Biotechnol Adv 18:481–497. https://doi.org/10.1016/S0734-9750(00)00048-3
57. Gill I, Valivety R (1997) Polyunsaturated fatty acids, part 1: occurrence, biological activities and applications. Trends Biotechnol 15:401–409. https://doi.org/10.1016/S0167-7799(97)01076-7
58. Řezanka T, Sigler K (2009) Odd-numbered very-long-chain fatty acids from the microbial, animal and plant kingdoms. Prog Lipid Res 48:206–238. https://doi.org/10.1016/j.plipres.2009.03.003
59. Jenkins B, West J, Koulman A (2015) A review of odd-chain fatty acid metabolism and the role of pentadecanoic acid (C15:0) and heptadecanoic acid (c17:0) in health and disease. Molecules 20:2425–2444. https://doi.org/10.3390/molecules20022425
60. Vlaeminck B, Fievez V, Cabrita ARJ et al (2006) Factors affecting odd- and branched-chain fatty acids in milk: a review. Anim Feed Sci Technol 131:389–417. https://doi.org/10.1016/j.anifeedsci.2006.06.017
61. Qin N, Li L, Wang Z, Shi S (2023) Microbial production of odd-chain fatty acids. Biotechnol Bioeng 120:917–931. https://doi.org/10.1002/bit.28308

Chapter 5
Macro-, Micro-, Trace, and Toxic Elements of Pollen

Pawel Pohl, Anna Dzimitrowicz, Piotr Jamroz, Anna Lesniewicz, Anna Szymczycha-Madeja, Maja Welna, and Krzysztof Greda

For years, herbalists have considered bee pollen as an extremely nutritious food and even still claim that it can be a remedy for some health problems. Unique properties of bee pollen are related to its chemical composition that includes carbohydrates, proteins, lipids, fiber, vitamins, and minerals, represented by different groups of elements, *i.e.,* macro-, micro-, and trace. Particularly, the presence of various physiologically important biogenic elements makes it an essential dietary supplement that may supplement deficiencies of elements in the daily human diet. On the other hand, the information on the content of selected elements causes, that the role of bee pollen goes also beyond its nutrition and supplementation effect and can be widespread for the indication of environmental pollution. Therefore, this chapter surveys the relevant literature devoted to the elemental composition of bee pollen, moreover, coming from different countries. The significance of bee pollen in human nutrition, illustrated by contributions of its intake to recommended dietary allowances (RDAs), is detailed. Additionally, the application of bee pollen for bioindication purposes (to monitor the occurrence of hazardous elements in the environment) as well as its role in the classification and/or the discrimination of samples is also highlighted.

P. Pohl (✉) · A. Dzimitrowicz · P. Jamroz · A. Lesniewicz · A. Szymczycha-Madeja
M. Welna · K. Greda
Faculty of Chemistry, Department of Analytical Chemistry and Chemical Metallurgy, Wroclaw University of Science and Technology, Wroclaw, Poland
e-mail: pawel.pohl@pwr.edu.pl

N. Ecem Bayram et al. (eds.), *Pollen Chemistry & Biotechnology*,
https://doi.org/10.1007/978-3-031-47563-4_5

5.1 Introduction

Bee pollen is a product of the agglutination of pollen grains collected by bee workers within their flight. These insects gather pollen grains from blooming plants by electrostatic attractions when visiting flowers of these plants to search for forages and other food supplies critical for colony development [1]. Using moistening combs and hairs of their hind legs as well as flower secretions and nectar (as a kind of a binder), bee workers form characteristic pollen pellets. Then, they stack pellets in the form of baskets on these hind legs, and transport pollen to beehive [1–3]. In the beehive, pollen pellets are unloaded and other workers, using saliva, mix them with nectar and/or honey, and stock in wax hexagonal brood cells. Cells are closed with wax, where pollen is fermented to produce bee bread, being a primary food source for the bee colony [4]. When beekeepers use special bottom-fitting pollen traps at the entrance to beehives, pollen pellets drop down from the hind legs of bees to trap-trays [3].

Bee pollen has a distinct chemical composition, including carbohydrates (18–84%), proteins (4–41%), lipids (0.4–14%), fibers (0.2–31%), and minerals (the content of ash changing from 0.5 to 8%). In addition, different valuable biologically active substances of plant origin are recognized to be present in bee pollen, including vitamins, carotenoids, and polyphenols [1, 2, 5, 6]. In the latter case, these secondary plant metabolites, *i.e.*, anthocyanins, flavonoids, flavonols, flavonones, and phenolic acids, give bee pollen a special biological activity in addition to unique regulatory properties, counting antioxidant, antibacterial, anti-inflammatory, anticarcinogenic, and antiallergic [4, 5]. For that reason bee pollen can be regarded as both i) a natural bee product for medicinal and therapeutic uses as well as ii) a natural food ingredient for food production and other manufacturing and processing uses [1–3]. Especially health-promoting and therapeutic properties of bee pollen are essential and unique, which increases the scientific interest in this bee product. Simultaneously, this goes hand in hand with numerous practical applications in pharmaceutical and food industries. Served in the form of tablets, capsules, powders, or as received, bee pollen products are particularly recommended as dietary supplements. In this case, their nutritional value certainly improves the body condition and health. However, regardless of how they are served, they must meet relevant hygienic, sanitary, and safety requirements [4]. In addition, bee pollen also contains various biogenic elements, including macro-, micro-, and trace elements, which levels are certainly higher than those established in bee honey [4, 7]. A primary role of the elements contained in bee pollen (also linked with its general chemical composition) is the nutrition and the nutritional supplementation of deficiencies of elements in the daily human diet [3, 7]. On the other hand, considering the primary source of elements in bee pollen and the fact that their content in this bee product is fairly constant and mostly related to some external conditions, the role of bee pollen certainly goes beyond its nutrition and supplementation importance [3]. Accordingly, the information on the content of selected elements in bee pollen is important and often applied for detecting and monitoring environmental pollution on a local scale, mostly limited to a specific area around the apiary. For similar reasons, exogenously and endogenously present elements in bee pollen are regarded as discriminators of its botanical and geographical provenience and used to compare and/or distinguish samples of this bee product of different origins.

The present chapter is devoted to the elemental composition of bee pollen and surveys the relevant literature published in this regard in more than two last decades. Important sources of different groups of elements, *i.e.*, macro-, micro-, and trace elements, are detailed, while the content of these individual groups of elements coming from different countries is given as well. The role of bee pollen as a functional food product and a supplement in the daily diet is illustrated by contributions of its intake to recommended dietary allowances (RDAs) of various physiologically and nutritionally important elements. The application of bee pollen for the indication of a contamination state that is a result of environmental pollution with hazardous elements in a territory close to the beehive or its use for the classification and/or the discrimination of samples is also highlighted.

5.2 Different Sources of Elements in Bee Pollen

The concentration of elements in different types of bee pollen changes a lot, and scientists dealing with bee products agree that this is the effect of both the geographical origin of flowering plants from which pollen is collected (extrinsic conditions) and the biological provenience of these plant species (intrinsic conditions) [7–31].

The most abundant macroelement in bee pollen is K, which concentration varies from 800 to 38,000 mg/kg. Other macroelements are Ca (100–6500 mg/kg), Mg (200–3600 mg/kg), P (1400–9200 mg/kg), and S (40–3600 mg/kg), although the two latter elements are rarely determined [3]. Such elements as Fe (3–520 mg/kg), Zn (0.1–340 mg/kg), Na (0.1–2400 mg/kg), Al (0.1–110 mg/kg) and Mn (0.1–370 mg/kg) are the most abundant microelements. Tables 5.1 (K, Ca, Mg, P, and S) and 5.2 (Fe, Zn, Na, Al, and Mn) give concentrations of both mentioned groups of elements in bee pollen coming from different countries all over the world. The content of these elements is quite differentiated even within the same country, indicating the effect of extrinsic and intrinsic conditions on the final content of elements in this bee product. Such elements as Ba, Cu, Ni, and Sr are also regarded as microelements (micronutrients). Finally, trace elements and contaminants of bee pollen and bee bread are As [30], Ba [19], Cd [7, 14, 19, 30], Co [7, 30], Cr [7, 30], Hg [30], Li [19], Ni [14], Pb [14, 19, 30], Sb [30], and V [19]. Sometimes Al, Ba, and Ni are included in the latter group of elements too [30]. Concentrations of other, less frequently determined microelements as well as trace elements, *i.e.*, As, Ba, Cd, Co, Cr, Cu, Ni, Pb, Sb, Sr, and V, are given in Tables 5.3 (European countries) and 5.4 (African, American and Asian countries).

The variation in the concentration of elements present in bee pollen, particularly of macro- (K, Ca, Mg, P, S) and micro- (Fe, Zn, Na, Al, Mn) elements, should be

Table 5.1 Concentrations (in g/kg) of macroelements (K, Ca, Mg, P and S) in bee pollen and bee bread originating from different countries

	K	Ca	Mg	P	S
Africa					
Nigeria					
[32]	0.8–0.9	0.1	0.04–0.07		
America					
Argentina					
[33]	4.2–7.1	0.8–1.7			
Brazil					
[8]	3.4–9.8	0.9–4.1	0.6–2.4		
[9]	4.8–5.4	0.8–1.2	0.7–0.8	6.7–7.1	
[10]	6.4–3.7	0.3–2.2	0.5–1.2		
[19]		0.6–0.7		4.4–5.2	
[31]	1.4–9.9	0.8–4.7	0.4–3.6	2.2–8.2	
[34]	5.9–13	1.9–3.4			
Asia					
China					
[17]	2.4–8.2	0.8–3.0	0.3–2.8	2.1–9.6	
[24]	4.2–6.0	0.2–1.0	1.1–1.9		
Jordan					
[11]			0.6–1.6		
Turkey					
[12]	1.0–2.9	0.5–1.5	0.3–1.2		
[35]	2.4–4.9	0.9–2.4	0.6–1.1		
[36]	3.8–5.8	1.0–2.4	0.7–1.2	3.0–5.3	1.7–2.2
Europe					
Greece					
[16]	2.7–12	0.4–4.5	0.4–2.3	1.4–9.2	
[37]		0.5–2.5	0.4–1.5		
Lithuania					
[27]	2.9–3.5	1.2–2.5	0.6–1.0	3.5–4.5	
Poland					
[14]			1.3–2.6		
[24]	2.8–4.8	0.5–1.1	0.7–1.7		
[38]	0.4–0.8	0.3–1.8	0.4–1.5	2.2–7.6	0.04–3.6
[39]	0.7	0.8–1.1	0.5–0.9		
Romania					
[21]	2.0–4.3	0.3–4.3	0.3–1.5		
[22]	2.5–7.6	0.6–2.8	0.2–3.6		
[23]	3.2–5.4	1.4–2.6	0.4–1.0		
[25]	3.6–3.9	0.8–1.3	0.6–2.3		
Serbia					
[7]	2.5–4.2	0.9–2.1	0.5–1.0		

(continued)

Table 5.1 (continued)

	K	Ca	Mg	P	S
[40]	6.5–7.4	3.3–6.5	2.4–2.7		
Slovakia					
[41]		1.8–2.0			
Spain					
[15]	5.0–5.4	0.5–0.8	0.4		
[42]	2.9–6.4	0.2–0.8	0.3–0.8		
Other					
Australia					
[20]	2.2–38	0.4–3.1	0.2–2.7		
Philippines					
[28]		3.8–5.2	1.8–2.4		

Table 5.2 Concentrations (in mg/kg) of microelements (Fe, Al, Zn, Na, and Mn) in bee pollen and bee bread originating from different countries

	Fe	Al	Zn	Na	Mn
Africa					
Egypt					
[44]	20–96		0.2–42		
Nigeria					
[32]	2.6–4.3	0.1–0.2	0.1–3.0	0.1	0.1–0.4
America					
Argentina					
[33]	59–122		27–72		14–37
Brazil					
[8]	46–1180		30–101	20–374	25–215
[9]	60–87		45–55	191–215	43–74
[10]	28–203		36–82	13–636	19–129
[19]	58–312		29–43		28–62
[29]	6.6–66		14–68		5.7–178
[30]		10–315			
[31]	11–552		5.1–76	<0.004–1470	12–211
[34]	16–34		36–71	ND	35–75
[45]			51–90		34–86
[46]	205–973	68–836	27–51		25–80
Chile					
[47]	147–175		63–64		52–63
[48]	43–76		82–87		42–63
Asia					
China					
[17]	75–208	94–218	28–65	274–846	8.7–357
[24]	59–182		24–38	1070–2450	13–430
Jordan					
[11]			25–77		

(continued)

Table 5.2 (continued)

	Fe	Al	Zn	Na	Mn
Turkey					
[12]	29–725		15–39		8.2–201
[35]	45–161	21–108	26–50	5500–6220	12–40
[36]	62–129	27–89	29–162	62–273	11–59
[49]	34–125	12–108	10–24		12–40
Europe					
Finland					
[50]	44–120		29–49		21–110
Greece					
[16]	32–230		24–90	102–633	11–152
[37]	49–199		41–94		14–64
Lithuania					
[27]	46–67		20–28	34–46	18–31
Poland					
[14]	106–139		75–159		
[24]	40–136		26–54	293–2190	13–60
[38]	24–229		23–56	40–202	7.1–367
[39]	33–40		27–31	24–138	14–25
Romania					
[22]	19–135		19–60		
[23]	27–123		32–40		
[25]	38–62		27–35		
[29]	22–151		20–60		
[51]	21–123		22–58		21–134
[52]	21–123		22–58		21–134
Serbia					
[7]	40–141	8.5–112	29–76	5.0–55	14–92
[40]	104–122		44–50		30–101
Slovakia					
[41]			37		
Spain					
[15]	20–30		16–20	57–65	17–38
[42]	11–80		19–81	614–2380	7.4–18
[43]					47–131
Other					
Australia					
[20]	14–520		16–340	16–480	5–110
Philippines					
[28]	70–142		45–56		43–267

Table 5.3 Concentrations (in mg/kg) of selected micro- and trace elements (As, Ba, Cd, Co, Cr, Cu, Ni, Pb, Sb, and Sr) in bee pollen and bee bread originating from different European countries

Europe								
Bulgaria	Finland	France	Greece	Italy	Lithuania	Poland	Serbia	Spain
As								
						<0.02–1.52 [18, 53]		0.05–0.54 [43]
Ba								
					0.65–1.06 [27]		0.38–2.64 [7]	
Cd								
0.02–0.03 [13]	0.01–0.15 [50]		<0.5 [37]	<0.02–15.40 [54, 55]	0.007–0.35 [27]	0.001–0.75 [14, 18, 39, 53]	0.03–0.23 [7]	0.06–0.26 [43]
Co								
					ND-0.10 [27]	0.2–0.4 [39]	0.01–0.10 [7]	
Cr								
			<0.3 [37]	<0.002–0.24 [54, 55]	ND [50]	0.09–0.95 [18, 39]	0.17–0.46 [7]	
Cu								
	6.8–19.0 [50]		4–22 [16, 37]		ND-5.90 [27]	1.8–23.9 [24, 38, 39]	4.4–10.7 [7]	4.1–15.7 [15, 42]
Ni								
						0.004–3.3 [14, 39]	0.23–2.23 [7]	
Pb								
0.45–0.49 [13]	<0.2–0.37 [50]	0.004–0.80 [56]	<0.5 [37]	<0.02–12.17 [54, 55]	0.24–0.38 [27]	0.007–3.90 [14, 18, 39, 53]		
Sb								
								<0.01 [43]
Sr								
					0.73–5.37 [27]		0.67–3.37 [7]	

ND Not detected

Table 5.4 Concentrations (in mg/kg) of selected micro- and trace elements (As, Ba, Cd, Co, Cr, Cu, Ni, Pb, Sb, and Sr) in bee pollen and bee bread originating from different African, American and Asian countries

Africa		America			Asia		
Egypt	Nigeria	Argentina	Brazil	Chile	China	Jordan	Turkey
As							
			<0.01–1.83 [30, 46, 57]		2.2–14.7 [17]	<0.02 [11]	0.006–8.5 [12, 49]
Ba							
			0.32–17.63 [19, 30, 46]				
Cd							
0.03–1.38 [44]			<0.001–0.23 [19, 30, 34, 46]	0.02–0.43 [47, 48]	0.15–0.49 [17]	<0.005 [11]	ND-0.18 [12, 36, 49]
Co							
			<0.01–1.22 [30, 46]		0.20–0.54 [17]		ND [49]
Cr							
			<0.01–42 [19, 30, 46]	0.01–0.51 [47, 48]	ND [17]		0.13–7.94 [12, 35, 36, 49]
Cu							
1.0–24.2 [44]	0.4–0.5 [32]	11.8–26.0 [33]	0.1–25.4 [9, 10, 19, 31, 34, 45, 56]	8.8–12.5 [47, 48]	3.2–25.1 [17, 24]	0.03–11.4 [11]	2.7–15.0 [12, 35, 36, 49]
Ni							
			0.02–6.85 [30, 46]		0.53–3.97 [17]	<0.01–2.81 [11]	0.002–2.56 [12, 35, 36, 49]
Pb							
	ND [32]		0.001–18.2 [19, 30, 45, 46, 57]	0.02–0.23 [47, 48]	ND-1.37 [17, 24]	<0.03–2.57 [11]	ND-0.60 [12, 36, 49]
Sb							
			<0.04–1.33 [30, 46]				
Sr							
							1.37–8.89 [35]

ND Not detected

seen in changes in the chemical composition and mineralogy of the soil on which flowering plants are grown and uptake chemical compound through their root systems. Many researchers believe so [7–10, 12, 14, 16, 19, 24–27, 29, 34, 35, 42] and connect this phenomenon with the so-called "effect of the geographical origin", where the location of a beehive, directly related to the soil geochemistry and growing conditions at the territory at which worker bees collect different food sources, significantly affect the concentration and the type of elements species available from the soil to flowering plants. However, even more important than the geographical origin of flowering plants and the overall growth and breeding conditions of these plants on their element content is the effect of the botanical origin of pollen grains (that are stuck to form bee pollen pellets) [7–9, 16, 19, 24–27, 34, 35, 41, 42, 44]. The natural seasonality of certain flowering and pollen-donating plants is closely related to their botanical origin [25]. In this case, at a given geographical location and flowering time, certain plants are available for bees which, depending on the plant species distribution, can bear monofloral (80% of leading pollen grains according to [58]) or multifloral pollen to the hive. Considering this inherent relationship between the location site of the beehive and botanical conditions of plants growing at a certain site, multifloral bee pollen, is not a good material to study the effect of the geographical or the botanical origin on the element content of this bee product [16]. The element composition of multifloral bee pollen pellets is more or less an average of the content of pollen grains of certain plant species that grow in a given area [7, 16].

When pollen sources of one plant species are collected by bees and then the resulting monofloral bee pollen is investigated, it is possible to assess the effect of the geographical origin on its element composition, and hence, to comprehensively classify and/or distinguish relevant bee pollen samples originating from different places. Indeed, such bee pollen has a more consistent element composition [7], which can vary a lot when it comes from different places, particularly in reference to macro- and microelements [16]. In this case, differences in the elemental composition of monofloral bee pollen are related to variances in the soil type, climatic conditions, and growth conditions only. Additionally, the element analysis of different monofloral pollen sources shows that the floral origin of pollen-donating plants, including their type and taxonomic specificity, has a higher impact on the element composition of pollen pellets than factors related to the soil composition and its characteristics [12, 16, 20–23, 43, 45]. This is likely because certain flowering plants selectively uptake the compounds of elements from the soil solution, transport them by the root system to above-ground parts of the plant and accumulate them in various organs, *i.e.*, stems, leaves, and flowers [7, 14, 22, 23, 45, 49]. In addition, flowering plants have a dissimilar tolerance to high concentrations of elements species present in the soil and a different ability to uptake them through the root system and next distribute them inside plant tissues [45]. One example of this behavior is described in ref. [45]. It appears that a certain flowering plant can accumulate relatively high amounts of Pb and Zn in the roots and aerial parts of this plant despite relatively low levels of these two elements in the soil [45]. An opposite phenomenon can be observed for other elements. For example, although a relatively high concentration of Mn is present in the soil, the transport rate of the element to the roots and aerial parts of this plant is quite low [45]. This example certainly proves that flowering plants can specifically uptake elements species from the soil. Very convenient are such flowering plants that preferentially translocate selected elements from the soil to their root systems to a very high degree but later on, translocate them from roots to stems, leaves, and flowers (and pollen) to a much lower extent [45].

A high variation in the concentration of elements, observed for bee pollen samples coming from a given territory or region, can also point out that, except for the predominating effect of the botanical origin of flowering plants, bees can enter different interactions before they carry collected pollen pellets to the hive [7, 23]. Since bee workers moisture then pollen grains with nectar to prepare pollen pellets before their transportation, a part of the resultant element composition of the bee pollen material can also originate from nectar [7, 20, 23].

Finally, except for the direct effect of the botanical origin of flowering plants, in which pollen grains are included in gathered pollen pellets, and the indirect effect of the geographical origin, the subsequent handling of bee pollen by beekeepers, *e.g.*, drying, freezing, lyophilizing, has also influence on its element composition [7, 12, 14, 15, 34].

The content of selected elements, *e.g.*, Ca, Fe, K, and Mg, is higher in bee bread than in the respective bee pollen [40]. This is likely due to the addition of honey and bee secretions during its transformation into brood cells. This trend, however, is not observed in all works, in which both bee products originating from the same apiary are sampled and analyzed [25, 27].

5.3 Role of Elements in Bee Pollen

5.3.1 Nutrition and Supplementation

It is recognized that bee pollen is of high quality and safety when it contains no trace elements that are documented to be harmful to human health, or these elements are present at acceptably low levels. These elements are Al, As, Ag, Cd, Co, Cr, Ge, Hg, Li, No, Pb, Sb, Sr, and V. In the case of a few of them, similarly as for other bee products, allowable concentrations are established and regulated by the low, *i.e.*, 0.03 mg/kg for Cd, 0.01 mg/kg for Hg, 0.5 mg/kg for Pb [58]. For others, permissible limits given for other food products should be taken into consideration. In general, the presence of some trace elements in bee pollen above their allowable limits may suggest that the area from which it comes can be exposed to environmental pollution. In this case, the quality of bee pollen should be of particular interest and significance for potential commercial buyers and consumers because this bee product should carry no health risks. A regular uptake of bee pollen should not be associated with a high contribution of elements listed above to their provisional weakly tolerance intakes (PWTIs), especially of As, Al, Cd, and Pb [17, 18, 30, 49, 53]. Although few or dozen percent contributions to PWTIs of selected elements, *e.g.*, 5–8% in the case of As [30], 19–27% in the case of Al [30], reached through the consumption of bee pollen portions, seem to be relatively safe, in a long term, the uptake of such bee pollen can be dangerous and may lead to the intoxication of the human body. It is also established that the increased concentrations of some harmful elements in bee pollen are responsible for its decreased antioxidant activity [48], likely due to the complexation of these elements by polyphenolic species. All quoted facts undoubtedly show how important the bee pollen analysis is; it should be routinely carried out to verify its quality, safety, and nutritional value. This is especially justified because this bee product is willingly applied to produce pharmaceuticals and dietary supplements, or is treated as functional food itself [12, 28, 30, 35, 48].

Considering the presence of various macro- and microelements (see their concentrations listed in Tables 5.1 and 5.2), bee pollen can be indeed regarded as an important food product, which element composition makes it quite important in human nutrition and supplementation. This is pointed out by many researchers [8–10, 14, 15, 17, 19, 21, 22, 24, 27–29, 31, 32, 36, 41, 58]. In reference to relatively high levels of Ca, Cu, Fe, Mg, P, and Zn, as well as Cr, Cu, Mn, Mo, and Se, it appears that a regular intake of bee pollen (20–40 g per day) can vastly contribute to RDIs of these elements (see Table 5.5).

That is why bee pollen may be treated as a valuable food supplement for all those, who need a balanced diet with a special relation between some elements, *e.g.*, a high K concentration over the Na concentration, or a high Cu concentration over the Zn concentration [8–11, 14–17, 19, 24, 27, 31, 34–36, 41, 42]. This unique element composition decides that it is a really valuable food and nutrition product [9–12, 16, 17, 31, 34, 36, 42]. Nevertheless, it should be considered that the sandwich structure of pollen grains, where the inner layer, primarily consisting of cellulose and pectins, and the outer layer, being composed of a highly decay-resistant compound, *i.e.*, sporopollenin, are distinguished, can hinder the bioaccessibility of elements [59]. The *in vitro* gastro-intestinal digestion study, mimicking conditions of the uptake of elements species in the digestive track,

Table 5.5 Realization (in %) of recommended daily intakes (RDIs) for selected elements through consumption of bee pollen dozes (20–40 g/day)

Macroelements	
Ca	up to 2% [12, 19], 4% [16], 6% [31], or 15% [8]
K	up to 1% [12], 2% [16], 3% [31], or 5% [8]
Mg	up to 6% [12], 7% [16], 11% [32], or 28% [8]
P	up to 8% [16], 9% [12], 17% [31], or 16–19% [19]
Microelements and Tracelements	
Cr	up to 36% [12], or 132–3020% [19]
Cu	up to 9% [16], 27% [31], 28% [12], 19–36% [19], or 100% [8]
Fe	up to 3.9% [29], 15% [9, 10, 31, 42], 20% [16], 36% [12], 10–56% [19], or 100% [8]
Mn	up to 28% [16], 32% [12], 30–67% [19], 70% [30], 72% [29], or 100% [8]
Mo	11–258% [19]
Se	up to 38% [31], or 46% [12]
Zn	up to 10% [12, 16], 10–15% [19], 15% [9, 10, 29, 42], 17% [31], or 77% [8]

shows that the bioaccessible fraction of selected elements (Ca, Cu, Fe, Mg, Mn, and Zn) is lower than 100%, hence, does not correspond to their total concentrations in bee pollen [60]. Accordingly, the contribution of this bioaccessible fraction, as determined based on the enzymatic digestion of bee pollen, varies in the following ranges: 41.8–95.3% (Ca, the average value of 70.4%), 33.6–55.4% (Cu, the average value of 42.7%), 21.1–59.0% (Fe, the average value of 37.7%), 76.2–99.1% (Mg, the average value of 84.8%), 19.8–51.2% (Mn, the average value of 33.9%), and 13.3–41.1% (Zn, the average value of 26.8%) [60]. The administration of bee pollen is more efficient when a bee pollen portion is poured with warm water and left to swallow overnight. In this case, the overall release of Ca, Cu, Fe, Mg, Mn, and Zn from such pretreated bee pollen portion, enzymatically digested latter on, increases by 19% (for Ca and Cu), 18% (for Fe), 14% (for Mg), 30% (for Mn) and 24% (for Zn) [60]. Water, resulting from this treatment, should be administrated as well because it contains relatively high concentrations of elements (in reference to their total concentrations in the treated bee pollen), *i.e.*, 25.2–54.5% (for Ca), 5.9–13.1% (for Cu), 1.6–25.0% (for Fe), 38.4–56.9% (for Mg), 30.6–47.5% (for Mn), and 24.8–41.5% for (Zn) [60].

Considering the results of the quoted *in vitro* gastro-intestinal digestion, it can be assumed that toxic elements can be partly trapped as mentioned above macro- and microelements. However, it should not be overlooked that the treatment of bee pollen before its administration is quite important, additionally, other components of food products intaken along with bee pollen can also modulate the bioaccessibility of these elements for the human body [59].

5.3.2 Bioindication of the Environmental Pollution

As it is mentioned above, markedly high concentrations of selected elements, *i.e.*, Al [7, 30, 46, 49], As [12, 18, 43, 46, 49, 53], Cd [12, 18, 19, 27, 30, 46, 48, 53, 55, 61], Cr [18, 27, 46, 49], Cu [11, 47–49], Fe [47, 51, 52], Mn [29, 51, 52], Ni [11, 49], Pb [12–14, 18, 27, 30, 44, 46, 48, 53–56, 61], Sb [46], Se [61], and Zn [11, 29, 43, 51, 52], discovered when analyzing the material originating from a certain territory, may be related to the environmental pollution in this territory. This can happen in the case of sites directly located in or being close to regions, where industrial or municipal activities are elevated and hence, accompanied by industrial and municipal emissions of gaseous, liquid, and solid wastes into the environment [11, 12, 14, 18, 30, 43, 46–49, 51, 52, 55, 61]. Considering rural regions and/or areas with intensive agricultural activity, the contamination of bee pollen is also noted due to the use of fertilizers and pests [11, 12, 14, 19, 29, 44, 49, 53].

In both mentioned here cases, bee pollen is regarded as a bioindicator of environmental pollution, being better for that aim than honey [11, 13, 14, 18, 29, 43, 46–49, 51–55, 61]. It seems that bee pollen can especially well reflect any accidental or seasonal exposure of bees and pollen sources to locally happening environmental contaminants [14, 46, 48, 49, 53, 61]. In general, this is true that in urban, industrial, or agricultural sites, where the exposure to different kinds of emissions is much greater, concentrations of a selected group of elements, *e.g.*, Al, As, Cd, Co, Cr, Hg, Ni, Pb, Sb, Zn, are higher in samples of bee pollen from these sites [30, 46, 49, 51, 53]. However, the botanical effect is also strong and linked with the differentiated tendencies of flowering plants to absorb and accumulate the elements at such sites. This is especially evident in areas where different plants are grown. Indeed, in this case, multifloral bee pollen is collected and loaded in beehives, hence, the compensation effect from some plants is observed. As such, it is difficult to find any correlation between the pollution rate of the environment in a given territory, as represented by elevated concentrations of some elements in the soil sampled at this territory, and the content of contaminating elements, *e.g.*, Cd, Pb [50, 55, 56], collected at this territory. That is why some researchers do not agree with the statement that bee pollen can be used as the proper bioindicator for the monitoring of environmental pollution [50, 55, 56].

Finally, in countries with a longer vegetation period and only two climatic seasons, a coincidence between the level of the contamination of bee pollen with some trace elements and the occurrence of a given season is noted [30, 31, 33, 43, 44, 46]. Accordingly, it is experimentally established that during the dry season, with a higher wind activity, the amount of dust and deposits is higher. As a result, pollen sources are contaminated with higher concentrations of Al, As, Cd, Co, Cr, Cu, Fe, Mn, Ni, Pb, and Zn [30, 43, 46].

5.4 Chemometric Analysis of Bee Pollen Due to the Element Composition

Since the source of elements in bee pollen are directly related to pollen-donating plant species and indirectly to the place from which these plants come from, concentrations of elements can be used as variables for the chemometric evaluation of bee pollen samples by using multivariate methods. In the latter case, it can be the data reduction, the grouping of objects, the classification of objects, and the modeling of relationships between objects and their botanical and geographical origin. Indeed, several reports show that data matrices, comprising concentrations of elements determined in bee pollen samples of different origins, can be useful for their visualization and grouping due to the geographical origin (location site, collection area) and a year season [7, 8, 30, 31, 35, 44, 46, 49, 62] or the botanical origin in the case of miofloral bee pollen [8, 15, 16, 20, 22, 63] or the material with leading pollen [35].

Essential multivariate chemometric methods that are used for such evaluation are principle component analysis (PCA) [7, 8, 15, 16, 30, 31, 35, 44, 46, 49, 63], canonic variate analysis (CVA) [20], or hierarchic cluster analysis (HCA) [8, 22, 31, 49, 62]. When concentrations of elements are the only variables taken to build the data matrix prior to the chemometric evaluation, bee pollen samples coming from different locations can be just partially discriminated [7, 8, 30, 31]. In this case, distinctive elements patterns for given samples are interfered by the effect related to the environmental pollution of location sites affected by urban and industrial contaminants. A more distinctive differentiation between samples of bee pollen can be achieved due to climatic changes and the year season [44, 46, 49, 62]. For that reason, the supplementary information is used in the form of other physicochemical parameters, including the content of ash [8], fatty acids [15], carboxylic acids [35], carbohydrates [15, 35], total carotenoids [15], total phenolic compounds [8, 15, 35], total flavonoids [8], individual phenolic compounds or antioxidant activity [8, 15, 35] or the antimicrobial activity [8].

In the same way, only partial discrimination of bee pollen samples can be obtained due to the botanical origin of pollen grains when using the information on the total content of elements [8, 15, 16, 20, 22, 63]. Again, much better discrimination is achieved when other physicochemical parameters are considered in the chemometric evaluation [35, 63]. Because of the intercorrelation between the botanical and the geographical origin effects on the actual content of elements in bee pollen only in the case of monofloral bee pollen, differences between samples collected from various areas can be established. In such an approach, for a given area, the botanical origin can be the descriptive variable for samples coming from different plant species [7, 16].

5.5 Conclusions and Future Trends

Bee pollen is of particular interest to consumers due to the potential health profits that come with its consumption. The elemental composition of this bee product and the presence of various physiologically important minerals makes it an important dietary supplement. Its use for environmental pollution monitoring purposes is also noteworthy because of the variability of its elements content due to changes in vegetation conditions of pollinating plants in a given area. Certainly, an explanation of the impact of the transfer of elements from the soil to pollinating plants and pollen collected by bees would allow in the future obtaining bee pollen that would not carry any risk of excessive accumulation of some hazardous elements, without compromising its nutritional properties. At the same time, quality criteria should be established for the selection of appropriate plant species, which bee pollen is best suited for human nutrition and mineral supplementation. To achieve this, a comprehensive database with information on the elements content in bee pollen derived from different pollinating plants, taking into account environmental and geographical factors, should be further developed and validated.

References

1. Thakur M, Nanda V (2020) Composition and functionality of bee pollen: a review. Trends Food Sci Technol 98:82–126. https://doi.org/10.1016/j.tifs.2020.02.001
2. Vegh R, Csoka M, Soros C, Sipos L (2021) Food safety hazards of bee pollen – a review. Trends Food Sci Technol 114:490–509. https://doi.org/10.1016/j.tifs.2021.06.016
3. Pohl P, Dzimitrowicz A, Greda K, Jamroz P, Lesniewicz A, Szymczycha-Madeja A, Welna A (2020) Element analysis of bee-collected pollen and bee bread by atomic and mass spectrometry – methodological development in addition to environmental and nutritional aspects. Trends Anal Chem 128:115922. https://doi.org/j.trac.2020.115922
4. Kieliszek M, Piwowarek K, Kot AM, Blazejak S, Chlebowska-Smigiel WI (2018) Pollen and bee bread as new health-oriented products: a review. Trends Food Sci Technol 71:170–180. https://doi.org/10.1016/j.tifs.2017.10.021
5. Denisow B, Denisow-Pietrzyk M (2016) Biological and therapeutic properties of bee pollen: a review. J Sci Food Agric 96:4303–4309. https://doi.org/10.1002/jsfa.7729
6. Margaoan R, Stranţ M, Varadi A, Topal E, Yucel B, Cornea-Cipcigan M, Campos MG, Vodnar DC (2019) Bee collected pollen and bee bread: bioactive constituents and health benefits. Antiox 8:568. https://doi.org/10.3390/antiox8120568
7. Kostic AZ, Pestic MB, Mosic MD, Dojcinovic BP, Natic MM, Trifkovic JD (2015) Mineral content of bee pollen from Serbia. Arch Hig Rada Toksikol 66:251–258. https://doi.org/10.1515/aiht-2015-66-2630
8. De-Melo AAM, Estevinho LM, Moreira MM, Delerue-Matos C, da Silva de Freitas A, Barth OM, de Almeida-Muradian LB (2018) A multivariate approach based on physicochemical parameters and biological potential for the botanical and geographical discrimination of Brazilian bee pollen. Food Biosci 25:91–110. https://doi.org/10.1016/j.fbio.2018.08.001
9. Carpes ST, Cabral ISR, Luz CFP, Capeletti JP, Alencar SM, Masson ML (2009) Palynological and physicochemical characterization of *Apis mellifera* L. bee pollen in the Southern region of Brazil. J Food Agric Environ 7:667–673. https://doi.org/10.1234/4.2009.2739
10. Carpes ST, Mourao GB, de Alencar SM, Masson ML (2009) Chemical composition and free radical scavenging activity of Apis mellifera bee pollen from Southern Brazil. Braz J Food Technol 12:220–229. https://doi.org/10.4260/BJFT2009800900016
11. Aldgini HMM, Al-Abbadi AA, Abu-Nameh ESM, Alghazeer RO (2019) Determination of metals as bio indicators in some selected bee pollen samples from Jordan. Saudi J Biol Sci 26:1418–1422. https://doi.org/10.1016/j.sjbs.2019.03.005
12. Altunatmaz SS, Tarhan D, Aksu F, Barutcu UB, Or ME (2017) Mineral element and heavy metal (cadmium, lead and arsenic) levels of bee pollen of Turkey. Food Sci Technol 37(Suppl):136–141. https://doi.org/10.1590/1678-457x.36016
13. Dinkov D, Stratev S (2016) The content of two toxic heavy metals in Bulgarian bee pollen. Int Food Res J 23:1343–1345
14. Formicki G, Gren A, Stawarz R, Zysk B, Gal A (2013) Metal content in honey, propolis, wax and bee pollen and implications for metal pollution monitoring. Pol J Environ Stud 22:99–106
15. Dominguez-Valhondo D, Gil DB, Hernandez MT, Gonzalez-Gomez D (2011) Influence of the commercial processing and floral origin on bioactive and nutritional properties of honeybee-collected pollen. Int J Food Sci Technol 46:2204–2211
16. Liolios V, Tananaki C, Papaioannou A, Kanelis D, Rodopoulou MA, Argena N (2019) Mineral content in monofloral bee pollen: investigation of the effect of the botanical and geographical origin. J Food Meas Charact 13:1674–1682. https://doi.org/10.1007/s11694-019-00084-w
17. Yang K, Wu D, Ye X, Liu D, Chen J, Sun P (2013) Characterization of chemical composition of bee pollen in China. J Agric Food Chem 61:708–718. https://doi.org/10.1021/j5804056b
18. Roman A, Popiela-Pleban E, Migdal P, Kruszynski W (2016) As, Cr, Cd, and Pb in bee products from Polish industrialized region. Open Chem 14:33–36. https://doi.org/10.1515/chem-2016-0007

19. Sattler JAG, Demelo AAM, do Nascimento KS, de ILP M, Mancini-Filho J, Sattler A, de Almeida-Muradian LB (2016) Essential minerals and inorganic contaminants (barium, cadmium, lithium, lead and vanadium) in dried bee pollen produced in Rio Grande do Sul State, Brazil. Food Sci Technol 36:505–509. https://doi.org/10.1590/1678-457x.0029
20. Somerville DC, Nicol HI (2002) Mineral content of honey bee-collected pollen from New South Wales. Aust J Exp Agric 42:1131–1136. https://doi.org/10.1071/EA01086
21. Spulber R, Dogaroglu M, Babeanu N, Popa O (2018) Physicochemical characteristics of fresh bee pollen from different botanical origins. Rom Biotechnol Lett 23:13357–13365
22. Stanciu OG, Marghitas LA, Dezmirean D, Campos MG (2012) Specific distribution of minerals in selected unifloral bee pollen. Food Sci Technol Lett 3:27–31
23. Stanciu OG, Marghitas LA, Dezmirean D, Campos MG (2011) A comparison between the mineral content of flower and honeybee collected pollen of selected plant origin (*Helianthus annuus* L. and *Salix* sp.). Rom Biotechnol Lett 16:6291–6296
24. Szczesna T (2007) Concentration of selected elements in honeybee-collected pollen. J Apicult Sci 51:5–13
25. Stanciu OG, Marghitas LA, Dezmirean D (2009) Macro- and oligo-mineral elements from honeybee-collected pollen and beebread harvested from Transylvania (Romania). Bull UASVM Anim Sci Biotechnol 66:276–281
26. Bakour M, Fernandes A, Barros L, Sokovic M, Ferreira ICFR, Iyoussi B (2019) Bee bread as a functional product: chemical composition and bioactive properties. Food Sci Technol 109:276–282. https://doi.org/10.1016/j.lwt.2019.02.008
27. Adaskeviciute V, Kaskoniene V, Kaskonas P, Barcauskaite K, Maruska A (2019) Comparison of physicochemical properties of bee pollen with other bee products. Biomol Ther 9(2019):819. https://doi.org/10.3390/biom9120819
28. Belina-Aldemita MD, Opper C, Schreiner M, D'Amico S (2019) Nutritional composition of pot-pollen produced by stingless bees (Tetragonula biroi Friese) from The Philippines. J Food Comp Anal 82:103215. https://doi.org/10.1016/j.jfca.2019.04.003
29. Siqueira JS, Pereira JB, Lemos MS, Filho HAD, Dantas KGF (2017) Optimization of a digestion method using diluted acid in bee pollen samples for determination of Fe, Mn, and Zn by flame atomic absorption spectrometry. Food Anal Methods 10:759–763. https://doi.org/10.1007/s12161-016-0625-0
30. Morgano MA, Teixeira Martins MC, Rabonato LC, Milani RF, Yotsuyanagi K, Rodriguez-Amaya DB (2010) Inorganic contaminants in bee pollen from southeastern Brazil. J Agric Food Chem 58:6876–6883. https://doi.org/10.1021/j4300433p
31. Morgano MA, Martins MCT, Rabonato LC, Milani RF, Yotsuyanagi K, Rodriguez-Amaya DB (2012) A comprehensive investigation of the mineral composition of Brazilian bee pollen: geographic and seasonal variations and contribution to human diet. J Braz Chem Soc 23:727–736
32. Odimba VO, Azu DEO, Oko EC (2016) Macro and micro mineral elements composition of bee pollen from rainforest zone of Nigeria. Int J Sci Res 5:413–415. https://doi.org/10.21275/art2016899
33. Basso IM, Lorenzo DS, Mouteira MC, Custo GS (2019) Mineral analysis of pollen by total reflection X-ray fluorescence. Appl Rad Isot 152:168–171. https://doi.org/10.1016/j.apradiso.2019.06.015
34. da Silva GR, da Natividade TB, Camara CA, da Silva EMS, dos Santos FAR, Silva TMS (2014) Identification of sugar, amino acids, and minerals from pollen of Jandaira stingless bees (*Melipona subnitida*). Food Nutr Sci 4:1015–1021. https://doi.org/10.4236/fns.2014.511112
35. Kalaycioglu Z, Kaygusuz H, Doker S, Kolayli S, Erim FB (2017) Characterization of Turkish honeybee pollens by principal component analysis based on their individual organic acids, sugars, minerals and antioxidant activities. Food Sci Technol 84:402–408. https://doi.org/10.1016/j.lwt.2017.06.003
36. Ozcan MM, Aljuhaimi F, Babiker EE, Uslu N, Ceylan DA, Ghafoor K, Ozcan MM, Dursun N, Ahmed IM, Jamiu FG, Alsawmahi ON (2019) Determination of antioxidant activity, phenolic

compound, mineral contents and fatty acid compositions of bee pollen grains collected from different locations. J Apic Sci 63:69–79. https://doi.org/10.2478/jas-2019-0004
37. Maragou NC, Pavlidis G, Karasali H, Hatjina F (2017) Major and minor element levels in Greek apicultural products. Global NEST J 19:423–429
38. Filipiak M, Kuszewska K, Asselman M, Denisow B, Stawiarz E, Woyciechowski M, Weiner J (2017) Ecological stoichiometry of the honeybee: pollen diversity and adequate species composition are needed to mitigate limitations imposed on the growth and development of bees by pollen quality. PLoS One 12:e0183236. https://doi.org/10.1371/journal.pone.0183236
39. Grembecka M, Szefer P (2013) Evaluation of honeys and bee products quality based on their mineral composition using multivariate techniques. Environ Monit Assess 185:4033–4047. https://doi.org/10.1007/s10661-012-2847-y
40. Andelkovic B, Jevtic G, Mladenovic M, Markovic J, Petrovic M, Nedic N (2012) Quality of pollen and honey bee bread collected in spring. J Hyg Eng Design 1:275–277
41. Fatrcova-Sramkova K, Nozkova J, Kacaniova M, Mariassyova M, Kropkova Z (2010) Microbial properties, nutritional composition and antioxidant activity of *Brassica napus* subsp. *napus* L. bee pollen used in human nutrition. Ecol Chem Eng A 17:45–54
42. Bonvehi JS, Jorda RE (1997) Nutrient composition and microbiological quality of honeybee-collected pollen in Spain. J Agric Food Chem 45:725–732. https://doi.org/10.1021/j5060265q
43. Alvarez-Ayuso E, Abad-Valle P (2017) Trace element levels in an area impacted by old mining operations and their relationship with beehive products. Sci Total Environ 599–600:671–678. https://doi.org/10.1016/j.scitotenv.2017.05.030
44. Al Naggar YA, Naiem EA, Seif AI, Mona MH (2010) Honey bees and their products as a bio-indicator of anvironmental pollution with heavy metals. Mellifera 13–26:10–20
45. Silva AS, Araujo SB, Souza DC, Silva FAS (2012) Study of the Cu, Mn, Pb and Zn dynamics in soil, plants, and bee pollen from the region of Teresina (PI), Brazil. Ann Braz Acad Sci 84:881–889. https://doi.org/10.1590/S0001-37652012005000065
46. Nascimento NO, Nalini HA, Ataide F, Abreu AT, Antonini Y (2018) Pollen storage by stingless bees as an environmental marker for metal contamination: spatial and temporal distribution of metal elements. Sociobiol 65:259–270. https://doi.org/10.13102/sociobiology.v65i2.2078
47. Mejias E, Montenegro G (2012) The antioxidant activity of Chilean honey and bee pollen produced in the Llaima volcano's zones. J Food Qual 35:315–322. https://doi.org/10.1111/j.1745-4557.2012.00460.x
48. Mejias E, Gomez CJ, Gareil P, Delaunay N, Montenegro G (2018) Characterization of phenolic profile alterations in metal-polluted bee pollen via capillary electrophoresis. Cien Inv Agr 45:51–63. https://doi.org/10.7764/rcia.v45i1.1890
49. Temizer IK, Guder A, Temel FA, Avci E (2018) A comparison of the antioxidant activities and biomonitoring of heavy metals by pollen in the urban environments. Environ Monit Assess 190:462. https://doi.org/10.1007/s1061-018-6829-6
50. Fakhimzadeh K, Lodenius M (2000) Honey, pollen and bees as indicator of metal pollution. Acta Univ Carolinae Environ 14:13–20
51. Popescu IV, Dima G, Dinu S (2010) The content of heavy metals in pollen from Dambovita region. J Sci Arts 10:171–174
52. Dima G, Popescu IV, Dinu S, Nitescu O, Stribescu R (2012) Heavy metals in pollen samples collected from Dambovita county analysed by EDXRF method. Rom J Phys 57:1411–1416
53. Roman A (2009) Concentration of chosen trace elements of toxic properties in bee pollen loads. Polish J Environ Stud 18:265–271
54. Conti ME, Botre F (2001) Honeybees and their products as potential bioindicators of heavy metals contamination. Environ Monit Assess 69:267–282. https://doi.org/10.1023/A:101071910
55. Satta A, Verdinelli M, Ruiu L, Buffa F, Salis S, Sassu A, Floris I (2012) Combination of beehive matrices analysis and ant biodiversity to study metal pollution impact in a post-mining area (Sardinia, Italy). Environ Sci Pollut Res 19:3977–3988. https://doi.org/10.1007/s11356-012-0921-1

56. Lambert O, Piroux M, Puyo S, Thorin C, Larhantec M, Delbac F, Pouliquen H (2012) Bees, honey and pollen as sentinels for lead environmental contamination. Environ Pollut 170:254–259. https://doi.org/10.1016/j.envpol.2012.07.012
57. de Oliveira FA, de Abreu AT, de Oliveira NN, Froes-Silva RES, Antonini Y, Nalini HA Jr, de Lena JC (2017) Evaluation of matrix effect on the determination of rare earth elements and As, Bi, Cd, Pb, Se and In in honey and pollen of native Brazilian bees (*Tetragonisca angustula* – Jatai) by Q-ICP-MS. Talanta 162:488–494. https://doi.org/10.1016/j.talanta.201610.058
58. Campos MGR, Bogdanov S, de Almeida-Muradian LB, Szczesna T, Mancebo Y, Frigerio C, Ferreira F (2008) Pollen composition and standardization of analytical methods. J Apicult Res Bee World 47:154–161
59. Kostic AZ, Dojcinovic B, Spirovic Trifunovic B, Milincic DD, Nedic N, Stanojevic S, Pesic M (2022) J Environ Sci Health B 57:568–575. https://doi.org/10.1080/03601234.2022.2079348
60. Pohl P, Dzimitrowicz A, Lesniewicz A, Welna M, Szymczycha-Madeja A, Cyganowski P, Jamroz P (2020) Room temperature solvent extraction for simple and fast determination of total concentration of Ca, Cu, Fe, Mg, Mn, and Zn in bee pollen by FAAS along with assessment of the bioacessible fraction of these elements using in vitro gastrointestinal digestion. J Trace Elem Med Biol 60:126479. https://doi.org/10.1016/j.jtemb.2020.126479
61. Golubkina NA, Sheshnitsan SS, Kapitalchuk MV, Erdenotsogt E (2016) Variations of chemical element composition of bee and beekeeping products in different taxons of the biosphere. Ecol Indic 66:452–457. https://doi.org/j.ecolind.2016.01.042
62. Topal E, Cakici N, Margaoan R, Takma C, Guney F, Kosoglu M, Cornea-Cipcigan M, Atmaca H (2022) Annual development performance of fixed honeybee colonies linked with chemical and mineral profile of bee collected pollen. Chem Biodivers 19:e202200468. https://doi.org/10.1002/cbdv.202200468
63. Asmae EG, Nawal EM, Bakour M, Lyoussi B (2021) Moroccan monofloral bee pollen: botanical origin, physicochemical characterization, and antioxidant activities. J Food Qual 2021:8877266. https://doi.org/10.1155/2021/8877266

Chapter 6
Phenolic Acids in Pollen

Aleksandar Ž. Kostić, Yusuf Can Gercek, and Nesrin Ecem Bayram

6.1 Introduction

Phenolic compounds present one of the most important groups of plant secondary metabolites. Among all other categorizations of this quite diverse group of compounds the simplest one classifies it as phenolic acids (usually contain only one aromatic ring), flavonoids (usually contain two or three aromatic rings substituted by different functional groups and/or substituents) and coumarins (characterized by naturally occurring benzopyrone structure) [1]. The most important biosynthetic way to produce these compounds in plants is shikimate pathway continuing with phenylpropanoid pathway and finishing with biosynthesis of simple phenolic acids (PAs) or other types of phenolics [1]. There are two important groups of PAs: (1) derivatives of benzoic acid (hydroxybenzoic acids, HBAc); (2) derivatives of *p*-coumaric acid (hydroxycinnamic acids, HCAs) [2]. The most important HBAs are: gallic acid, vanillic acid, *p*-hydroxybezoic acid, salicylic acid, etc. while HCAs are: chlorogenic acid, ferulic acid, *p*-coumaric acid, caffeic acid, etc. [2].

Phenolic compounds, including PAs, have several important roles in plants including protecting (from herbivores, UV lights and other abiotic stresses), attracting (to pollinators), color role, allelopathic, microbiota attracting, etc. [1]. Among

A. Ž. Kostić
Faculty of Agriculture, Chair of Chemistry and Biochemistry, University of Belgrade, Belgrade, Serbia
e-mail: akostic@agrif.bg.ac.rs

Y. C. Gercek
Faculty of Science, Department of Biology, Istanbul University, Istanbul, Turkey
e-mail: yusuf.gercek@istanbul.edu.tr

N. E. Bayram (✉)
Department of Food Processing, Bayburt University, Aydıntepe Vocational College, Bayburt, Turkey

N. Ecem Bayram et al. (eds.), *Pollen Chemistry & Biotechnology*,
https://doi.org/10.1007/978-3-031-47563-4_6

all plant parts pollen grain is particularly sensitive since it represents the male gametophyte cell that must be protected from any adverse impact. Accordingly, it was observed that *p*-coumaric acid can play a very important role in the protection of genetic material in *Pinus* sp. pollen grain from deterioration effect of UV-B light [3]. Similarly, exposure of pollen to low temperature induced biosynthesis of phenolic compounds in *Citrus* pollen [4].

Bees collect pollen during the flowering season in order to provide the most important source of nutrients for their brood and larvae development. More specifically, pollen is a crucial source of proteins for bees [5]. During collection, they mix pollen with saliva, nectar and honey, thus making bee-collected pollen. By doing so, pollen is enriched mostly with sugars and some enzymes. However, phenolic compounds remain mostly the same as they originate from plants. It can be concluded that bee-collected pollen can be easily linked to the plant of origin through determination of the phenolic profile. Thus, phenolic compounds can be used as important chemotaxonomic markers in bee-collected pollen [6–8].

As a source of important antioxidants [9, 10] such as phenolic compounds but also carotenoids [11–13] and vitamins [14–16] bee-collected pollen has gained a lot of attention over the past decades as a potential functional food ingredient [17] with health benefits for humans. Due to all this, detailed information of different phenolic compounds present in bee-collected pollen, including their diversity and abundance, can provide us with a broader view of exactly how important bee-collected pollen is as a potential source of bioactive compounds for humans. Flower pollen is quite limited for this purpose due to shorter time of stability as bees and their digestive system can preserve pollen. Consequently, this chapter aimed to summarize all available data about identification and quantification of phenolic acids identified and determined in different pollen samples (both floral and bee-collected) to establish this natural product as an important carrier of bioactivity. Flavonoids in pollen are not elaborated on in this chapter since they will be discussed in detail in the next part of this book.

6.2 Phenolic Acids in Pollen

In addition to carotenoids and different vitamins (A,C,E and B group) phenolic compounds are the most important bioactive compounds found in pollen. Moreover, the most recent research focused on several more "specific" compounds in bee-collected pollen such as phenylamides [18, 19] and glucosinolates [20, 21].

On average, phenolic compounds in (bee) pollen make up 1.6% of the chemical composition with a usual share of 1.4% of different flavonoids and derivatives and 0.2% of phenolic acids [22]. However, diversity of compounds found in pollen is quite impressive and depends on the botanical origin of the samples as well as agro-ecological conditions and geographical location where the samples were collected.

In general, it is assumed that phenolic compounds (including phenolic acids) are part of the exine, outer wall of pollen grain where they are linked to sporopollenin

[23]. The main forms are conjugates of hydroxycinnamic and benzoic acids present in pollen of several different higher plants, in particular esters of *p*-coumaric acid [23] as well as ferulic acid, vanillic acid and *p*-hydroxybenzoic acid identified in *Pinus* pollen after saponification of sporopollenin [24]. Similarly, *p*-coumaric and ferulic acids were determined as components of sporopollenin from pollen of several plant species (*Vicia faba*, *Pinus sylvestris*, *Helleborus foetidus*, *Betula pendula*) with an important role in protection from UV-B radiation [25]. However, data about the phenolic acids content for floral pollen are quite limited. Recently, *p*-coumaric acid was determined (7.5 μg/g) in pine floral pollen (*Pinus yunnanensis*) from Shandong Province (China) [26]. It was also observed that exposing pollen to stressful factors such as postharvest irradiation provoked increased content of this compound (~ 13 μg/g) which is in line with previously stated information that *p*-coumaric acid protects pollen grain from unfavorable radiation in the environment [3, 25]. Interestingly, based on the data from UV spectra obtained through HPLC analysis it was determined that 94% of free (extractable) phenolic compounds in *Pinus* and *Betula* floral pollen were phenolic acids [27] . In both types, *trans*-ferulic acid was found (22% of total phenolics in *Betula* i.e. 57% in *Pinus* pollen) while there was a difference regarding the abundance of other phenolic acids and it was related to the botanical origin of pollen. The most predominant phenolic acids in *Betula* pollen samples were vanillic (42%) and chlorogenic (30%) acids while in case of *Pinus*, apart from *trans*-ferulic acid, there was gallic acid (13%) as the second one. In addition, 3,4-dihydroxybenzoic acid was detected and quantified in traces.

Unlike floral pollen, data about phenolic profiles of different bee-collected pollen samples are considerable and can help us get a full picture about the great diversity of these compounds. All available data are summarized in Table 6.1. It includes data about both qualitative and quantitative analyses of PAs in bee-collected pollen samples depending on the availability of previously presented analyses.

Based on the data collected from literature and presented in Table 6.1 a significant diversity of phenolic acids and their derivatives was observed. In total, 23 different PAs were identified/quantified in the pollen samples with predominance of the following: caffeic acid (23 reports), ferulic acid (17 reports) and *p*-coumaric acid (15 reports + 4 reports without precise data about the type of coumaric acid). On the other hand, *cis*-coutaric acid, 3-hydroxycinnamic acid (both research studies from China without the data on botanical origin) and *m*-coumaric acid (Poland, no botanical origin data) were identified/quantified in only one report while 2,5-dihydroxybenzoic acid (Turkey, polyfloral samples and *Rhododendron ponticum* monofloral sample), and benzoic acid were reported in two articles. Moreover, several different PA derivatives (esters, ethers, amides) were identified in the samples. Figure 6.1 presents summarized data about the ranges for several most studied PAs in published articles.

However, it should also be pointed out that the data about so-called "bound" or "conjugate" or "hydrolyzable" fraction of phenolics found in pollen are still missing. It is well known that some plants contain a significant part of phenolic compounds, in particular PAs, attached to some cellular structure or biomacromolecules like pectin, cellulose, structural proteins, etc. [60, 61]. It usually counts from 20 to

Table 6.1 The content/presence of phenolic acids and derivatives in bee-collected pollen samples according to the available literature

Geographical origin	Botanical origin	Compound/quantity	Analytical method of determination	Reference
China	Monofloral samples (MoS): Pine Buckwheat Maize Rape Papaver Camellia Basswood Chinese gooseberry	vanillic acid (μg/g): 46.4 (pine); 4985.5 (rape); 3087.0 (papaver) gallic acid (μg/g): 460.9 (camellia) protocatechuic acid (μg/g): 309.5 (pine)	A high-performance capillary electrophoresis	[28]
Mexico	MoS: *Prosopis julifolora* (Leguminosae)	cinnamic acid derivative: n.q.*	HPLC-DAD	[29]
Sonar Desert (USA)	MoS: Mesquite Turpentine bush Mimosa	methyl benzoylformate (mg/g fw*): 0.51 (Mimosa); 0.35 (Turpentine bush); 0.21 (Mesquite)	SPME coupled with GC-MS	[30]
Croatia	Polyfloral sample (PoS): *Quercus ilex* *Cystus incanus* L. *Quercus* sp. *Asphodelus* sp. Brassicaceae	caffeic acid: 0.38 μmol/g (nonhydrolized); 0.7 μmol/g (hydrolized)	HPLC/UV/Vis detector	[31]
Taiwan	MoS: Tea (*Camellia sinensis*)	gallic acid: 0.69–2.55 mg/g dw* *p*-coumaric acid: 0.34–0-41 mg/g dw chlorogenic acid: 0.12–0.48 mg/g dw caffeic acid: 0.41–0.45 mg/g dw ferulic acid: 0.25–0.45 mg/g dw	RP-HPLC	[32]
Poland	PoS without detailed botanical data	salicylic acid: 0.119–2.570 mg/100 g *p*-coumaric acid: 0.014–0.224 mg/100 g caffeic acid: 0.004–0.036 mg/100 g	HPLC-DAD	[33]

(continued)

Table 6.1 (continued)

Geographical origin	Botanical origin	Compound/quantity	Analytical method of determination	Reference
Serbia	MoS: Sunflower (*Helianthus annuus* L.)	caffeic acid: 2.16–4.28 mg/kg dw ferulic acid: 4.7 mg/kg dw *p*-coumaric acid: 1.10 mg/kg dw chlorogenic acid: 1–2.54 mg/kg dw protocatechuic acid: 0.62 mg/kg dw	UHPLC/MS-MS Orbitrap	[34]
Türkiye and Russia	No data for botanical origin	gallic acid: 6.17–101.77 mg/100 g 3,4-dihydroxybenzoic acid: 17.09–94.74 mg/100 g syringic acid: 5.56–23.77 mg/100 g caffeic acid: 5.84–23.86 mg/100 g *p*-coumaric acid: 1.39–10.46 mg/100 g *trans*-ferulic acid: 3.25–42.24 mg/100 g *trans*-cinnamic acid: 1.57–31.13 mg/100 g	HPLC-PDA	[35]

(continued)

Table 6.1 (continued)

Geographical origin	Botanical origin	Compound/quantity	Analytical method of determination	Reference
Italy	MoS: *Hedera helix* L. *Cistus* L. *Cistus incanus* *Brassica* sp. *Gleditsia triacanthos* L. *Hedysarum coronarium* L. *Trifolium pretense* *Trifolium alexandrinum* L. *Castanea sativa* Miller *Lamium* gr. *Fraxinus ornus* L. *Papaver rhoeas* L. *Prunus* L. *Rubus ulmifolius* Schott. *Daucus* and *Corinadrum* gr. PoS: *Helianthus annuus* L. (predominant type) *Taraxacum* and *Cichorium* sp. (predominant type) *Trifolium pretense* gr. (predominant type) *Magnolia* sp. (predominant type) *Crataegus monogyna* Jacq. (predominant type)	Phenolic acids/derivatives represented 17% (in average) of total phenolic compounds Identified phenolic compounds were: syringic acid caffeic acid gallic acid ethyl ester caffeic acid-4-*O*-glucoside caffeoyl glucose feruloyl glucose	UHPLC-ESI-QTOF	[36]
Serbia	MoS: Artichoke (*Cynara scolymus*)	caffeic acid: 0.416 mg/kg dw (extractable fraction ferulic acid: 30.393 mg/kg dw (alkaline hydrolyzable fraction)	HPLC DAD/ MS-MS	[37]

(continued)

Table 6.1 (continued)

Geographical origin	Botanical origin	Compound/quantity	Analytical method of determination	Reference
Türkiye	PoS: Apiaceae, Brassicaceae, Lamiaceae (major) *Centaurea* sp. (the accompanying) *Castanea sativa* (the accompanying) *Cistus* sp. (the accompanying) *Crepis* sp. (minor) *Lotus* sp. (minor) Ranunuculaceae (minor) Rosaceae (minor)	caffeic acid: 12.46–56.17 μg/100 g 2,5-dihydroxybenzoic acid: 2.69–9.85 μg/100 g chlorogenic acid: 4.37–36.09 μg/100 g gallic acid: 35.31–157.84 μg/100 g *p*-coumaric acid: 36.28–78.47 μg/100 g protocatechuic acid: 1.11–61.3 μg/100 g salicylic acid: 19.27–65.20 μg/100 g *trans*-ferulic acid: 23.23–107.86 μg/100 g	LC-MS/MS	[38]
Lithuania, Poland, Sweden, Denmark, Slovakia, The Netherlands, Malta, Italy, Spain	No data for botanical origin	gallic acid: 2.11 (Spain)-31.42 (Lithuania) μg/g benzoic acid: 0.54 (Malta)-5.51 (Lithuania) μg/g chlorogenic acid: 1.84 (Italy)-5.51 (Slovakia) μg/g vanillic acid: 1.14 (Italy only) μg/g caffeic acid: 1.92 (Netherlands)-2.53 (Italy) μg/g syringic acid: 1.54 (Malta)-8.81 (Lithuania) μg/g salicylic acid: 6.71 (Malta)-17.71 (Lithuania) μg/g coumaric acid: 0.31 (Italy)-3.61 (Lithuania) μg/g ferulic acid: 7.91 (Malta)-38.14 (Lithuania) μg/g ellagic acid: 2.71 (Malta)-28.12 (Lithuania) μg/g	UPLC-DAD	[39]

(continued)

Table 6.1 (continued)

Geographical origin	Botanical origin	Compound/quantity	Analytical method of determination	Reference
Romania	PoS: Originated from spontaneous plant mixtures including *Cornus* sp., *Prunus* sp., *Rosa* sp., *Salix* sp., *Taraxacum officinale*	gallic acid: 0.015–2.31 μg/g 3,4-dihydroxybenzoic acid: 0.21–0.55 μg/g 4-dihydroxybenzoic acid: 2.68–19.77 μg/g chlorogenic acid: 0.27–46.94 μg/g caffeic acid: 0.27–1.17 μg/g syringic acid: 0.09–0.32 μg/g *p*-coumaric acid: 1.04–1.26 μg/g ferulic acid: 0.96–2.98 μg/g cinnamic acid: 0.22–2.24 μg/g	UHPLC-DAD-ESI/MS	[40]
Chile	Commercial samples without data for botanical origin	chlorogenic/caffeic acid: 5.10–19.67 mg/L sinapic acid: 4.97–23.86 mg/L ferulic acid: 3.46–16.29 mg/L syringic acid: 4.73–7.92 mg/L *p*-coumaric acid: 7.28–52.17 mg/L cinnamic acid: 3.00–5.77 mg/L	HPLC-DAD	[41]

(continued)

Table 6.1 (continued)

Geographical origin	Botanical origin	Compound/quantity	Analytical method of determination	Reference
Poland	Commercial samples without data for botanical origin	*m*-coumaric acid: 0.00 (free)-1.86 (conjugated) µg/g *m*-hydroxybezoic acid: 0–00 (free)-0.62 (conjugated) µg/g chlorogenic acid: 0–00 (free)-0.52 (conjugated) µg/g sinapic acid: 0.00 (free)-1.62 (conjugated) µg/g protocatechuic acid: 0.00 (free)-7.73 (conjugated) µg/g *p*-hydroxybezoic acid: 0–00 (free)-3.13 (conjugated) µg/g caffeic acid: 0.00 (free)-1.11 (conjugated) µg/g ferulic acid: 0.00 (free)-0.71 (conjugated) µg/g	HPLC–TOF–MS/MS	[42]
Italy	MoS: Ivy (*Hedera helix*)	caffeic acid: 0.2–1.5 mg/100 g *p*-coumaric acid: 1.7–5.6 mg/100 g ferulic acid: 1.1–1.3 mg/100 g	LC-ESI-MS/MS	[43]
China	Commercial samples without data for botanical origin	57 different derivatives of phenolic acids were identified (much more in fermented samples) such as: 3-aminosalicylic acid, caffeic acid, 3-hydroxycinnamic acid, brevifolin carboxylic acid, 3,4-dihydroxybenzoic acid ethyl ester, (S)-2-hydroxy-3-(4-hydroxyphenyl) propanoic acid, cis-coutaric acid, arbutin, picein, acetovanillone and 2,4,6-trihydroxybenzoic acid methyl ester	UPLC-ESI-MS	[44]

(continued)

Table 6.1 (continued)

Geographical origin	Botanical origin	Compound/quantity	Analytical method of determination	Reference
Tuscany, Italy	MoS: Brassicaceae *Rubus* sp. *Viburnum* sp. *Erica* sp. *Trifolium pretense* Asteraceae *Eucalyptus* sp. *Rosa* sp. PoS. *Prunus* sp.	Not quantified phenolic acids - Brassicaceae sample: contained predominantly vanillic, ferulic, caffeic, and *p*-coumaric acids - *Erica* sample: 4-hydroxybenzoic acid (predominant signal) followed with vanillic acid and *p*-coumaric acid - All other samples contained predominantly *p*-coumaric and caffeic acids	Front-Face Fluorescence Spectroscopy	[45]
Morocco	MoS: *Reseda luteola* PoS: *Thymus vulgaris* (predominant type) *Cystus* sp. (predominant type)	*o*-coumaric acid: 13.9–50.4 mg/kg dw ferulic acid: 17.8–22.2 mg/kg dw cinnamic acid: 4.8–261.1 mg/kg dw rosmarinic acid: 68.2–167.7 mg/kg dw chlorogenic acid: 14.6–16.8 mg/kg dw ellagic acid: 12.9–138.6 mg/kg dw	UPLC-DAD	[46]
Türkiye	PoS with no data about botanical origin	*p*-coumaric acid: 120.25–321.66 µg/100 g *trans*-ferulic acid: 27.33–157.40 µg/100 g chlorogenic acid: 41.50–113.83 µg/100 g syringic acid: 29.48 µg/100 g (one sample) caffeic acid: 133.95–401.82 µg/100 g salicylic acid: 10.24–20.66 µg/100 g protocatechuic acid: 40.81–414.63 µg/100 g gallic acid: 96.90–392.95 µg/100 g	LC-MS/MS	[47]

(continued)

Table 6.1 (continued)

Geographical origin	Botanical origin	Compound/quantity	Analytical method of determination	Reference
Portugal	PoS: *Echium plantagineum* (predominant type) *Cystus ladanifer* (predominant type) *Quercus* sp. (predominant type)	caffeoyl di-hexoside: n.q coumaroyl hexose: n.q caffeoyl hexose: 0.056 μg/g dw	HPLC/DAD/ESI-MS^n	[48]
Romania	PoS: *Helianthus annuus* L. (predominant type) *Robinia pseudoacacia* (predominant type) *Tilia* sp. (predominant type) Pinaceae (predominant type) *Quercus* sp. (predominant type) *Prunus* sp. (predominant type) *Zea mays* (predominant type) *Taraxacum* sp. (predominant type) *Crataegus monogyna* (predominant type)	protocatechuic acid: 88.93 (Pinaceae sp.) mg/kg *p*-hydroxybenzoic acid: 21.02 (Pinaceae sp.) mg/kg vanillic acid: 21.19 (*H. annuus*) mg/kg chlorogenic acid: 0.78 (*H. annuus*) -3.82 (polyfloral mg/kg *p*-coumaric acid: 2.92 (*C. monogyna*) -238.97 (*R. pseudoacacia*) mg/kg rosmarinic acid: 2.01 (*H. annuus*) -85.14 (*Z. mays*) mg/kg	HPLC-DAD	[49]

(continued)

Table 6.1 (continued)

Geographical origin	Botanical origin	Compound/quantity	Analytical method of determination	Reference
Chile	PoS: *Brassica rapa* L. (predominant type) *Lotus* sp. (predominant type) *Pedunculatus* cav. (predominant type) *Ulex europaeus* L. (predominant type)	sinapic acid: 1.28 (*B. rapa; U. europaeus*) – 62 13 (*B. rapa*) mg/100 g cinnamic acid: 3.30 (*Caldcluvia paniculata*) – 29.5 (*B. rapa*) mg/100 g abscisic acid: 2.56 (PoS) – 35.52 (*B. rapa*) mg/100 g ferulic acid: 5.82 (*B. rapa*) – 43.6 (*Lotus pedunculatus*) mg/100 g coumaric acid: 0.03 (*B. rapa*; *L. pedunculatus*) – 4.31 (*L. pedunculatus*) mg/100 g syringic acid: 4.68 (*B. rapa*; *L. pedunculatus*) – 18.85 (*B. rapa*) mg/100 g caffeic acid: 1.72 (*L. pedunculatus*; *U. europaeus*) – 4.48 (*B. rapa*) mg/100 g chlorogenic acid: 0.38 (*B. rapa*) – 5.07 (*B. rapa; U. europaeus*) mg/100 g	HPLC- DAD	[50]
Portugal	PoS: *Jasione montana* (predominant type) *Rubus* sp. (predominant type) *Eucalyptus* sp. (predominant type)	different caffeoyl, coumaroyl and feruloyl spermidines: 0.18 mg/g (N^1,N^5-di-*p*-coumaroyl-N^{10}-caffeoylspermidine; *Eucalyptus* sp. and *Salix* sp.) – 11.94 mg/g (N^1,N^5,N^{10}-tri-*p*-coumaroylspermidine isomer; *Rubus* sp.)	LC/DAD/ESI-MS^n	[51]

(continued)

Table 6.1 (continued)

Geographical origin	Botanical origin	Compound/quantity	Analytical method of determination	Reference
Türkiye	MoS: *Rhododendron ponticum*	caffeic acid: 1137.09 µg/100 g chlorogenic acid: 931.8 µg/100 g *p*-coumaric acid: 945.95 µg/100 g abscisic acid: 408.66 µg/100 g *trans*-ferulic acid: 147.73 µg/100 g gallic acid: 83.17 µg/100 g protocatechuic acid: 59.79 µg/100 g 2,5-dihydroxybezoic acid: 18.81 µg/100 g salicylic acid: 37.92 µg/100 g	LC-MS/MS	[15]
Brazil	*Melipona seminigra* pot-pollen without data about botanical origin	coumaric acid: n.q. dicoumaroyl-caffeoyl spermidine: n.q. N′, N″, N‴-tris-*p*-coumaroylspermidine: n.q. N′, N″, N‴-tris-*p*-feruloylspermidine: n.q.	UPLC-ESI-QToF-MS	[52]

(continued)

Table 6.1 (continued)

Geographical origin	Botanical origin	Compound/quantity	Analytical method of determination	Reference
Chile	PoS: *Brassica rapa* (predominant type) *Acacia caven* (predominant type) *Eschscholzia californica* (predominant type) *Cryptocarya alba* (predominant type) *Azara celastrina* (predominant type)	chlorogenic acid: 15.92 (*Azara celastrina*) – 26.21 (*C. alba*) mg/100 g caffeic acid: 0.93 (*B. rapa*; *C. alba*) and 5.67 (*B. rapa*; *E. californica*) mg/100 g syringic acid: 1.35 (*B. rapa*; *C. alba*) – 13.64 (*A. celastrina*) mg/100 g coumaric acid: 0.2 (*B. rapa*; *E. californica*) – 0.94 (*B. rapa*; *C. alba*) mg/100 g sinapic acid: 0.98 (*B. rapa*; *C. alba*) – 19.05 (*C. alba*; *Anthemis cotula*) mg/100 g ferulic acid: 1.03 (*B. rapa*; *E. californica*) – 20.2 (*A. celestina*) mg/100 g cinnamic acid: 2.81 (*B. rapa*; *E. californica*) – 21.42 (*B. rapa*; *C. alba*) mg/100 g	HPLC-DAD	[53]

(continued)

Table 6.1 (continued)

Geographical origin	Botanical origin	Compound/quantity	Analytical method of determination	Reference
Türkiye	No data for botanical origin	ferulic acid: 36.83–230.55 μg/100 g *p*-coumaric acid: 34.16–127.85 μg/100 g gallic acid: 8.34–18.59 μg/100 g protocatechuic acid: 4.73–19.77 μg/100 g *p*-hydrohybenzoic acid: 2.74–122.68 μg/100 g chlorogenic acid: 14.64–75.08 μg/100 g vanillic acid: 22.96–87.02 μg/100 g caffeic acid: 10.88–98.03 μg/100 g syringic acid: 10.55–259.53 μg/100 g benzoic acid: 46.87–1077.64 μg/100 g *o*-coumaric acid: 2.63–42.23 μg/100 g abscisic acid: 21.04–288.70 μg/100 g *tert*-cinnamic acid: 6.82–56.38 μg/100 g	HPLC- UV	[54]
Portugal	MoS: *Eucalyptus globulus* *Salix atrocinerea* Brot.	derivatives (sperimidine) of cinnamic acid-1 and 2 (*E. globulus*): n.q. derivatives of caffeic acid (*S. atrocinerea*): n.q.	HPLC-DAD	[55]
Lithuania	PoS: *Brassica napus* (predominant type) *Trifolium repens* (accompanying type)	kaempferol 3-o-α-l-(2″-e-*p*-coumaroyl-3″-z-p-coumaroyl)rhamnoside (cinnamate ester): 2.67 mg/g	LC-MS	[56]

(continued)

Table 6.1 (continued)

Geographical origin	Botanical origin	Compound/quantity	Analytical method of determination	Reference
India	MoS: Coconut Coriander Rapeseed PoS: Maize and pearl millet (predominant type) Onionweed, Pigeon pea and Cotton (accompanying type)	gallic acid (mg/100 g): 0.81 (coconut); 7.83 (coriander); 2.65 (rapeseed); 1.76 (PoS) caffeic acid (mg/100 g): 11.65 (coriander), 2.39 (rapeseed); 1.52 (PoS) syringic acid (mg/100 g): 9.47 (coriander); 1.01 (rapeseed) cinnamic acid (mg/100 g): 0.74 (coconut); 3.25 (coriander); 5.36 (rapeseed)	UHPLC-DAD-MS/MS	[57]
Morocco	MoS: *Coriandrum* and *Daucus* sp. *Olea europea* *Raphanus* sp. *Ononis spinosa*/*Astralagus* sp. PoS: *Brassica* sp. (predominant type) *Sinapis* sp. (accompanying type) *Helianthemum* sp. (predominant type) *Anthemis* sp. (accompanying type) Asteraceae (mixture)	caffeic acid (mg/g): 0. 15 (*O. spinosa*/*Astralagus* sp.) caffeic acid hexoside (mg/g): 0.14 (*Raphanus* sp.) i.e. 0.36 (*Coriandrum* and *Daucus* sp.) *p*-coumaric acid hexoside (mg/g): 0.16 (*Brassica* sp. and *Sinapis* sp.) i.e. 0.22 (*Coriandrum* and *Daucus* sp.) several spermidine and spermine derivatives (mg/g): 0.14 (tetracoumaroyl spermine isomer) – 10.52 (N10-tri-*p*-coumaroylspermidine isomer)	LC/DAD/ESI-MS^n	[58]

(continued)

Table 6.1 (continued)

Geographical origin	Botanical origin	Compound/quantity	Analytical method of determination	Reference
Morocco	PoS: *Pinus sylvestris*, *Ceratonia siliqua*, *Rosmarinus officinalis*, *Juniperus oxycedrus*, *Thymus vulgaris*, *Bupleurum spinosum*, *Globularia alypum*, *Cistus landanferus*	ferulic acid: 17.7 mg/kg fw cinnamic acid: 46.01 mg/kg fw *o*-coumaric acid: 27.10 mg/kg fw rosmarinic acid: 127.3 mg/kg fw gallic acid: 32.54 mg/kg fw vanillic acid: 6.13 mg/kg fw ellagic acid: 13.02 mg/kg fw	UPLC-DAD	[59]

* *n.q.* not quantified, *dw* dry weight, *fw* fresh weight

60% of total phenolics in different plant foodstuffs, especially abundant in cereals, pulses and legumes [61]. These phenolics could not be extracted by simple treatment with some solvent such as methanol, ethanol, and water but must be detached from these structures by acidic or alkaline treatment. However, these treatments are time-consuming and that is why most authors actually only monitored so-called "free" or "extractable" fractions. Reports about the bound phenolic fraction in pollen are scarce and quite novel. For instance, authors reported "alkaline hydrolyzable" fraction monitored in artichoke bee-collected pollen from Serbia and determined a significant presence of ferulic acid (30.39 mg/kg dw) which was absent in the free fraction [37]. The observed result is quite logical since ferulic acid is one of the most common bound phenolic compounds/acids in plants [60–62]. Moreover, in the bee-collected pollen sample from Poland authors recently determined a significant presence of conjugated phenolic compounds released after acidic hydrolysis including several PAs: *m*-coumaric, *m*-hydroxybenzoic, chlorogenic, sinapic, protocatechuic, *p*-hydroxybenzoic, caffeic and ferulic acids [42] in different ranges.

6.3 Importance and Bioaccessibility of Phenolic Acids and Derivatives in Human Nutrition

In order to determine true accessibility of some nutrients/phytochemicals it is not enough to establish their content in some foodstuffs but to monitor their behavior during digestion process in human body. Currently one of the most reported topic in literature is the application of a standardized *in vitro* digestion model to determine bioaccessibility of different compounds in food [63]. However, the next step is the

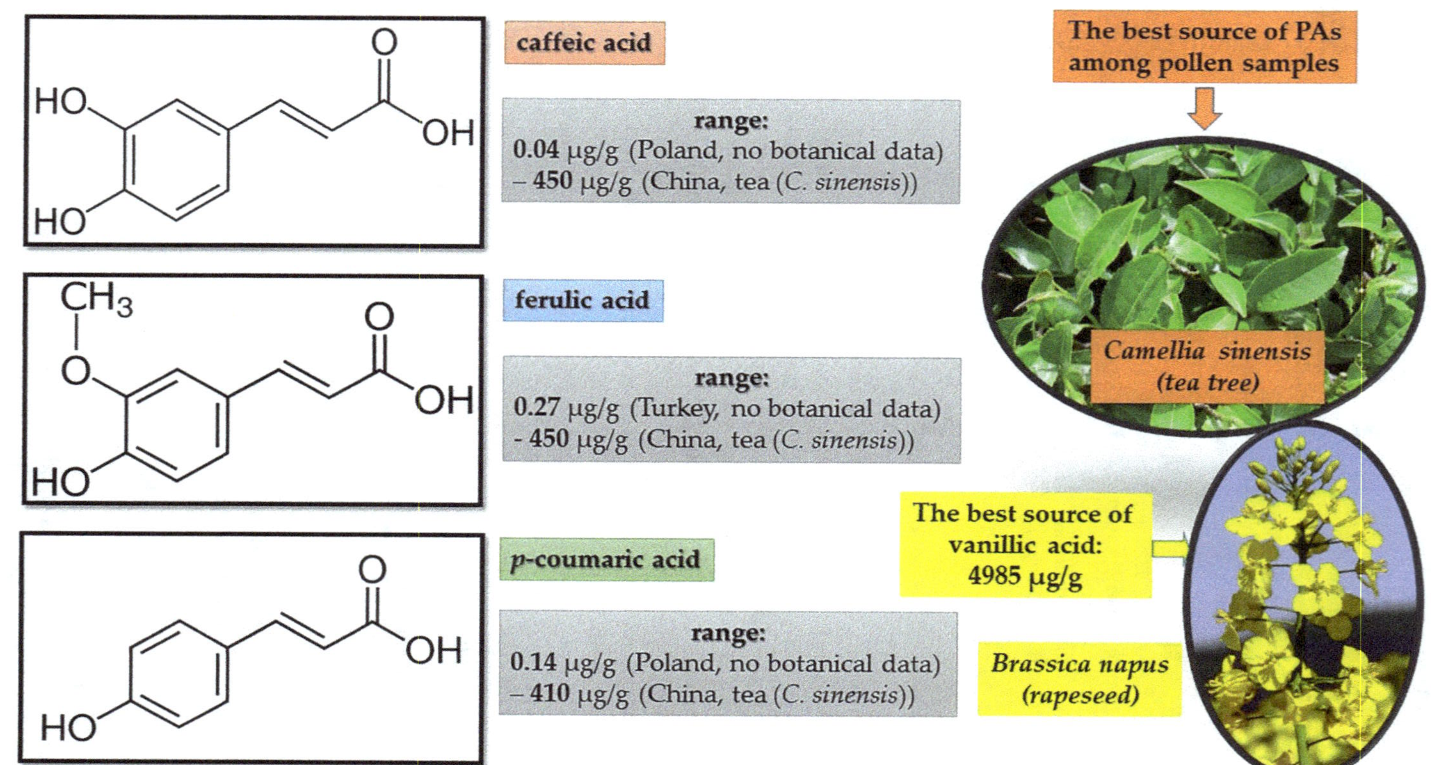

Fig. 6.1 The most representative phenolic acids (PAs) in pollen samples with associated ranges

determination of nutrients/phytochemicals' bioavailability which is much more expensive and demanding since it requires *in vivo* models. However, currently there are quite good correlations between *in vitro* and *in vivo* obtained data [64] allowing us to rely on the data obtained through *in vitro* artificial models. Following this method, the bioaccessibility of different phenolic compounds from bee-collected pollen samples has recently been monitored and published [65]. It was observed that the content of different phenolics decreased after oral digestion phase in a wide range depending on the pollen sample- from 7% to 92%. The most sensitive compounds under digestion conditions were phenylamides, an important group of hydroxycinnamic acid derivatives. Total phenolic bioaccessibility was determined to be 31% [65]. In addition, bioaccessibility of different phenolics was determined in functional food product obtained by enrichment of skimmed goat milk with monofloral sunflower pollen [66]. For several phenolic acids authors have determined a strong influence of a specific compound nature on the final bioaccessibility. To this effect, chlorogenic acid as well as ferulic acid were completely absent in the digested sample while *p*-coumaric, *p*-hydroxybenzoic and vanillic acids had bioaccessibility in the range from 48.6 to 53.6% [66]. Apart from possible degradation of compounds during digestion process there is also the probability of making strong interactions between phenolics and some macronutrients in foodstuffs such as proteins, lipids and sugars [67] which can „mask" the monitored compound making it unavailable for detection and this shouldn't be overlooked. Regardless, it is clear that a great diversity of PAs and derivatives, as well as other phenolics, determined in pollen samples make it such a perfect source of these bioactive ingredients that are important for human diet. This is why some authors define pollen as „the almost perfect food" [68].

6.4 Concluding Remarks and Future Perspectives

Natural sources of bioactive compounds are important nowadays as part of functional food, very popular among modern consumers. Pollen can serve as an excellent source of bioactivity including different phenolic compounds such as phenolic acids. Although there are extensive data about pollen phenolic acids, especially in bee-collected samples, there are still several tasks for further investigation:

- to continue monitoring the composition of monofloral samples with much more botanical diversity in order to determine possible chemotaxonomic significance
- to expand the examination of phenolics from the bound fraction in pollen and to determine how relevant it is for the general phenolic profile
- to include modern statistical methods and models in the research and to try to determine whether it is possible to establish any regularities between botanical and geographical origin with the representation of certain types of phenolic compounds.

References

1. Marchiosi R, dos Santos WD, Constantin RP et al (2020) Biosynthesis and metabolic actions of simple phenolic acids in plants. Phytochem Rev 19:865–906. https://doi.org/10.1007/s11101-020-09689-2
2. Goleniowski M, Bonfill M, Cusido R, Palazón J (2013) Phenolic acids. In: Natural products. Springer Berlin Heidelberg, Berlin, Heidelberg, pp 1951–1973
3. Seddon AWR, Festi D, Nieuwkerk M et al (2021) Pollen-chemistry variations along elevation gradients and their implications for a proxy for UV-B radiation in the plant-fossil record. J Ecol 109:3060–3073. https://doi.org/10.1111/1365-2745.13720
4. Mohammadrezakhani S, Hajilou J, Rezanejad F (2018) Evaluation of phenolic and flavonoid compounds in pollen grains of three *Citrus* species in response to low temperature. Grana 57:214–222. https://doi.org/10.1080/00173134.2017.1358763
5. Frias BED, Barbosa CD, Lourenço AP (2016) Pollen nutrition in honey bees (*Apis mellifera*): impact on adult health. Apidologie 47:15–25. https://doi.org/10.1007/s13592-015-0373-y
6. Almaraz-Ab N, Campos MG, Delgado-Al EA et al (2008) Pollen flavonoid/phenolic acid composition of four species of Cactaceae and its taxonomic significance. Am J Agric Biol Sci 3:534–543. https://doi.org/10.3844/ajabssp.2008.534.543
7. Campos M, Markham KR, Mitchell KA, da Cunha AP (1997) An approach to the characterization of bee pollens via their flavonoid/phenolic profiles. Phytochem Anal 8:181–185. https://doi.org/10.1002/(SICI)1099-1565(199707)8:4<181::AID-PCA359>3.0.CO;2-A
8. Tomás-Barberán FA, Tomás-Lorente F, Ferreres F, Garcia-Viguera C (1989) Flavonoids as biochemical markers of the plant origin of bee pollen. J Sci Food Agric 47:337–340. https://doi.org/10.1002/jsfa.2740470308
9. Aylanc V, Falcão SI, Ertosun S, Vilas-Boas M (2021) From the hive to the table: nutrition value, digestibility and bioavailability of the dietary phytochemicals present in the bee pollen and bee bread. Trends Food Sci Technol 109:464–481. https://doi.org/10.1016/j.tifs.2021.01.042
10. Mărgăoan R, Stranţ M, Varadi A et al (2019) Bee collected pollen and bee bread: bioactive constituents and health benefits. Antioxidants 8:568. https://doi.org/10.3390/antiox8120568
11. Mărgăoan R, Mărghitaş LA, Dezmirean DS et al (2014) Predominant and secondary pollen botanical origins influence the carotenoid and fatty acid profile in fresh honeybee-collected pollen. J Agric Food Chem 62:6306–6316. https://doi.org/10.1021/jf5020318
12. Salazar-González CY, Rodríguez-Pulido FJ, Stinco CM et al (2020) Carotenoid profile determination of bee pollen by advanced digital image analysis. Comput Electron Agric 175:105601. https://doi.org/10.1016/j.compag.2020.105601
13. Schulte F, Mäder J, Kroh LW et al (2009) Characterization of pollen carotenoids with *in situ* and high-performance thin-layer chromatography supported resonant Raman spectroscopy. Anal Chem 81:8426–8433. https://doi.org/10.1021/ac901389p
14. de Melo ILP, de Almeida-Muradian LB (2010) Stability of antioxidants vitamins in bee pollen samples. Quim Nova 33:514–518. https://doi.org/10.1590/S0100-40422010000300004
15. Ecem Bayram N (2021) Vitamin, mineral, polyphenol, amino acid profile of bee pollen from *Rhododendron ponticum* (source of "mad honey"): nutritional and palynological approach. J Food Measur Character 15:2659–2666. https://doi.org/10.1007/s11694-021-00854-5
16. de Arruda VAS, Pereira AAS, de Freitas AS et al (2013) Dried bee pollen: B complex vitamins, physicochemical and botanical composition. J Food Compos Anal 29:100–105. https://doi.org/10.1016/j.jfca.2012.11.004
17. Kostić AŽ, Milinčić DD, Barać MB et al (2020) The application of pollen as a functional food and feed ingredient—the present and perspectives. Biomol Ther 10:84. https://doi.org/10.3390/biom10010084
18. Qiao J, Feng Z, Zhang Y et al (2023) Phenolamide and flavonoid glycoside profiles of 20 types of monofloral bee pollen. Food Chem 405:134800. https://doi.org/10.1016/j.foodchem.2022.134800

19. Zhang X, Yu M, Zhu X et al (2022) Metabolomics reveals that phenolamides are the main chemical components contributing to the anti-tyrosinase activity of bee pollen. Food Chem 389:133071. https://doi.org/10.1016/j.foodchem.2022.133071
20. Ares AM, Redondo M, Tapia J et al (2020) Differentiation of bee pollen samples according to their intact-glucosinolate content using canonical discriminant analysis. LWT 129:109559. https://doi.org/10.1016/j.lwt.2020.109559
21. Ares AM, Tapia JA, González-Porto AV et al (2022) Glucosinolates as markers of the origin and harvesting period for discrimination of bee pollen by UPLC-MS/MS. Foods 11:1446. https://doi.org/10.3390/foods11101446
22. Komosinska-Vassev K, Olczyk P, Kaźmierczak J et al (2015) Bee pollen: chemical composition and therapeutic application. Evid Based Complement Altern Med 2015:1–6. https://doi.org/10.1155/2015/297425
23. Wiermann R, Gubatz S (1992) Pollen Wall and Sporopollenin. pp 35–72
24. Schulze Osthoff K, Wiermann R (1987) Phenols as integrated compounds of sporopollenin from *Pinus* pollen. J Plant Physiol 131:5–15. https://doi.org/10.1016/S0176-1617(87)80262-6
25. Rozema J, Broekman RA, Blokker P et al (2001) UV-B absorbance and UV-B absorbing compounds (*para*-coumaric acid) in pollen and sporopollenin: the perspective to track historic UV-B levels. J Photochem Photobiol B 62:108–117. https://doi.org/10.1016/S1011-1344(01)00155-5
26. Cheng Y, Quan W, He Y et al (2021) Effects of postharvest irradiation and superfine grinding wall disruption treatment on the bioactive compounds, endogenous enzyme activities, and antioxidant properties of pine (*Pinus yunnanensis*) pollen during accelerated storage. LWT 144:111249. https://doi.org/10.1016/j.lwt.2021.111249
27. Kerienė I, Šaulienė I, Šukienė L et al (2023) Patterns of phenolic compounds in *Betula* and *Pinus* pollen. Plan Theory 12:356. https://doi.org/10.3390/plants12020356
28. Chu Q, Tian X, Jiang L, Ye J (2007) Application of capillary electrophoresis to study phenolic profiles of honeybee-collected pollen. J Agric Food Chem 55:8864–8869. https://doi.org/10.1021/jf071701j
29. Almaraz-Abarca N, da Graça CM, Ávila-Reyes JA et al (2007) Antioxidant activity of polyphenolic extract of monofloral honeybee-collected pollen from mesquite (*Prosopis juliflora*, Leguminosae). J Food Compos Anal 20:119–124. https://doi.org/10.1016/j.jfca.2006.08.001
30. LeBlanc BW, Davis OK, Boue S et al (2009) Antioxidant activity of Sonoran Desert bee pollen. Food Chem 115:1299–1305. https://doi.org/10.1016/j.foodchem.2009.01.055
31. Šarić A, Balog T, Sobočanec S et al (2009) Antioxidant effects of flavonoid from Croatian *Cystus incanus* L. rich bee pollen. Food Chem Toxicol 47:547–554. https://doi.org/10.1016/j.fct.2008.12.007
32. Kao YT, Lu MJ, Chen C (2011) Preliminary analyses of phenolic compounds and antioxidant activities in tea pollen extracts. J Food Drug Anal 19. https://doi.org/10.38212/2224-6614.2177
33. Waś E, Szczęsna T, Rybak-Chmielewska H et al (2017) Application of HPLC-DAD technique for determination of phenolic compounds in bee pollen loads. J Apic Sci 61:153–162. https://doi.org/10.1515/jas-2017-0009
34. Kostić AŽ, Milinčić DD, Gašić UM et al (2019) Polyphenolic profile and antioxidant properties of bee-collected pollen from sunflower (*Helianthus annuus* L.) plant. LWT 112:108244. https://doi.org/10.1016/j.lwt.2019.06.011
35. Özcan MM, Aljuhaimi F, Babiker EE et al (2019) Determination of antioxidant activity, phenolic compound, mineral contents and fatty acid compositions of bee pollen grains collected from different locations. J Apic Sci 63:69–79. https://doi.org/10.2478/jas-2019-0004
36. Rocchetti G, Castiglioni S, Maldarizzi G et al (2019) UHPLC-ESI-QTOF-MS phenolic profiling and antioxidant capacity of bee pollen from different botanical origin. Int J Food Sci Technol 54:335–346. https://doi.org/10.1111/ijfs.13941
37. Kostić AŽ, Milinčić DD, Nedić N et al (2021) Phytochemical profile and antioxidant properties of bee-collected artichoke (*Cynara scolymus*) pollen. Antioxidants 10:1091. https://doi.org/10.3390/antiox10071091

38. Bayram NE, Gercek YC, Çelik S et al (2021) Phenolic and free amino acid profiles of bee bread and bee pollen with the same botanical origin – similarities and differences. Arab J Chem 14:103004. https://doi.org/10.1016/j.arabjc.2021.103004
39. Adaškevičiūtė V, Kaškonienė V, Barčauskaitė K et al (2022) The impact of fermentation on bee pollen polyphenolic compounds composition. Antioxidants 11:645. https://doi.org/10.3390/antiox11040645
40. Ilie C-I, Oprea E, Geana E-I et al (2022) Bee pollen extracts: chemical composition, antioxidant properties, and effect on the growth of selected probiotic and pathogenic bacteria. Antioxidants 11:959. https://doi.org/10.3390/antiox11050959
41. Velásquez P, Montenegro G, Giordano A et al (2019) Bioactivities of phenolic blend extracts from Chilean honey and bee pollen. CyTA – J Food 17:754–762. https://doi.org/10.1080/19476337.2019.1646808
42. Sawicki T, Ruszkowska M, Shin J, Starowicz M (2022) Free and conjugated phenolic compounds profile and antioxidant activities of honeybee products of polish origin. Eur Food Res Technol 248:2263–2273. https://doi.org/10.1007/s00217-022-04041-8
43. Filannino P, di Cagno R, Gambacorta G et al (2021) Volatilome and bioaccessible phenolics profiles in lab-scale fermented bee pollen. Foods 10:286. https://doi.org/10.3390/foods10020286
44. Zhang H, Lu Q, Liu R (2022) Widely targeted metabolomics analysis reveals the effect of fermentation on the chemical composition of bee pollen. Food Chem 375:131908. https://doi.org/10.1016/j.foodchem.2021.131908
45. Barbieri D, Gabriele M, Summa M et al (2020) Antioxidant, nutraceutical properties, and fluorescence spectral profiles of bee pollen samples from different botanical origins. Antioxidants 9:1001. https://doi.org/10.3390/antiox9101001
46. Laaroussi H, Ferreira-Santos P, Genisheva Z et al (2023) Unveiling the techno-functional and bioactive properties of bee pollen as an added-value food ingredient. Food Chem 405:134958. https://doi.org/10.1016/j.foodchem.2022.134958
47. Çelik S, Kutlu N, Gerçek YC et al (2022) Optimization of ultrasonic extraction of nutraceutical and pharmaceutical compounds from bee pollen with deep eutectic solvents using response surface methodology. Foods 11:3652. https://doi.org/10.3390/foods11223652
48. Gonçalves AC, Lahlou RA, Alves G et al (2021) Potential activity of Abrantes pollen extract: biochemical and cellular model studies. Foods 10:2804. https://doi.org/10.3390/foods10112804
49. Oroian M, Dranca F, Ursachi F (2022) Characterization of Romanian bee pollen—An important nutritional source. Foods 11:2633. https://doi.org/10.3390/foods11172633
50. Bridi R, Echeverría J, Larena A et al (2022) Honeybee pollen From Southern Chile: phenolic profile, antioxidant capacity, bioaccessibility, and inhibition of DNA damage. Front Pharmacol 13. https://doi.org/10.3389/fphar.2022.775219
51. Aylanc V, Ertosun S, Russo-Almeida P et al (2022) Performance of green and conventional techniques for the optimal extraction of bioactive compounds in bee pollen. Int J Food Sci Technol 57:3490–3502. https://doi.org/10.1111/ijfs.15672
52. Rebelo KS, Cazarin CB, Iglesias AH et al (2021) Nutritional composition and bioactive compounds of *Melipona seminigra* pot-pollen from Amazonas, Brazil. J Sci Food Agric 101:4907–4915. https://doi.org/10.1002/jsfa.11134
53. Oyarzún JE, Andia ME, Uribe S et al (2020) Honeybee pollen extracts reduce oxidative stress and steatosis in hepatic cells. Molecules 26:6. https://doi.org/10.3390/molecules26010006
54. Ulusoy E, Kolayli S (2014) Phenolic composition and antioxidant properties of Anzer bee pollen. J Food Biochem 38:73–82. https://doi.org/10.1111/jfbc.12027
55. Campos MG, Frigerio C, Bobiş O et al (2021) Infrared irradiation drying impact on bee pollen: case study on the phenolic composition of *Eucalyptus globulus* Labill and *Salix atrocinerea* Brot. pollens. Processes 9:890. https://doi.org/10.3390/pr9050890
56. Pukalskas A, Kazernavičiūtė R, Balžekas J (2016) Evaluation of antioxidant activity and flavonoid composition in differently preserved bee products. Czech J Food Sci 34:133–142. https://doi.org/10.17221/312/2015-CJFS

57. Thakur M, Nanda V (2021) Screening of Indian bee pollen based on antioxidant properties and polyphenolic composition using UHPLC-DAD-MS/MS: a multivariate analysis and ANN based approach. Food Res Int 140:110041. https://doi.org/10.1016/j.foodres.2020.110041
58. Aylanc V, Larbi S, Calhelha R et al (2023) Evaluation of antioxidant and anticancer activity of mono- and polyfloral moroccan bee pollen by characterizing phenolic and volatile compounds. Molecules 28:835. https://doi.org/10.3390/molecules28020835
59. Laaroussi H, Bakour M, Ousaaid D et al (2020) Effect of antioxidant-rich propolis and bee pollen extracts against D-glucose induced type 2 diabetes in rats. Food Res Int 138:109802. https://doi.org/10.1016/j.foodres.2020.109802
60. Shahidi F, Yeo J (2016) Insoluble-bound phenolics in food. Molecules 21:1216. https://doi.org/10.3390/molecules21091216
61. Shahidi F, Hossain A (2023) Importance of insoluble-bound phenolics to the antioxidant potential is dictated by source material. Antioxidants 12:203. https://doi.org/10.3390/antiox12010203
62. Yoshida A, Sonoda K, Nogata Y et al (2010) Determination of free and bound phenolic acids, and evaluation of antioxidant activities and total polyphenolic contents in selected pearled barley. Food Sci Technol Res 16:215–224. https://doi.org/10.3136/fstr.16.215
63. Minekus M, Alminger M, Alvito P et al (2014) A standardised static *in vitro* digestion method suitable for food – an international consensus. Food Funct 5:1113–1124. https://doi.org/10.1039/C3FO60702J
64. Bohn T, Carriere F, Day L et al (2018) Correlation between *in vitro* and *in vivo* data on food digestion. What can we predict with static in vitro digestion models? Crit Rev Food Sci Nutr 58:2239–2261. https://doi.org/10.1080/10408398.2017.1315362
65. Aylanc V, Tomás A, Russo-Almeida P et al (2021) Assessment of bioactive compounds under simulated gastrointestinal digestion of bee pollen and bee bread: bioaccessibility and antioxidant activity. Antioxidants 10:651. https://doi.org/10.3390/antiox10050651
66. Kostić AŽ, Milinčić DD, Stanisavljević NS et al (2021) Polyphenol bioaccessibility and antioxidant properties of *in vitro* digested spray-dried thermally-treated skimmed goat milk enriched with pollen. Food Chem 351:129310. https://doi.org/10.1016/j.foodchem.2021.129310
67. Jakobek L (2015) Interactions of polyphenols with carbohydrates, lipids and proteins. Food Chem 175:556–567. https://doi.org/10.1016/j.foodchem.2014.12.013
68. Li Q-Q, Wang K, Marcucci MC et al (2018) Nutrient-rich bee pollen: a treasure trove of active natural metabolites. J Funct Foods 49:472–484. https://doi.org/10.1016/j.jff.2018.09.008

Chapter 7
Flavonoids in Pollen

Milica Kalaba, Živoslav Tešić, and Stevan Blagojević

7.1 Introduction

Among the wide variety of foods available today, the way they are prepared and consumed still reduces their nutritional value. Therefore, the increasing number of products that are considered functional food could compensate for that. The concept of "functional food" has been introduced due to the appearance of another function in addition to serving food. Functional food has a preventive effect on many diseases as it contains functional compounds, such as flavonoids, among others. Due to the presence of these phytochemicals, functional food usually exhibits a high antioxidant capacity, which is important for the human body in which free radicals are constantly being created and which is additionally exposed to pollution in the environment.

Pollen can be considered the cell of life, as it ensures the reproduction of plant species, which additionally contributes to biodiversity, as well as the continuous creation of oxygen necessary for life on earth. Each pollen particle has its own morphological differences, owing to which it is possible to determine both botanical and geographical origin. Moreover, due to the specification of each pollen grain, as well as their resistance to chemical attacks [1], they could be valuable in forensic applications.

Pollen collected by bees, among other bee products, represents a functional food. It has shown a wide range of therapeutic properties [2]. The possibility of consuming food without prior preparation by humans facilitates the intake of

M. Kalaba (✉) · S. Blagojević
Institute of General and Physical Chemistry, Belgrade, Serbia
e-mail: milicaffh@yahoo.com

Ž. Tešić
University of Belgrade – Faculty of Chemistry, Belgrade, Serbia
e-mail: ztesic@chem.bg.ac.rs

N. Ecem Bayram et al. (eds.), *Pollen Chemistry & Biotechnology*,
https://doi.org/10.1007/978-3-031-47563-4_7

phytochemicals into the organism even more. Therefore, pollen as a functional food, although difficult for humans to digest when it is hand-collected, is easily accessible to humans thanks to bees. Bees compensate for the reduced possibility of humans consuming natural pollen grains from flowers (hand-collected pollen). The bees additionally increase the digestibility of pollen grains, which is important for human consumption. Due to the cell wall structure of pollen, which leads to poor digestibility, fermentation or enzymatic hydrolysis methods are very effective since they release valuable pollen compounds [2]. One of the essential constituents of pollen are flavonoids, which, due to the influence of bee enzymes, are further facilitated for better utilization by the human body. Consuming bee-collected pollen, honey or other bee products is a very useful way for humans to ingest flavonoids. In addition, these bee products really contribute to the prevention and treatment of many diseases. However, pollen can cause allergic reactions in some people, so it should be consumed with caution at first.

Flavonoids as most important subgroup of polyphenols are well-known phytochemicals with various health benefits [3, 4], which are based on their antioxidant activity. Today, the term "antioxidant" is misused in many spheres of life. It is considered that taking a certain amount of product rich in antioxidants is enough to protect organisms from harmful effects or the appearance of diseases. The processes of formation of free radicals as well as the mechanisms of action of antioxidants are rarely mentioned by the general public. The possibility that a sufficiently large quantity of antioxidants can really prevent the occurrence of diseases is imprecisely interpreted. Namely, it is not always possible to consume that quantity of phytochemicals, such as polyphenols, as well as flavonoids, in order to stop a harmful process in the organism. It is much more important to focus on moderate consumption of different foods, with an emphasis on functional food, which must also be varied. Emphasis should also be put on the synergistic effect of flavonoids, as well as other phytochemicals, which can enhance their effect as antioxidants and bioavailability.

The consumption of pollen, as well as other bee products, could be without prior human preparation. Nevertheless, there are potential risks of consuming unprepared food due to the possible contamination by pesticides or toxic metals from the environment. Various mycotoxins also have a major impact on food contamination, among which pollen is highly exposed [5]. One way to avoid this is to isolate biologically active substances from these products and proceed to further possible combinations with other therapeutic drugs. Based on well-described methods for extraction of many phytochemicals from pollen, as well as their identification and quantification [6], the contribution of pollen composition to its health properties could be analyzed. Due to the complex content of pollen particles, as well as the presence of many flavonoids in it, and their mutual interaction with other phytochemicals, a deeper physicochemical analysis of pollen could provide markers as additional criteria for successful chemotaxonomy, as well as a beneficial effect on health.

7.2 Flavonoids – Structure and Chemistry

Flavonoids are phytochemicals with a high positive impact on human health. They represent a large subgroup of polyphenols. They have a great ability to reduce and prevent oxidative stress that occurs due to various metabolic processes, as well as the excessive production of free radicals. The negative effect of free radicals affects the functioning of cells, as well as the whole organism, so free radicals are considered the main cause of diseases for a reason. On the other hand, due to flavonoids' effect on free radicals through several mechanisms, they are considered as good antioxidants [3]. Flavonoids, as well as other polyphenols could activate the defense system and affect free radicals by translating them into other types. The structure of polyphenols, including flavonoids, affects their antioxidant activity and causes them to manifest it in different ways. Thus, the antioxidant activity of flavonoids depends on the number of hydroxyl groups, as well as on the saturation of the heterocyclic C ring. However, the quantity of flavonoids and their relation, as well as the expression of synergism, could have a great influence on the final antioxidant capacity. Flavonoids' solubility depends primarily on the presence of side groups attached to the aromatic rings [6].

In general, phenols are biomolecules characterized by the presence of one or more hydroxyl groups. In case of a higher number of hydroxyl groups and aromatic rings, they are called polyphenols. There are several divisions of the phenolic compounds. Some of authors classified them into two major classes: flavonoids and non-flavonoids [6]. Other authors emphasized the important group such as flavonoids, and the phenolic acids group among other groups [4, 7]. These groups are also present in pollen, which is stated by many authors [2, 6, 8] as valuable for human consumption.

The structure of the flavonoids is characterized by a heterocyclic system of flavonoid nuclei of two benzenes (ring A and B), and the γ-pyrone part (C ring). The basic structure of flavonoids is presented in Fig. 7.1. Further classification of flavonoids (Fig. 7.1) is based on the position of the aromatic B ring, as well as the substitutions on rings A or B.

Flavones include some of the important aglycones such as apigenin, luteolin, etc. They differ from other flavonoids as they have a double bond and no substitution at the C3 position. In their native forms, they are found as both *O*- and *C*-glycosides. Flavones from plants are typically conjugated as 7-*O*-glycosides, while flavone *C*-glycosides are commonly found as 6-*C*- and 8-*C*-glucosides [9]. Well-known isoflavones are genistein and daidzein. Flavonones include naringin, naringenin, eriodictyol, pinocembrin, etc. The flavonol group includes kaempferol, quercetin, myricetin. The most important group of flavanols is flavan-3-ol, which includes both monomeric and dimeric units, such as catechin, epicatechin, then epicatechin gallate, epigallocatechin. Procyanidins are compounds consisting of (epi)catechin units, while the presence of (epi)gallocatechin is characteristic of prodelphinidins. Depending on the bonds between the monomer units, B-type procyanidins are characterized by an interflavone bond between the catechin and epicatechin units between the C4 and C8 carbon atoms (for dimers B1, B2, B3 and B4), or the C4 and

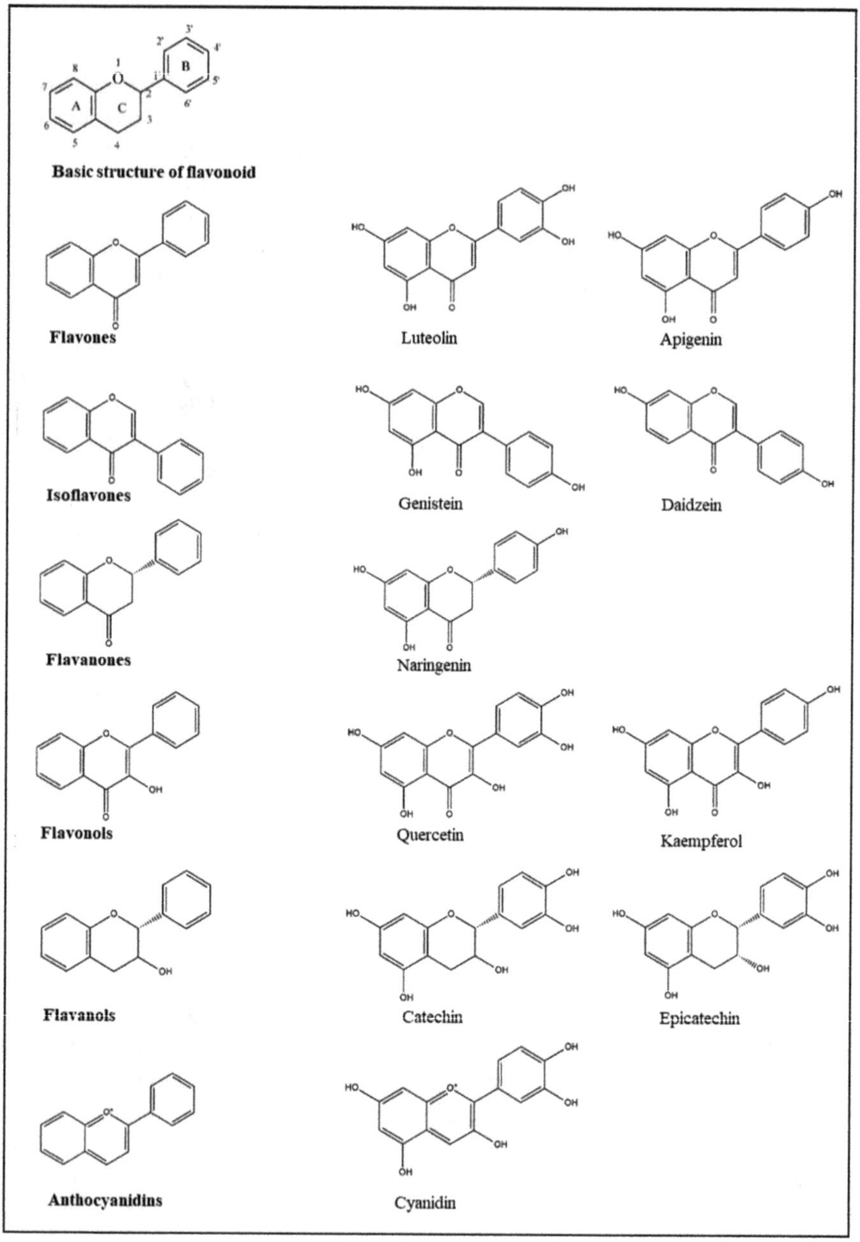

Fig. 7.1 The structure of the main flavonoid groups, and their most common representatives

C6 carbon atoms (dimers B5, B6, B7 and B8). Anthocyanidins, i.e. anthocyanin glycosides and their compounds have a positive charge on B ring. Thus, it can be observed that flavonoids differ greatly in the way they are formed [10], which contributed to their wide distribution.

7.3 Transport of Flavonoids Through the Plant to Pollen Grains

Flavonoids are secondary plant metabolites. They are formed by a direct continuation of the primary metabolic processes through the phenylpropanoid, shikimate and polyketide pathways. The processes of secondary metabolism are very different. Moreover, the similarity of the mechanisms of these processes in different plants is very rare. Plants contribute to the specificity of metabolic processes, which leads to a great variety of secondary metabolites. Primarily, flavonoids serve as catalysts in the light phase of photosynthesis [11]. In addition, secondary metabolites can be considered as a mechanism of plant species' adaptation to external influences, such as various types of stress. Flavonoids are transferred through the plant and they could be present in different quantities and ratios in all plant parts [4, 11–13]. In plants, flavonoids are rarely found as aglycones, but rather as glycosides (bonded to sugars moiety). They also represent a group of plant pigments that give color to leaves, flowers, and fruits. Considering the great influence of other phytochemicals, as well as many external factors on the biosynthesis pathway of polyphenols and their further transport through the plant, their presence in many fruit products [14, 15] and different plant parts [13, 16] is very important for research regarding their health properties as well as their consumption.

If the plant is observed from the root to the stem, and then to other plant organs, as well as to the leaf where photosynthesis takes place to the greatest extent, it can be said that the pollen grains represent the last organ of the plant to which flavonoids reach, and of course, in which they are also synthesized. Pollination by insects is, beside wind and self-reproduction, the most important way of increasing biodiversity. Studies of hand-collected pollen are very rare. However, some authors investigated pollen directly from flowers. For example, Bakour et al. [8] showed different phenolic profiles of several hand-collected pollens. Moreover, it was shown that pollen grains were promising sources of antioxidant substances, mostly flavonoids such as flavonols and flavones [8], which is in accordance with the properties of bee-collected pollen [17]. Many studies, however, focus more on bee-collected pollen for several reasons. Firstly, the bees collect more pollen grains, which are free from other flower parts. During the visits to flowers, the bees could collect hundreds to thousands of pollen grains, by using an electrostatic field that is generated between the flower and their body [18]. Secondly, from the point of view of essential substances for humans, bee-collected pollen can be considered more bioavailable than hand-collected pollen. The reason for this is the enzymes that bees insert into the pollen grains [19] during transport and storage in the comb. Pollen grains have cell walls that limit digestibility in the human organism. Thus, it was shown that the fermentation process increases the digestibility, as well as the polyphenol content in pollen [20].

7.4 Flavonoids – A Remarkable Constituents of Pollen

Some studies [3, 4] presented a different group of flavonoids and their appearance in various foods. However, many of these flavonoids were also present in pollen and other bee products. Pollen particles are rich in many phytochemicals, where, in addition to flavonoids, phenolic acids, essential amino acids, vitamins, minerals and fatty acids stand out [21]. The nutritional and functional properties, as well as flavonoids composition of pollen, are greatly influenced by plant species [21, 22]. In addition, many authors showed variations between the chemical compositions of beepollen depending on origin or season [2, 17, 23]. Examination of the hand-collected pollen [8] revealed that it is a good source of flavonoids and that they certainly contribute to overall bioactivity. Furthermore, flavonoids contribute to the health benefits of pollen [2]. Knowing that pollen is difficult for collecting, bee-collected pollen, as well as honey and other bee products, represents an easier way for humans to consume pollen. Hence, considering bee products as an available source of flavonoids for humans, more attention is devoted to the examination of pollen collected by bees.

One of the most abundant flavonoids in pollen is quercetin [24]. In addition to quercetin, major flavonoids in pollen are kaempferol, isorhamnetin, apigenin or luteolin and their derivatives, mainly glycosides [6]. According to Tomás-Barberán et al. [25], the list of flavonoids such as quercetin, luteolin, 8-methoxykaempferol, kaempferol, apigenin, isorhamnetin, myricetin, myricetin 3-methyl ether that originate from pollen could be also classified as pollen-nectar flavonoids. In addition, these authors also suggested that riboflavin derivate should originate from pollen or nectar. In one study it was noted that flavonoids such as quercetin, rutin, luteolin, and naringenin were similar in several bee-collected pollen samples with different geographical, as well as botanical origins [20]. It was also reported that in pollen from buckwheat were found flavonoids such as myricetin, vitexin, quercetin derivatives, as well as flavan-3-ols [16]. Moreover, it was stated [16] that flavan-3-ols and procyanidins are proposed as buckwheat markers. Flavonoids such as quercetin, quercetin 3-*O*-rhamnoside, rutin, isorhamnetin, isorhamnetin 3-*O*-glucoside, kaempferol, galangin, lutelolin, apigenin, acacetin, genkwanin, eriodictyol, naringenin, aesculetin, phloretin, narcissin were quantified in sunflower pollen [26] (Table 7.1).

De Melo et al. [23] analyzed several different pollen samples and found the presence of rutin, naringenin, naringin, kaempferol, catechin, and epicatechin. On the other hand, in one study that analyzed commercial bee pollen from several plant species there were found some unusual compounds such as quercetin 3-*O*-rhamnosyl-galactoside, quercetin 3-*O*-xylosyl-glucuronide, as well as isorhamnetin 3-*O*-glucoside 7-*O*-rhamnoside, quercetin 3-*O*-rutinoside [27]. Various content of flavonoids in bee pollen of different botanical origins was found in Chilean bee pollen samples, where the majority of samples corresponded to nonnative (*Brassica rapa* and *Eschscholzia californica* predominated) multifloral and monofloral and native (*Cryptocarya alba* (peumo) and *Acacia caven* (espinillo) predominated)

Table 7.1 Summarized results from literature of identified flavonoids, their quantitative values or range, and antioxidant activity expressed through different parameters in pollen samples

Pollen type	Flavonoids	Quantified values/Range	Antioxidant activity	References
Pollen samples from 14 different plants collected in Morocco	Quercetin-3-*O*-glycoside	/	PC = (9.20–71.20) mg GAE/g of pollen	[8]
	Luteolin-derivative	/	FFC = (1.27–15.44) mg QE/g of pollen	
	Luteolin- 7-*O*-derivative	/	TAC = (17.84–99 54) mg AAE/g pollen	
	Quercitin-3-*O*-derivative	/	DPPH = (0.009–0.86) mg/mL expressed as IC_{50}	
	Luteolin	/	ABTS = (0.003–0.89) mg/mL expressed as IC_{50}	
	Isorhamnetin-3-*O*-derivative	/	RP = (0.04–0.32) mg/mL expressed as EC_{50}	
	Isorhamnetin-3-*O*-glycoside	/		
	Quercetin	/		
	Apigenin derivative	/		

PC phenolic content, *FFC* flavones and flavonol content, *TAC* phenolics, flavones and flavonol content, *DPPH* scavenging of the 2, 2-diphenyl-1-picrylhydrazyl free radical, *ABTS* azino-bis(3-ethylbenzothiazoline-6-sulphonic acid free radical-scavenging activity, *RP* reducing power, *GAE* gallic acid equivalents, *QE* quercetin equivalents

(continued)

Table 7.1 (continued)

Pollen type	Flavonoids	Quantified values/Range	Antioxidant activity	References
Pollen sample from buckwheat plant collected in Serbia	B type procyanidin dimer gallate	/		[16]
	Methyl-B type prodelphinidin dimer	/		
	Epicatechin	/		
	B type procyanidin dimer gallate isomer 1	/		
	(Epi)catechin gallate	/		
	Methyl-(epi)gallocatechin gallate	/		
	B type procyanidin dimer gallate isomer 2	/		
	Luteolin 6-*C*-hexoside	/		
	Apigenin 8-*C*-hexoside (Vitexin)	/		
	Myricetin	/		
	Quercetin 3-*O*-(6″- rhamnosyl)-hexoside (Rutin)	/		
	Quercetin 3-*O*-galactoside	/		
	Quercetin 3-*O*-rhamnoside	/		
	Kaempferol 3-*O*-rhamnoside	/		
	Quercetin	/		
	Kaempferol	/		
Bee pollen samples (56 of them) collected in Brazil	Catechin	(n.d – 20.0) mg/100 g	TPC = (6.5–29.2) mg GAE/g	[23]
	Epicatechin	(n.d – 2.9) mg/100 g	TFC = (0.3–17.5) mg QE/g	
	Naringin	(n.d – 10.0) mg/100 g		
	Rutin	(n.d – 237.0) mg/100 g		
	Naringenin	(n.d – 143.0) mg/100 g		
	Quercetin	(n.d – 63.0) mg/100 g		
	Kaempferol	(n.d. – 101) mg/100 g		

TPC total phenolic content, *TFC* total flavonoid content, *n.d.* not detected

Bee pollen samples from sunflower collected in Serbia		Methanolic extract/ Ethanolic extract	Methanolic extract/Ethanolic extract	[26]
	Quercetin	(26.14/38.96) mg/kg dw	TPC = (3816/2907) mg/kg GAE of dw	
	Quercetin 3-*O*-galactoside	(112.86/128.64) mg/kg dw	TFC = (843/865) mg/kg QE of dw	
	Quercetin 3-*O*-rhamnoside	(7.00/7.76) mg/kg dw		
	Quercetin 3-*O*-rutinoside (Rutin)	(8.64/2.72) mg/kg dw		
	Isorhamnetin	(13.62/14.36) mg/kg dw		
	Isorhamnetin 3-*O*-glucoside	(14.46/19.50) mg/kg dw		
	Isorhamnetin 3-*O*-rutinoside (Narcissin)	(3.40/2.90) mg/kg dw		
	Kaempferol	(2.42/6.12) mg/kg dw		
	Galangin	(0.24/7.28) mg/kg dw		
	Luteolin	(1.14/1.22) mg/kg dw		
	Apigenin	(0.58/1.06) mg/kg dw		
	Acacetin	(n.d./0.40) mg/kg dw		
	Genkwanin	(n.d./0.42) mg/kg dw		
	Eriodictyol	(1.22/1.36) mg/kg dw		
	Naringenin	(0.46/n.d.) mg/kg dw		
	Taxifolin	(2.60/n.d.) mg/kg dw		
	Phloretin	(n.d./0.66 mg/kg dw		
	Aesculin	(0.48/n.d. mg/kg dw		

dw dry weight

(continued)

Table 7.1 (continued)

Pollen type	Flavonoids	Quantified values/Range	Antioxidant activity	References
Bee pollen from Greece	Quercetin 3-*O*-rhamnosyl-galactoside	/	AC = (75. 70–86. 81) %	[27]
	Quercetin 3-*O*-xylosyl-glucuronide	/	Δ[DPPH] = (53.70–71.67) %	
	Isorhamnetin 3-*O*-glucoside 7-*O*-rhamnoside	/		
	Quercetin 3-*O*-rutinoside	/		
AC antioxidant capacity; Δ[*DPPH*] % decrease in free radical concentration				
Bee pollen samples (28 of them) of different plant species (classified as first, second and third predominant species) collected in Chile		(Quantifies values of six pollen samples)		[28]
	Epicatechin	(0–38.42) mg/100 g pollen	TP = (102–8379) mg GAE/100 g	
	Rutin	(2.74–74.25) mg/100 g pollen	Flavonoids = (62–504) mg QE/100 g	
	Myricetin	(28.51–775.89) mg/100 g pollen	FRAP = (19–194) μmol TE/g	
	Quercetin	(6.59–98.70) mg/100 g pollen	ORAC FL = (109–492) μmol TE/g	
	Apigenin	(0–767.24) mg/100 g pollen	[QE] = (4.29–409.94) mg/100 g	
	Rhamnetin	(0–80.37) mg/100 g pollen		
	Catechin	(0–17.96) mg/100 g pollen		
TP total phenols, *ORAC FL* oxygen radical absorbance capacity, fluorescein, [*QE*] quercetin concentration; *TE* Trolox equivalent				
Bee pollen samples (13 of them) collected in Turkey	Quercetin	(55.94–499.20) μg/100 g pollen	TP = (44.07–124.10) mg/g pollen	[29]
	Rutin	(25.59–692.85) μg/100 g pollen	FRAP TEAP = (11.77–105.06) μmol Trolox/g	
	Catechin	(22.96–87.02) μg/100 g pollen	DPPH SC_{50} = (0.65–8.20) mg/mL	
	Epicatechin	(n.d. – 520.02) μg/100 g pollen	CUPRAC TEAC = (33.1–91.8) μmol Trolox/g	

FRAP ferric reducing antioxidant power, *TEAP* Trolox equivalent antioxidant power, SC_{50} the concentration of the extract (mg/mL) required to inhibit 50% of the free radical-scavenging activity, *CUPRAC* cupric reducing antioxidant capacity, *TEAC* Trolox equivalent antioxidant capacity

Monofloral bee pollen from sunflower collected in Slovakia	Quercetin	10.19 mg/kg	Polyphenols = 763.67 mg/kg	[30]
	Luteolin	63.62 mg/kg	DPPH = 48.43%	
	Kaempferol	lod		
	Apigenin	32.01 mg/kg		

Lod below limit of detection

Bee pollen (10 samples) with different botanical origin (5 samples) and different geographical origin (5 samples) collected in China	Rutin	(0.152–2.237) mg/100 g		[31]
	Isoquercitrin	(0.439–0.841) mg/100 g		
	Quercitrin	(0.456–0.717) mg/100 g		
	Naringenin	(n.d. – 20.510) mg/100 g		
	Kaempferol	(n.d. – 4.784) mg/100 g		
	Luteolin	(n.d. – 0.445) mg/100 g		

(continued)

Table 7.1 (continued)

Pollen type	Flavonoids	Quantified values/Range	Antioxidant activity	References
Bee-collected artichoke pollen collected in Serbia		EF/AHF	EF/AHF	[32]
	Quercetin 3-*O*-glucoside (Isoquercetin)	(1.836/n.d.)	TPC = (5314.2/497.9) mg GAE/kg dw	
	Rutin	(3.662/0.655)	TFC = (812.0/957.0) mg QE/kg dw	
	Isorhamnetin 3-*O*-glucoside	(49. 171/4.613)	TCD = (1064.7/n.d.) mg CGAE/kg dw	
	Kaempferol	(1.527/n.d.)	TC = (5.00/5.00 mg/kg dw	
	Kaempferol 3-*O*-glucoside (Astragalin)	(2.508/0.189)	EF/AHF/LF	
	Apigenin	(2.633/0.373)	TAC = (21912.4/813.0/7830.9) mg AAE kg dw	
	Apigenin 7-*O*-glucoside (Apigetrin)	(1.332/0.274)	FRP = (468.6/115.1/n.d.) mg AAE kg dw	
	Aesculetin	(0.438/n.d.)	ABTS = (81.48/14.74/4.80) % of inhibition	
			DPPH = (38.30/82.23/22.32) % of inhibition	
			FCC = (13.31/14.25/n.a.) % of chelating ability	

EF Extractable Fraction, *AHF* Alkaline Hydrolyzable Fraction, *LF* Lipid Fraction, *TCD* total dihydroxycinnamic and derivative content, *TC* total carotenoid content, *CGAE* chlorogenic acid equivalents, *TAC* total antioxidant capacity, *FRP* ferric reducing capacity, *FCC* ferrous chelating capacity, *AAE* ascorbic acid equivalents, *n.a.* not applicable

Pollen type	Flavonoids	Quantified values/Range	Antioxidant activity	References
Fresh bee pollen collected from five different beehives in Turkey	Catechin	(n.d. – 76.73)		[33]
	Isorhamnetin	(238.42–337.76)		
	Kaempferol	(112.94–768.04)		
	Luteolin	(21.91–3325.73)		
	Myricetin	(20.46–244.78)		
	Naringin	n.d.		
	Phlorizin	(n.d. – 78.60)		
	Quercetin	(381.10–3631.93)		
	Resveratrol	(n.d. – 122.81)		
	Rutin	(1728.76–5845.77)		

monofloral bee pollen [28]. Table 7.1 shows summarized polyphenols in pollen reported by many authors. Some of them were quantified, as well as expressed in different units. However, based on their results, the most abundant flavonoids in many pollen samples were quercetin, rutin, than kaempferol, luteolin, apigenin, etc. Different quantified values are influenced by many factors, which also contribute to different antioxidant capacity of these pollen samples (Table 7.1).

7.4.1 *Pollen Flavonoids as Potential Maker for Assessment of Geographical and Botanical Origins*

Flavonoids in pollen are well-known as reliable chemotaxonomic markers [17]. The botanical origin of pollen is characterized by the microscopic examination of grains. Due to the specific characteristic of pollen grains, such as shape, size, and surface [6], it could define which pollen type it is. However, this method is a highly specialized procedure with the participation of an experienced melissopalynologist and a comprehensive content pollen atlas. Contrarily, polyphenolic profiles have been found to be species-specific [17].

Pollen analysis should classify pollen as monofloral or polyfloral, which is also an important marker the assessment of geographical and botanical origins. Primarily, the geographical origin of polyfloral pollen sample should be identified, as it ensures the representation of specific plant species in that area [6]. Additionally, the procedure for identification of a good marker or a combination of markers for each taxon that can be used is well described through the standard methods of pollen analysis [6].

One study by Turkish authors [29] reported that rutin as the main flavonoid in pollen samples could be possible marker for the Anzer pollen, among quantified phenolic acids. In another study, it was shown that geographical origin had a strong effect on the qualitative differences of the phytochemical composition of pollen [20]. Authors distinguished bee-collected pollen samples from different regions in Europe by the presence or absence of some flavonoids [20]. For example, coumarin was detected in Danish, Swedish and Italian pollen, hesperidin in Lithuanian, Dutch, Maltese pollen [20]. Moreover, these authors point out that the quantitative composition of flavonoids and other polyphenols is related to the geographical origin of pollen samples. Quantification of the flavonoids in bee pollen showed that different areas have a great influence on flavonoids content. However, Chinese scientists [31] were reported that five bee pollen samples with different botanical origins show more similarities than samples with different geographical origins, by evaluating characteristic common peaks using the chromatographic method. Thus, the fact that the mentioned pollen samples were also of different botanical origins should not be excluded. It is necessary to additionally highlight the influence of botanical origin on the composition of analyzed pollen samples.

Compounds such as apigenin and luteolin derivatives (flavones), as well as quercetin, kaempferol, and isorhamnetin (flavonols) are mainly found in pollen samples

[6]. A similar observation can be observed from the studies that are described in Table 7.1. Throughout the mentioned studies presented in Table 7.1, pollen samples from species that belong to the Asteraceae family dominate. The next were species from Fabaceae, Brassicaceae and then Papaveraceae family. Based on the detail analysis of pollen samples, some flavonoids could be pointed out. Thus, apigenin [26, 28, 30, 32] and quercetin 3-*O*- and isorhamnetin 3-*O*- derivatives [8, 26, 27, 32] were present in different pollen samples (such as *Anacyclus radiatus*, *Calendula officinalis*, *Helianthus annuus*, *Cynara scolymus*) from the same Asteraceae family. Another study stated that flavones were dominant in vegetables in Asteraceae family [9]. Furthermore, mentioned flavones were characteristic compounds for marigold (quercetin and isorhamnetin derivatives) and chamomile (apigenin), which are famous due to their beneficial effect on human health [34, 35]. In that view, these flavonoids or their combination could be possible taxonomic characteristics of the Asteraceae species.

Besides kaempferol and quercetin, in samples from analyzed studies (Table 7.1) rutin prevails as present in a larger number of pollen samples that belong to Brassicaceae family [23, 27, 28, 31]. Other authors also listed vegetables from Brassicaceae family, among other food products, with high content of kaempferol and quercetin [36]. In the pollen samples (Table 7.1) that had different percentage of pollen particles that belong to several families (Asteraceae, Brassicaceae, Fabaceae or Papaveraceae), besides rutin, quercetin, and myricetin, was also found catechin and/or epicatechin [28]. In another study, the same trend was also found in pollen species from Fabaceae family [23]. Accordingly, it has been singled out that a plant from the Fabaceae family (*Bauhinia monandra*) is a species rich in rutin [4], while tea (which could be from different plants) is a source of catechin and epicatechin [3, 4, 12]. These observations are important for further investigations and determination of the assessment of pollen botanical and geographical origins. However, further and more detailed studies (which will also include statistical analysis) are needed to provide additional confirmation for all these observations.

In addition to the similarities noted for the families Asteraceae, Brassicaceae and Papaveraceae, another common feature is their positive influence against several bacteria stated by Denisow et al. [2]. Moreover, the antibacterial activity of rutin [4, 10] or other effects of quercetin, myricetin, and catechins are already mentioned in the literature through their positive influence on human health [3, 4, 10, 12]. In addition to highlighting these compounds, the antioxidant activity of plants is also based on many other phytochemicals.

Thus, the strong influence that the phytochemical composition of pollen has is an important marker for the assessment of the geographical and botanical origins of many products, including bee products. Bearing in mind that honey is a widely used bee product, in the following text several examples of the influence of pollen, or more precisely flavonoids from pollen, on the botanical origin of honey be will listed. Bees primarily need pollen as a source of necessary proteins that they cannot obtain through collected nectar. Therefore, pollen and flavonoids from pollen (partially transformed or not) are also naturally found in honey. Pollen particles are important for honey distinction. Hence, the generally accepted method of honey classification is pollen analysis. Most studies distinguished honey samples by melissopalynological analysis [37, 38] among other techniques [39]. However, the similarity of the flavonoid profiles of honey and pollen, and honey and nectar, could be different. Our previous study showed more similarity between the flavonoid profiles of honey and nectar than between honey and pollen [16]. Moreover, the study shows that in honey with different content and percentage of pollen grains, there was no significant variation between the flavonoid profiles of the analyzed honey samples [16]. In addition to pollen, nectar and propolis were confirmed as origins of some flavonoids [25]. Furthermore, flavonoids in honey could be specific for certain regions as well as for the harvest season of honey [37]. According to Gašić et al. [7], several examples of polyphenols that were have been proposed to be markers of different types of honey are listed. Thus, quercetin was reported as a marker for sunflower honey (*Helianthus annuus,* Asteraceae) [7], which is in accordance with the previously mentioned presence of this flavonoid in the analyzed pollen from the Asteraceae family [26, 30]. Naringenin and kaempferol have been proposed as markers for lavender (*Lavandula angustifolia,* Lamiaceae) and rosemary (*Rosmarinus officinalis*, Lamiaceae) honey, respectively, as well as in *Leonurus* pollen (Lamiaceae family) [31]. Furthermore, kaempferol prevails in analyzed *Leonurus* pollen [31] among the other four flavonoids (rutin, quercitrin, isoquercitrin, and naringenin).

With this in mind, as well as many studies that investigating polyphenol profiles [7, 14–16, 26, 30], flavonoid content, synergism and influence of other compounds are valuable for further chemotaxonomy of many products. Besides the phytochemical composition of flavonoids, the evaluation of the markers is influenced by cultivars and external factors.

7.4.2 Pollen Flavonoids Important for Antioxidant Activity

The intake requirements of important phytochemicals for humans are far higher than what is found in pollen [2]. Regardless, the essential content of these antioxidant substances and their effects should not be overlooked. It is well known that flavonoids could show antibacterial, anti-inflammatory, anti-allergic, antithrombotic and vasodilator effects, as well as antitumor activities [4, 10, 12]. Furthermore, there are stuidies that support many medicinal effects of flavonoids from bee pollen [2, 28]. All these properties are based on their antioxidant activity.

The antioxidant activity of flavonoids depends on the number and distribution of hydroxyl groups. For example, flavones with their structure stand out. With more OH groups (e.g. flavonols quercetin, kaempferol and myricetin compared to chrysin, apigenin), they have a lower redox potential which characterizes them as more effective antioxidants. Also, the saturation of the heterocyclic C ring stabilizes the radical and reduces the antioxidant activity of flavonoids (e.g. flavanes, flavanols, flavanones have lower antioxidant activity compared to flavones and flavonols).

The differences between hand-collected and bee-collected pollen could be defined by the influence of the fermentation process that occurs during the bee's transport of pollen grains to the honeycomb. Furthermore, the increase of the antioxidant activity in the range of around 1.2–3.1 times, expressed through the total phenolic content, total flavonoid content, and radical scavenging activity was shown in the study of fermentation influence on pollen from various regions of Europe [20].

Flavonoids, as well as other polyphenols, and many phytochemicals, contribute to the final antioxidant activity of pollen. However, due to the well-known antioxidant capacity of flavonoids, Table 7.1 shows the parameters from the literature that are the results of the combined effects of many compounds that contribute to the total antioxidant activity of pollen.

7.5 Summary

Primarily, this chapter provides a comprehensive overview of the main subgroup of polyphenols, i.e. flavonoids in pollen, both hand-collected, which is rare in the studies, and bee-collected pollen. The importance of the presence of flavonoids in pollen is emphasized, which contributes to classifying this food as a functional food, which is also common for other bee products. Also, the connection of flavonoids with the assessment of the botanical and geographical origins of pollen and other products, and therefore with taxonomy, is highlighted. And finally, among others, the antioxidant activity of the pollen, which is manifested due to the presence of flavonoids in it, is also shown.

In the end, this chapter is only a small part of the whole story about pollen, as well as about a small part of the most important and the largest group of polyphenols, i.e. flavonoids. The quantity of flavonoids in pollen should not be excluded, which is often not enough for human consumption. However, this is a very important topic that can be used for further research in order to achieve more markers valuable for taxonomic characteristics and to confirm their influence on the level of free radicals and any diseases. Further developments will provide newer insights into pollen flavonoids and will certainly lead to new pharmaceutical agents for the treatment of many diseases.

Author Contributions Conceptualization, writing, and visualization M.K; supervision, review and editing Ž.T.; funding acquisition, Ž.T., S.B. All authors have read and agreed to the published version of the manuscript. ORCID Nos of authors: 0000-0003-0989-7398 (M.K.); 0000-0002-5162-3123 (Ž.T.); 0000-0002-0451-5912 (S.B.).

Funding The authors acknowledge the financial support from the Ministry of Education, Science and Technological Development of the Republic of Serbia (Contracts Nos: 451-03-68/2022-14/200168 (Ž.T.); 451-03-68/2022-14/200051 (M.K.).

Conflicts of Interest The authors declare no conflict of interest.

References

1. Alotaibi SS, Sayed SM, Alosaimi M, Alharthi R, Banjar A, Abdulqader N, Alhamed R (2020) Pollen molecular biology: applications in the forensic palynology and future prospects: a review. Saudi J Biol Sci 27(5):1185–1190. https://doi.org/10.1016/j.sjbs.2020.02.019
2. Denisow B, Denisow-Pietrzyk M (2016) Biological and therapeutic properties of bee pollen: a review. J Sci Food Agric 96(13):4303–4309. https://doi.org/10.1002/jsfa.7729
3. Fraga CG, Croft KD, Kennedy DO, Tomás-Barberán FA (2019) The effects of polyphenols and other bioactives on human health. Food Funct 10(2):514–528. https://doi.org/10.1039/C8FO01997E
4. Guven H, Arici A, Simsek O (2019) Flavonoids in our foods: a short review. JBACHS 3(2):96–106. https://doi.org/10.30621/jbachs.2019.555
5. Kostić AŽ, Milinčić DD, Petrović TS, Krnjaja VS, Stanojević SP, Barać MB, Tešić ŽLJ, Pešic MB (2020) Mycotoxins and mycotoxin producing fungi in pollen. Toxins 11(2):64. https://doi.org/10.3390/toxins11020064
6. Campos MG, Anjos O, Chica M, Campoy P, Nozkova J, Almaraz-Abarca N, Barreto LMRC, Nordi JC, Estevinho LM, Pascoal A, Paula VB, Chopina A, Dias LG, Tešić ŽLJ, Mosić MD, Kostić AŽ, Pešić MB, Milojković-Opsenica DM, Sickel W, Ankenbrand MJ, Grimmer G, Steffan-Dewenter I, Keller A, Förster F, Tananaki CH, Liolios V, Kanelis D, Rodopoulou M-A, Thrasyvoulou A, Paulo L, Kast C, Lucchetti MA, Glauser G, Lokutova O, de Almeida-Muradian LB, Szczęsna T, Carreck NL (2021) Standard methods for pollen research. J Apic Res 60(4):1–109. https://doi.org/10.1080/00218839.2021.1948240
7. Gašić UM, Milojković-Opsenica DM, Tešić ŽLJ (2017) Polyphenols as possible markers of botanical origin of honey. J AOAC Int 100(4):852–861. https://doi.org/10.5740/jaoacint.17-0144
8. Bakour M, Campos MDG, Imtara H, Lyoussi B (2020) Antioxidant content and identification of phenolic/flavonoid compounds in the pollen of fourteen plants using HPLC-DAD. J Apic Res 59(1):35–41. https://doi.org/10.1080/00218839.2019.1675336
9. Hostetler GL, Ralston RA, Schwartz SJ (2017) Flavones: food sources, bioavailability, metabolism, and bioactivity. Adv Nutr 8(3):423–435. https://doi.org/10.3945/an.116.012948
10. Dewick PM (2002) Medicinal natural products: a biosynthetic approach, 3rd edn. Wiley, Chichester
11. Palma-Tenango M, Soto-Hernández M, Aguirre-Hernández E (2017) Flavonoids in agriculture. In: Justino GC (ed) Flavonoids-from biosynthesis to human health. IntechOpen, pp 189–201. https://doi.org/10.5772/intechopen.68626
12. Kumar S, Pandey AK (2013) Chemistry and biological activities of flavonoids: an overview. Sci World J 2013:162750. https://doi.org/10.1155/2013/162750
13. Nešović M, Gašić U, Tosti T, Horvacki N, Nedić N, Sredojević M, Blagojević S, Ignjatović LJ, Tešić Ž (2021) Distribution of polyphenolic and sugar compounds in different buckwheat plant parts. RSC Adv 11(42):25816–25829. https://doi.org/10.1039/D1RA04250E
14. Fotirić Akšić MF, Nešović M, Ćirić I, Tešić Ž, Pezo L, Tosti T, Gašić U, Dojčinović B, Lončar B, Meland M (2022) Polyphenolics and chemical profiles of domestic Norwegian Apple (*Malus x domestica* Borkh.) cultivars. Front Nutr 9:9414. https://doi.org/10.3389/fnut.2022.941487
15. Fotirić Akšić M, Nešović M, Ćirić I, Tešić Ž, Pezo L, Tosti T, Gašić U, Dojčinović B, Lončar B, Meland M (2022) Chemical fruit profiles of different raspberry cultivars grown in specific norwegian agroclimatic conditions. Horticulturae 8(9):765. https://doi.org/10.3390/horticulturae8090765
16. Nešović M, Gašić U, Tosti T, Horvacki N, Šikoparija B, Nedić N, Blagojević S, Ignjatović LJ, Tešić Ž (2020) Polyphenol profile of buckwheat honey, nectar and pollen. R Soc Open Sci 7(12):201576. https://doi.org/10.1098/rsos.201576
17. Campos M, Markham KR, Mitchell KA, da Cunha AP (1997) An approach to the characterization of bee pollens via their flavonoid/phenolic profiles PCA. Int J Chem Biochem 8(4):181–185. https://doi.org/10.1002/(SICI)1099-1565(199707)8:4<181::AID-PCA359>3.0.CO;2-A

18. Clarke D, Morley E, Robert D (2017) The bee, the flower, and the electric field: electric ecology and aerial electroreception. J Comp Physiol 203:737–748. https://doi.org/10.1007/s00359-017-1176-6
19. Campos MG, Bogdanov S, de Almeida-Muradian LB, Szczesna T, Mancebo Y, Frigerio C, Ferreira F (2008) Pollen composition and standardisation of analytical methods. J Apic Res 47(2):154–161. https://doi.org/10.1080/00218839.2008.11101443
20. Adaškevičiūtė V, Kaškonienė V, Barčauskaitė K, Kaškonas P, Maruška A (2022) The impact of fermentation on bee pollen polyphenolic compounds composition. Antioxidants 11(4):645. https://doi.org/10.3390/antiox11040645
21. Somerville DC (2001) Nutritional value of bee collected pollens. A report for the Rural Industries Research and Development Corporation. Report, Rural Industries Research and Development Corporation, Project No DAN-134A
22. Thakur M, Nanda V (2020) Composition and functionality of bee pollen: a review. Trends Food Sci Technol 98:82–106. https://doi.org/10.1016/j.tifs.2020.02.001
23. de Melo AAM, Estevinho LM, Moreira MM, Delerue-Matos C, de Freitas ADS, Barth OM, de Almeida-Muradian LB (2018) A multivariate approach based on physicochemical parameters and biological potential for the botanical and geographical discrimination of Brazilian bee pollen. Food Biosci 25:91–110. https://doi.org/10.1016/j.fbio.2018.08.001
24. Algethami JS, El-Wahed AAA, Elashal MH, Ahmed HR, Elshafiey EH, Omar EM, Naggar YA, Algethami AE, Shou Q, Alsharif SM, Xu B, Shehata AA, Zhiming G, Khalifa SA, Wang K, El-Seedi HR (2022) Bee pollen: clinical trials and patent applications. Nutrients 14(14):2858. https://doi.org/10.3390/nu14142858
25. Tomás-Barberán FA, Ferreres F, García-Vignera C, Tomás-Lorente F (1993) Flavonoids in honey of different geographical origin. Z Lebensm Unters Forsh 196(1):38–44. https://doi.org/10.1007/BF01192982
26. Kostić AŽ, Milinčić DD, Gašić UM, Nedić N, Stanojević SP, Tešić ŽLJ, Pešić MB (2019) Polyphenolic profile and antioxidant properties of bee-collected pollen from sunflower (*Helianthus annuus* L.) plant. LWT-Food Sci Technol 112:108244. https://doi.org/10.1016/j.lwt.2019.06.011
27. Karabagias I, Karabagias V, Gatzias I, Riganakos K (2018) Bio-functional properties of bee pollen: the case of "bee pollen yoghurt". Coatings 8:423. https://doi.org/10.3390/coatings8120423
28. Oyarzún JE, Andia ME, Uribe S, Núñez Pizarro P, Núñez G, Montenegro G, Bridi R (2020) Honeybee pollen extracts reduce oxidative stress and steatosis in hepatic cells. Molecules 26(1):6. https://doi.org/10.3390/molecules26010006
29. Ulusoy E, Kolayli S (2014) Phenolic composition and antioxidant properties of Anzer bee pollen. J Food Biochem 38(1):73–82. https://doi.org/10.1111/jfbc.12027
30. Fatrcová-Šramková K, Nôžková J, Máriássyová M, Kačániová M (2016) Biologically active antimicrobial and antioxidant substances in the *Helianthus annuus* L. bee pollen. J Environ Sci Health B 51:176–181. https://doi.org/10.1080/03601234.2015.1108811
31. Duan H, Dong Z, Li H, Li WR, Shi SX, Wang Q, Cao W, Fang X, Fang A, Zhai KF (2019) Quality evaluation of bee pollens by chromatographic fingerprint and simultaneous determination of its major bioactive components. Food Chem Toxicol 134:110831. https://doi.org/10.1016/j.fct.2019.110831
32. Kostić AŽ, Milinčić DD, Nedić N, Gašić UM, Špirović Trifunović B, Vojt D, Tešić ŽLJ, Pešić MB (2021) Phytochemical profile and antioxidant properties of bee-collected artichoke (*Cynara scolymus*) pollen. Antioxidants 10(7):1091. https://doi.org/10.3390/antiox10071091
33. Bayram NE, Gercek YC, Çelik S, Mayda N, Kostić AŽ, Dramićanin AM, Özkök A (2021) Phenolic and free amino acid profiles of bee bread and bee pollen with the same botanical origin–similarities and differences. Arab J Chem 14(3):103004. https://doi.org/10.1016/j.arabjc.2021.103004
34. Olennikov DN, Kashchenko NI, Chirikova NK, Akobirshoeva A, Zilfikarov IN, Vennos C (2017) Isorhamnetin and quercetin derivatives as anti-acetylcholinesterase principles of marigold (*Calendula officinalis*) flowers and preparations. Int J Mol Sci 18(8):1685. https://doi.org/10.3390/ijms18081685

35. Dai Y-L, Li Y, Wang Q, Niu F-J, Li K-W, Wang Y-Y, Wang J, Zhou C-Z, Gao L-N (2023) Chamomile: a review of its traditional uses, chemical constituents, pharmacological activities and quality control studies. Molecules 28:133. https://doi.org/10.3390/molecules28010133
36. Dabeek WM, Marra MV (2019) Dietary quercetin and kaempferol: bioavailability and potential cardiovascular-related bioactivity in humans. Nutrients 11(10):2288. https://doi.org/10.3390/nu11102288
37. Bugeja DA, Nešović M, Šikoparija B, Radišić P, Tosti T, Trifković J, Luigi R, Everaldo A, Tešić Ž, Gašić U (2022) Melissopalynology analysis, determination of physicochemical parameters, sugars and phenolics in Maltese honey collected in different seasons. J Serb Chem Soc 87(9):983–995. https://doi.org/10.2298/JSC211214033B
38. Nedić N, Nešović M, Radišić P, Gašić U, Baošić R, Joksimović K, Pezo L, Tešić Ž, Vovk I (2022) Polyphenolic and chemical profiles of honey from the Tara Mountain in Serbia. Front Nutr 9:941463. https://doi.org/10.3389/fnut.2022.941463
39. Nešović M, Gašić U, Tosti T, Trifković J, Baošić R, Blagojević S, Ignjatović LJ, Tešić Ž (2020) Physicochemical analysis and phenolic profile of polyfloral and honeydew honey from Montenegro. RSC Adv 10(5):2462–2471. https://doi.org/10.1039/C9RA08783D

Chapter 8
Carotenoids and Vitamins of Pollen

Rodica Mărgăoan and Mihaiela Cornea-Cipcigan

8.1 Introduction

The term "functional food" was created to develop and provide novel and healthy sources of dietary constituents (i.e., carotenoids, fatty acids, vitamins, polyphenols, proteins etc.) that possess beneficial health effects [1]. Bee collected pollen (BP) varies in color, shape, texture, biological activity, and physicochemical properties [2]. This variation is due to regional, seasonal, and genetic differences in flower species visited by bees, each of which has distinct properties [3]. Based on the plant that bees visit, BP can be extremely diverse in color, from yellow, orange, brown or yellowish brown to yellowish blue [4]. The botanical origin of BP is important since it may serve as an indication of biological activity, phytochemicals, dietary contents, and may impact industrial quality [5].

In contrast to mixed pollen loads from different flowers, which are referred to as polyfloral, BP is categorized as monofloral when a single botanical classification (>80% of one grain type) describes the pollen loads while maintaining the plants physical and biochemical features [6]. BP is mostly sold commercially as polyfloral.

The color of BP is strongly correlated to plant pigments such as carotenoids and anthocyanins that are naturally present in varying amounts and can be used to generally assess the quality of BP [7].

Carotenoids, a class of isoprenoids, account for the color of fruits and flowers, and are key-roles in human health. Among them, β-carotene represents the main source of vitamin A, and its consumption may reduce the risk of several cancer types or cardiovascular diseases [8]. The total carotenoid content of saponified BP samples revealed a broad range of concentrations, mainly 49.90–425.32 μg/g in

R. Mărgăoan (✉) · M. Cornea-Cipcigan
Advanced Horticultural Research Institute of Transylvania, University of Agricultural Sciences and Veterinary Medicine, Cluj-Napoca, Romania
e-mail: rodica.margaoan@usamvcluj.ro; mihaiela.cornea@usamvcluj.ro

N. Ecem Bayram et al. (eds.), *Pollen Chemistry & Biotechnology*,
https://doi.org/10.1007/978-3-031-47563-4_8

fresh BP with the highest value in *Taraxacum officinale* BP [9]. *T. officinale* proves to be a natural source of β-carotene in high amount [10], exceeding the levels found in broccoli, carrots, and spinach [11]. *Filipendula ulmaria* BP also has high levels of β-carotene (36.3 ± 0.17 μg/g) [12].

Vitamins are a class of organic substances with numerous metabolic functions and high antioxidant activities [13]. Although vitamins are found in low amounts in human diets, they are essential micronutrients for health development and maintenance [14]. Vitamins are found naturally in foods, particularly fruits and vegetables, and have the role to maintain and regulate metabolic and cellular energy processes. The health benefits are proportional to daily consumption. Vitamin deficiency or excess disturb various metabolic processes, resulting in a variety of illnesses [15]. As a result, optimal consumption of all vitamins is required for cell and tissue function. Several studies have shown that BP is rich in vitamins, with an optimal concentration that meets this requirement [16, 17]. BP mostly includes vitamins from the B complex, vitamins C, D and E, together with carotenoids; however there is little research on the association between plant origin and vitamin quantification. BP samples from Spain presented high levels of carotenes with *Cistus* sp. as predominant pollen [17], whereas processed BP samples from southern Brazil lacked vitamin C or β-carotene [18]. These vitamins may have been lost during processing, or their absence may be due to botanical origin, as every pollen has its unique species-specific characteristics.

8.2 Carotenoids

8.2.1 Identification and Concentration in BP

Carotenoids are a class of pigmented components that are produced by plants (roots, flowers). They protect plants from photo damage and oxidative stress [19]. Roughly 100 dietary carotenoids are found in human diets as colored foods [20]. BP is rich in carotenoids accounting for its many hues, that vary depending on its botanical origin [21]. The most abundant and common carotenoids identified in BP are β-carotene, lutein, and β-cryptoxanthin [9]. After saponification, the main carotenoids identified in BP are lutein and β-cryptoxanthin, with lower or trace amounts of β-carotene detected. The quantity of carotenoids found in BP ranges from 5.0 and 1233.0 μg/g [7].

Romanian BP samples contain a broad range of carotenoid concentrations, mainly up to 425.32 μg/g in fresh BP and 527.27 μg/g in dry BP. Samples with predominant BP in *Taraxacum officinale* presented the highest carotenoid coencentration, with the lowest in *Prunus spinosa* BP [9]. Conversely, in *Filipendula ulmaria* BP samples the carotenoid content varied between 295.40 and 59.30 μg/g fresh weight (fw). The significantly difference in carotenoid content is due to the secondary pollen in *Calluna* sp. (P3) with lower content in carotenoids, compared with *Hypericum* sp. (P2 and P9) with significantly higher levels. Individual carotenoids present in BP samples can be separated, identified, and quantified using HPLC-PDA analysis of carotenoids. Lutein and β-cryptoxanthin were found in all tested samples, the most abundant being lutein with concentrations between 44.52 and 392.52 μg/g fw (57.04 and 476.30 μg/g dw), whereas β-cryptoxanthin ranged between 1.02 and 24.96 μg/g fw. β-carotene was identified in trace amounts in few samples. A broad range of total carotenoid concentration was observed in 10 Brazilian BP samples, ranging from "traces" to 451.50 μg/g fw (489.20 μg/g dw). Although the BP samples were marked as monofloral, the palynological analysis revealed the presence of 17 plant species. It was demonstrated that the botanical origin is closely related to the carotenoid content, with the maximum value achieved in Vernonia and Senecio BP. The absence of β-carotene may be due to the drying procedure [18]. The total carotenoid concentrations in fresh Brazilian BP ranged between 27.08 and 344.6 μg/g. The authors demonstrated that freeze-stored BP lost an average of 11–12% of β-carotene content after half and 1 year, demonstrating that this is the best method to preserve BP when its carotenoid content is taken into account. Thus, β-Carotene concentration in fresh BP ranged from 3.77 to 99.27 μg/g [16].

Romanian BP samples correspond with values of Brazilian fresh BP for total carotenoid concentration, but with lower levels of β-carotene in Romanian samples (up to 13.2 μg/g fresh BP). This may be due to differences in flora and environmental conditions amongst the samples.

Colombian BP revealed the presence of xanthophylls in high amount, mainly β-cryptoxanthin, zeaxanthin and its isomers, with the highest levels present in BP collected in March and September [22]. Colombian BP is the richest in zeaxanthin and its isomers and with lower levels in phytoene and lutein [22, 23]. Conversely, Chestnut and Willow BP presented the highest amount in anteraxanthin, lutein, α-carotene and zeaxanthin [24], whereas BP from Marmara (Turkey) presented higher levels of zeaxanthin and lutein, and those from the Black Sea region of α- and β-carotene [25]. Out of these, Chestnut BP revealed similar carotenoid levels with foods, such as watermelon and tomato, whereas Willow BP presented comparable levels to sweet potato and pumpkin [26]. Pacheco et al. (2014) revealed that α-carotene levels in vegetable matrices ranged from 3.0 to 121.0 mg/g and β-carotene levels ranged between 4.1 and 331.0 mg/g. These levels are consistent with the Black Sea Region BP samples that possess higher levels of α- and β-carotene compared to carrots (57.1 mg/g for β- and 45.0 mg/g for α-carotene). A teaspoon of chestnut BP (equivalent to 15 g) contains a high level of carotenoids and can supply the human with more than 100% of its daily need. Botanical sources of carotenoids found in BP and the quantification range are presented in Table 8.1.

Table 8.1 Botanical sources of carotenoids found in BP

Carotenoids	Bee product origin	Botanical source (Predominant BP)	Range quantification (μg/g)	Fresh or dry weight	Reference
β-Carotene	Brazil	Brassicaceae (*Brassica* sp.)	1.9–13.5	dw	[7]
		Mimosaceae (*Mimosa* sp.)	4.9 ± 0.4		
		Fabaceae (*Machaerium* sp.)	3.6–4.1		
		Asteraceae (*Piptocarpha* sp.)	112.7 ± 12.8		
		Asteraceae (*Eupatorium* sp.)	4.6 ± 0.2		
	Spain	Cistaceae (*Cistus ladanifer*)	4.31 ± 1.09	fw	[27]
		Cistaceae (*Cistus ladanifer*)	2.77 ± 0.78	dw	
	Italy	Willow (*Salix alba*)	8.98	fw	[24]
	Germany	Sallow (*Salix caprea*)	4.7 ± 0.8	fw	[28]
		Prunus mahaleb (Mahaleb cherry)	6.8 ± 1.4	fw	
		Tilia platyphyllus (Large-leaved linden	17.5 ± 0.8	fw	
	Turkey (Bursa)	Chestnut	8.39 ± 0.17	fw	[29]
	Turkey (Marmara)		2.42–294.42	fw	[25]
	Turkey (Black Sea Region)		19.31–395.02	fw	
	Romania	*Filipendula ulmaria*	13.98–18.18	dw	[9]
		Prunus sp.	3.7		
		Taraxacum officinale	0.72		
		Fabaceae (*Anthilis* sp.)	17.18		
		Fabaceae (*Trifolium repens*)	11.22		
		Fabaceae (*Robinia pseudoacacia*)	5.07		
		Salicaceae (*Salix* sp.)	2.71		

Lutein	Italy	Chestnut	6.4 (μg/g)	fw	[24]
		Willow	54.23		
	Germany	White horse chestnut (*Aesculus hippocastanum*)	2.5 ± 1.2		[28]
		Sallow	3.5 ± 0.7	fw	
		Mahaleb cherry	4.3 ± 0.4		
	Turkey (Bursa)	Chestnut	5.91 ± 0.09	fw	[29]
	Turkey (Marmara)		7.22–33.42	fw	[25]
	Turkey (Black Sea Region)		3.88–18.59	fw	
	Romania	*C. monogyna*	64.17	dw	[9]
		F. ulmaria	65.80–329.67		
		M. domestica	229.62		
		Prunus sp.	57.04		
		R. canina	72.47		
		C. officinalis	346.9		
		T. officinale	476.3		
		Brassica sp.	66.53		
		C. vulgaris	130.28		
		Anthilis sp.	72.47		
		T. repens	235.88		
		R. pseudoacacia	116.29		
		Salix sp.	85.33		
α-carotene	Brazil	*Machaerium* sp.	9.9–10.0		[7]
		Piptocarpha sp.	324.7 ± 49.7	dw	
		Eupatorium sp.	4.5 ± 0.3		
	Italy	Willow	17.65	fw	[24]
	Turkey (Marmara)	Chestnut	0.00–6.63		[25]
	Turkey (Black Sea Region)		6.20–152.42	fw	

(continued)

Table 8.1 (continued)

Carotenoids	Bee product origin	Botanical source (Predominant BP)	Range quantification (μg/g)	Fresh or dry weight	Reference
Pro-vitamin A	Brazil	*Brassica* sp.	15.8–112.1 (retinol μg/100 g)	dw	[7]
		Mimosa sp.	40.4		
		Machaerium sp.	71.0–75.6		
		Piptocarpha sp.	2292		
		Eupatorium sp.	57.1		
		Myrtaceae (*eucalyptus*)	0.0–6.5		
α-tocopherol	Colombia	*Brassica* sp.	15.50–88.71		[22]
		Asteraceae (*Hypochaeris radiata*)	47.32–82.02	dw	
		Eucalyptus sp.	25.49–30.45		
		T. repens	112.35 ± 3.31		
		Euphorbiaceae (*Acalypha diversifolia*)	33.95 ± 0.47		
		Polygonaceae (*Muehlenbeckia tamnifolia*)	65.17 ± 1.44		
	Italy	Chestnut	4.73	fw	[24]
		Willow	7.95		
	Colombia	*Brassica* sp.	65.15–631.20	dw	[22]
		Hypochaeris radiata	438.74–1084.64		
		Eucalyptus sp.	345.01–355.94		
		T. repens	682.51 ± 75.86		
		A. diversifolia	393.69 ± 39.33		
		M. tamnifolia	945.09 ± 114.6		
Zeaxanthin	Colombia	*Brassica* sp.	35.29–181.02		
		H. radiata	127.71–308.89		
		Eucalyptus sp.	115.43–117.18		
		T. repens	201.58 ± 21.64		
		A. diversifolia	123.93 ± 17.21		
		M. tamnifolia	243.41 ± 27.48		

	Italy	Willow	10.74	fw	[24]
	Germany	*A. hippocastanum*	3.9 ± 0.9		[28]
		Sallow	7.0 ± 0.2	fw	
		Mahaleb cherry	6.4 ± 0.8		
	Turkey (Bursa)	Chestnut	6.97 ± 0.14	fw	[29]
	Turkey (Marmara)		15.82–36.38	fw	[25]
	Turkey (Black Sea Region)		4.88–27.20	fw	
9,13-di-Z-zeaxanthin	Colombia	*Brassica* sp.	18.88–109.33	dw	[22]
		H. radiata	88.37–223.7		
		Eucalyptus sp.	73.01–78.11		
		T. repens	122.64 ± 14.35		
		A. diversifolia	81.02 ± 8.28		
		M. tamnifolia	181.10 ± 24.51		
9,15-di-Z-zeaxanthin	Colombia	*Brassica* sp.	18.23–101.90		
		H. radiata	81.89–189.17		
		Eucalyptus sp.	58.57–69.23		
		T. repens	110.49 ± 12.94		
		A. diversifolia	110.49 ± 12.94		
		M. tamnifolia	164.64 ± 20.36		
β-cryptoxanthin	Germany	Sallow	9.6 ± 2.5	fw	[28]
		Mahaleb cherry	4.3 ± 1.3	fw	
		Large-leaved linden	7.9 ± 1.2	fw	
	Turkey (Bursa)	Chestnut	8.39 ± 0.17	fw	[29]
	Turkey (Marmara)		7.81–44.67	fw	[25]
	Turkey (Black Sea Region)		7.55–34.21	fw	

(continued)

Table 8.1 (continued)

Carotenoids	Bee product origin	Botanical source (Predominant BP)	Range quantification (μg/g)	Fresh or dry weight	Reference
	Romania	*C. monogyna*	1.31	dw	[9]
		F. ulmaria	2.28–18.30		
		M. domestica	7.09		
		Prunus sp.	3.19		
		R. canina	4.62		
		C. officinalis	19.79		
		T. officinale	16.15		
		Brassica sp.	1.38		
		C. vulgaris	7.14		
		Anthilis sp.	28.78		
		T. repens	10.73		
		R. pseudoacacia	5.58		
		Salix sp.	2.76		

8.2.2 *Health Effects of Carotenoids*

Carotenoids have a significant role in cell proliferation, cellular malfunction management, and the prevention of frequent ailments such as cancer [30]. Pre-clinical studies listed in Table 8.2 have established the use of carotenoids against different disorders. Previously, Zhang et al. (2016) evaluated the use of β-carotene in combination with 5-fluorouracil (a commonly used cancer treatment) as a treatment against human oesophageal cancer [31]. Carotenoids, such as α- and β-carotene, lutein, zeaxanthin, and their isomers protect the skin against cell oxidation [32]. Supplementation with an antioxidant complex for 7 weeks composed of carotenoids, vitamins, and selenium enhanced epidermal resistance against UV-induced damage [33]. Lutein and zeaxanthin act as photo-protective agents against UVR (ultraviolet radiation), making them effective in the prevention or treatment of melasma (dark skin discoloration) and in eye protection [34, 35], whereas β-carotene supplementation (30 mg/day) for 90 days improved skin elasticity and reduced wrinkles [36]. Intake of vitamin C, E and β-carotene supplements in persons aged 55–80 years old led to a 25% lesser progression of macular aging [37].

Serum carotenoids (i.e. lutein and zeaxanthin) were demonstrated to lower the risk of neurodegenerative diseases by brain function improvement in elderly [38, 39]; patients with Alzheimer's disease have lower concentrations in vitamin A and β-carotene in plasma and serum, which increase with dietary supplementation of carotenoids [40].

Carotenoids are also an effective treatment against heart diseases, as low serum β-carotene levels increase the incidence of cardiac failure [41]. Research on coronary artery disease reported considerably reduced levels of β-carotene, along with higher level in inflammatory marker IL-6, indicating that β-carotene may have a preventive impact on atherosclerosis via inflammatory suppression [42, 43]. More details can be seen in Table 8.2.

8.3 Vitamins

8.3.1 *Vitamins Identification and Concentration in BP*

Vitamins are a class of chemical molecules that are required for growth and health maintenance in humans [70]. These chemicals are classified based on their solubility. Lipo-soluble vitamins (e.g., A, K, and E) are a class of chemical compounds that have various structures but are soluble in organic solvents. They accumulate as body fat and can be harmful if taken in large quantities. Hydro-soluble vitamins (B_1 (thiamine), B_2 (riboflavin), B_3 (niacin), B_5 (pantothenic acid), B_6 (pyrodoxal), B_9 (folic acid), B_{12} (cobalamin), and C) are found stored in significantly lower amounts in the body, so they are required on a daily basis [70].

Table 8.2 *In vivo* preclinical studies of different carotenoids present in BP

Carotenoids	Disorders	Study model	Treatment scheme	Duration	Biological activity	Reference
β-Carotene	Cadmium-induced (Cd) toxicity	Male SD rats	G1: control; G2: β-Carotene (10 mg/kg bw); G3: vit. E (100 mg/kg); G4- β-Carotene+ vit. E; G5: $CdCl_2$ (5 mg/kg); G6: $CdCl_2$ + β-Carotene; G7: $CdCl_2$+ vit. E; G8: $CdCl_2$ + β-Carotene+vit. E	30 days	↓TBARS levels in liver, plasma, brain and testes in G2 and G6 ↑ GST levels in plasma and liver in G2 and G6 ↓ GST levels in brain and testes ↓ plasma and testes AST and ALT ↑ liver AST, ALT, AIP ↑ plasma and brain AchE and protein ↑ sperm concentration and motility in G2 and G6	[44]
	Oxidative stress; diabetes	Female SD rats	G1: control; G2: β-carotene; G3: diabetic control; G4: Diabetic+ β-carotene	14 days	↑ γ -glutamyl transpeptidase in kidney G4 compared to G3 ↓ γ -glutamyl transpeptidase in liver ↑ glutathione reductase in kidney, liver and heart ↑ GP-x in liver and heart ↑ SOD in liver and ↓ SOD in heart and kidney in G4 compared to G3 ↑ CAT in liver and kidney ↓ CAT in heart ↓ TBARS in liver, kidney, and heart in G4	[45]
	Non-alcoholic fatty liver disease (NAFLD)	Male SD rats	G1: control; G2: NAFLD-HFD; G3: NAFLD-RD; G4: NAFLD-HFD + RSV; G5: NAFLD-RD + RSV; G6: NAFLD-HFD + β-carotene (10 mg/kg/day); G7: NAFLD-RD + β-carotene; G: NAFLD-HFD + RSV + β-carotene; G9: NAFLD-RD + RSV + β-carotene	4 weeks	↑ triglycerides and total cholesterol in alone and combination β-carotene groups on liver functions ↓ HDL, VLDL ↑LDL in liver ↑ GSH and HFD and RD groups with β-carotene ↓ SOD in HFD and RD groups with β-carotene ↑ TBARs in HFD+ β-carotene ↓ TBARS in RD + β-carotene group	[46]

	Methotrexate-induced testicular injury	Male Wistar albino rats	G1: control (saline i.p.) G2: β-carotene (10 mg/kg/day i.p.); G3: MTX (20 mg/kg i.p.) on day 21; G4: β-carotene + MTX	24 days	↓ testicular damage and MDA levels in G2 and G4 ↑ CAT, SOD, GP-x levels in G2 and G4	[47]
	Dextran sulfate sodium (DSS)-stimulated ulcerative colitis	Male SD rats	G1: control; G2: β-carotene (oral gavage, 50 mg/kg bw/day); G3: DSS polymers in drinking water 7-day; G4: DSS+ β-carotene	7 days	↓ inflammatory protein expressions in the serum (TNF-α, IL-6, IL-1β, IFN-γ, LPS and D-Lactate protein expression) ↓ inflammatory gene expression in colon ↓ NF-κB and MAPK protein expression levels pathway in G2 and G4	[48]
	Spinal cord injury	Male SD rats	G1: control; G2: SCI + saline; G3-G6: SCI + 10, 20, 40, and 80 mg/kg β-carotene i.p.	72 h	↑ functional recovery at 24 h post-SCI with increased β-carotene dosage (G4-G6 groups) ↓ ROS production, MDA and NO levels in β-carotene groups ↑ SOD, protein expression of Nrf2 and HO-1 at 72 h post-SCI ↓ Pro-inflammatory cytokines and astrocyte activation	[49]
Lutein	Cataract induced by diabetes	Male Wistar rats	G1: control; G2: lutein (0.5 mg/kg orally); G3: untreated diabetic; G4 (diabetic): insulin (50 mUI/g, subcutaneous); G5 (diabetic): lutein; G6 (diabetic): insulin+lutein	12 weeks	↓ blood glucose and glycated hemoglobin in G4 and G6 ↑ blood glucose in G3 & G5 ↓ bw in G3 ang G5 ↓ cataract incidence in lutein-treated groups	[50]
	Non-alcoholic fatty liver disease	Male SD rats	G1: ND; G2: HFD; G3: HFD + lutein 12.5 mg/kg bw; G3: HFD + lutein 25; G4: HFD + lutein 50	45 days	↓ abdominal and perinephric fat index in lutein-treated groups ↓ total cholesterol, LDL-C in G3; ↑ total cholesterol, LDL-C in serum, and tryglicerides in liver in G4; ↓ tryglicerides in serum and total cholesterol in liver; ↓ glutamic pyruvic transaminase and lipid accumulation in lutein-fed groups	[51]

(continued)

Table 8.2 (continued)

Carotenoids	Disorders	Study model	Treatment scheme	Duration	Biological activity	Reference
	Osteoporosis	SD Wistar female rats	G1: control; G2: lutein (50 mg/kg bw); G3: OVX; G4: OVX + lutein	4 weeks	↓ TBARS and oxidative stress in femur tissue in G4; ↓ IL expression in G4; ↑ NRF-2-dependent protein expression and suppression of osteoclast-specific markers	[52]
	Acrolein-induced ototoxicity	Albino Wistar male rats	G1: control; G2: acrolein (3 mg/kg daily); G3: acrolein+lutein (1 mg/kg by oral gavage)	30 days	↓ MDA, TOS, TNF-α, NF-κB, and IL-1β in vestibulocochlear nerve tissue in G1 and G3; ↑ GSH and TAS levels in G1 and G3	[53]
	Excessive alcohol to ameliorate reproductive damage	Seven-week-old male rats	G1: control; G2: lutein (24 mg/kg); G3: alcohol model group (12 mL/(kg.bw.d) 56% ethanol); G4: alcohol+ lutein (12 mg/kg.bw); G5: alcohol+ lutein (24 mg/kg.bw): G6: alcohol+ lutein (48 mg/kg.bw)	12 weeks	↑ testis and epididymis weight in G4-G6 by increased dosage; ↓ sperm count, motility and abnormality in G4-G6; ↑ marker enzyme levels in the testis; ↓ MDA, IL-6, TNF-α levels in testiculat tissue	[54]
	Spinal cord ischemia-reperfusion injury	Thirty-five male rats	G1: intact; G2: sham; G3: I-R; G4: I-R + 0.2 mg/kg lutein; G5: I-R + 0.4 mg/kg lutein	72 h	↓ MDI, plasma MDA in G4 and G5 compared with G3 ↑ plasma antioxidant capacity and number of normal motor neurons in G4 and G5	[55]
	Light-induced retinal damage	Male SD rats	G1: control; G2: water+light; G3: corn oil+light: G4: light+lutein (25 mg/kg bw); G5: lutein (50 mg/kg bw); G6: lutein (100 mg/kg bw)	30 days	↑ SOD, GST, CAT in lutein groups compared with G2 and G3 ↓ MDA, IL-6, IL-1β, and TNF-α in G4-G6 ↑ protein expression of BCO2 and thickness of the outer nuclear layer in G4-G6	[56]

α-Tocopherol	Diabetic retinopathy	Male Wistar rats	G1: control; G2: alloxan-induction; G3: retinol for 7 days; G4: α-tocopherol for 7 d; G5: retinol + α-tocopherol for 7 d; G6: retinol for 14 d; G7: α-tocopherol for 14 d; G8: retinol + α-tocopherol for 14 d	14 days	↑ retinal ganglion cell density in G7 compared with the others; ↓ apoptosis in photoreceptor cells in G4 and G7	[57]
	Ischemia/reperfusion injury	Male C57BL/6 mice	G1: control (vehicle, i.p.); G2: α-tocopherol (2.5 mg/kg bw)	3 days	↓ expression of seven inflammatory cytokines and receptors in G2 ↓ oxidized lipids in the infarcted area of the myocardium in G2 compared with control ↓ ROS production in G2	[58]
	Non-alcoholic fatty liver disease	Male C57BL/6 J mice	G1: control; G2: HFD; G3: HFD+ 50 mg/kg α-tocopherol; G4: HFD+ 200 mg/kg α-tocopherol	8 weeks	↓ liver weight and serum ALT levels in G3 compared with G2 and G4; ↓ triglyceride contents in the liver in G3 acompared with G4; ↑ serum α-tocopherol and PAO levels in G4 compared with G3; ↑ hepatic function in G3 compared with G4	[59]
	Alzheimmer's disease	5XFAD female mice	G1 and G2: control, 10 mg/kg etodalac i.p.; G3 and G4: control, 10 mg/kg/day of α-tocopherol; G5 and G6: control, etodalac+α-tocopherol	1 month	↑ BBB intactness in G6 compared to the other groups ↑ protein expression in mice brain microvessels in G6; ↑ SOD in G4 and G6; ↓ astrocytic marker GFAP in G4 and G6	[60]

(continued)

Table 8.2 (continued)

Carotenoids	Disorders	Study model	Treatment scheme	Duration	Biological activity	Reference
Zeaxanthin	Retinal pigment epithelial (RPE) atrophy	Sod2$^{flox/flox}$-VMD2-cre mice	G1: control; G2: 55 to 60 mg/kg of zeaxanthin	4 months	↑ antioxidant gene expression in the RPE/choroid in G2 compared with G1; ↓ protein nitrosylation in the RPE in G2; ↑ RPE structure and thickness in G2 compared to G1	[61]
	Uveal Melanoma Cell Line-inoculated eyes	Athymic nude mice	G1: control; G2: 57 μg of zeaxanthin; G3: 114 μg of zeaxanthin	21 days	↓ melanoma extension in G2 and G3 compared with control; ↓ tumor mass and size in G2 and G3 with increased zeaxanthin dosage	[62]
	Streptozotocin-induced diabetic	Male SD rats	G1: control (sterile saline, 2 ml/kg); G2: diabetic control (sterile saline, 2 ml/kg); G3: met (100 mg/kg) + HFHSD+STZ; G4: Zea (200 mg/kg) + HFHSD+STZ; G5: Zea (400 mg/kg) + HFHSD+STZ	4 weeks	↓ plasma glucose in G4 and G5; ↓ triglycerides, total cholesterol, LDL-C and HDL-C in G5 followed by G4; ↓ blood urea nitrogen, IL-2, IL-6, TNF-α and NF-κB in G4 and G5	[63]
	Cognitive decline	Male Wistar rats	G1: control; G2: amyloid-β peptide (20 μg/200 μl i.g.); G3: 60 mg/kg Zea daily; G4: Aβ1–42+ Zea	14 days	↓escape latency in G3 and G4 comapred with G2; ↑ target crossing in G3 and G4; ↓ ET-1 and Aβ1–42 expression levels in plasma and IL-1β in cerebrovascular tissue in G3 and G4	[64]
	Ovalbumin-induced allergic asthma	BALB/c nude mice	G1: control; G2: 3.75 μg ovalbumin i.p.; G3: 50 mg/kg Zea i.g.; G4: ovalbumin+Zea	26 days	↓ inflammatory cells infiltration in G3 and G4; ↓ IgE and pro-inflammatory cytokines in G3 and G4; ↑ SOD, GST and GSH in lung tissues in G3 and G4 compared to G2	[65]

β-cryptoxanthin	Bone loss	Female ddY mice	G1-3: sham, 1 mg/L β-cryptoxanthin, 10 mg/L β-cryptoxanthin; G4-6: OVX, OVX+ 1 mg/L β-cryptoxanthin, OVX+ 10 mg/mL β-cryptoxanthin	28 days	↑ inhibition of OVX-induced bone loss and OVX-induced osteoclastic activation in G5 and G6	[66]
	Cd-induced testicular injury	Male SD rats	G1: control (corn oil, 100 μl i.p.); G2: 10 μg/kg β-cryptoxanthin; G3: 2 mg Cd/kg bw; G4: β-cryptoxanthin+Cd	24 h	↑ testis weight, testicular T in G2 and G4 comapred with G3; ↑ SOD, CAT, GSH in G2 and G4; ↓ LPO and MDA in G2 and G4	[67]
	HFD	Male SD rats	G1: control (standard diet); G2: standard diet+β-cryptoxanthin (2.5 mg/kg bw daily); G3: HFD; G4: HFD + β-cryptoxanthin	12 weeks	↓ body and liver weight and visceral fat in G2 and G4 compared with G3; ↓ glucose, insulin, leptin, triglycerides, LDL-C, HDL-C and free fatty acids in G2 and G4; ↑ serum TAC, SOD, CAT, GSHp in G2 and G4	[68]
	Cyclopho sphamide-induced lung injury	Male albino rats	G1: control (corn oil); G2: β-cryptoxanthin (4 mg/kg daily); G3: Cyclophosphamide (single dose 200 mg/kg, i.p.) on day 7; G4: β-cryptoxanthin + cyclophosphamide	7 days	↓ MDA and MPO activity in lung tissues in G2 and G4 compared with G3; ↓lung tissue wet/dry weight ratios in G2 and G4; ↑body weight, GSH, GSH-px, SOD levels in G2 and G4	[69]

Note: acetylcholinesterase *AChE*, alkaline phosphatase *AlP*, alanine aminotransferase *ALT*, aspartate aminotransferase *AST*, Basso, Beattie and Bresnahan *BBB* score test; catalase *CAT*, dimethyl sulfoxide *DMSO*, glutathione S-transferase *GST*, glutathione peroxidase *GP-x*, high-fat diet *HFD*, high-density lipoprotein cholesterol *HDL-C*, high-fat, high-sucrose diet *HFHSD*, intragastric administration *i.g.*, interleukin 1-β beta *IL-1β*, ischemia-reperfusion *I-R*, low-density lipoprotein cholesterol *LDL-C*, malondialdehyde *MDA*, motor deficit index *MDI*, metformin hydrochloride *Met*, myeloperoxidase *MPO*, non-alcoholic fat liver diet-induced by high fat diet *NAFLD-HFD*, non-alcoholic fat liver diet with normal diet *NAFLD-RD*, normal diet *ND*, nuclear factor kappa B *NF-κB*, ovariectomized *OVX*, rosuvastatin *RSV*, superoxide dismutase *SOD*, streptozotocin *STZ*, total antioxidant status *TAS*, thiobarbituric acid reactive substances *TBARS*, total gluthatione *tGSH*, tumour necrosis factor alpha *TNF-α*, total oxidant status *TOS*

BP is a rich source of both hydro-soluble (0.3%) and lipo-soluble (0.1%) vitamins, including B complex, C, E, and pro-vitamin A. BP contains various amounts of these vitamins strongly influenced by its botanical origin and processing or storage conditions [7, 16, 71, 72].

Natural source-derived vitamin E, such as from BP is solubilised into micelles by bile salts and amphipathic lipids before entering the lymphatic system [73]. Many variables influence vitamin E absorption, including bile acids and pancreatic fluid [74]. Furthermore, depending on their type, lipids play a crucial function in vitamin E absorption [73]. Analysis of fresh and dry BP identified several vitamins, mainly B_1, B_2, vitamers of B_6 and of PP (nicotinic acid and nicotinamide). The concentrations of B_1 in fresh BP were from 0.59 to 1.09 mg/100 g (0.64 and 1.01 mg/100 g in dry BP). B_2 vitamin presented higher values in fresh (1.73–2.23 mg/100 g) and dry samples (1.77–2.56 mg/100 g) [72]. Vitamin PP (mg/100 g) with three vitamers (nicotinic acid, nicotinamide and niacin) in fresh BP varied between 2.42–4.61, 2.69–10.59, and 6.43–15.34, whereas B_6 (mg/100 g) presented lower values 0.50–0.79 in fresh BP (0.33 and 0.77 in dry BP) [72]. The vitamins content in *Rhododendron* BP was 315 µg/100 g for thiamin, 735 µg/100 g for riboflavin and 1940 µg/100 g for pantothenic acid [75]. Vitamin B_3 ranged between 0.04 mg/100 g in *Cucurbita pepo* BP and 0.77 mg/100 g in *Helianthus* BP [76]. Oliveira et al. (2009) revealed that floral sources such as *Eucalyptus* sp., *Macroptilium* sp., *Mimosa* sp. and *Raphanus* sp., are correlated to higher vitamin E concentrations; whereas *Arecaceae* type, *Anadenanthera* sp., and *Philodendron* sp. are related to high contents in vitamin C [17]. Table 8.3 contains further information.

Table 8.3 Botanical sources of vitamins found in BP

Vitamins	Bee product origin	Botanical source	Range quantification	Fresh or dry weight	Reference
Lipo-soluble					
Provitamin A (retinol µg/100 g)	Brazil	*Brassica* sp.	15.8–112.1	dw	[7]
		Mimosa sp.	40.4		
		Machaerium sp.	2292–4945.2		
		Piptocarpha sp.	71		
		Eupatorium sp.	57.1		
		Eucalyptus sp.	6.5		
		Fabaceae (*Andira* sp.)	12.5		
	Turkey	Ericaceae (*Rhododendron ponticum*)	22	fw	[75]
Vit. E (α-Tocopherol)	Brazil	*Eucalyptus* sp. (µg/g)	27.5–37.5	dw	[82]
		Eupatorium sp. (µg/g)	27.2–53.7		

(continued)

Table 8.3 (continued)

Vitamins	Bee product origin	Botanical source	Range quantification	Fresh or dry weight	Reference
	Thailand	*Z.mays*	6.21 (mg/100 g)	dw	[79]
	Brazil	*Mimosa caesapiniaefolia*	16.27–42.5 (μg/g)	dw	[17, 71]
		Mimosa scabrella	32.27 (μg/g)	dw	[71]
		Eucalyptus sp.	13.5–32.3 (μg/g)	dw	[17]
		Fabaceae (*Macroptilium sp.*)	13.1(μg/g)		
		Myrcia sp.	13.6 (μg/g)		
Water-soluble					
B_1 (mg/100 g)	Brazil	Arecaceae	0.68–0.80	dw	[77]
		Myrcia sp.	1.01		
		Piperaceae (*Piper* sp.)	0.74		
		Solanaceae (*Cestrum* sp.)	0.65–0.66		
	Brazil	*Eucalyptus sp.*	0.9–1.3	dw	[82]
B_2	Brazil	Arecaceae	1.92–2.03	dw	[77]
		Myrcia sp.	2.49		
		Piper sp.	2.05		
		Cestrum sp.	1.77–1.79		
	Thailand	*Z. mays*	0.5	dw	[79]
	Turkey	*R. ponticum*	0.735	fw	[75]
	Brazil	*Eucalyptus sp.*	0.4–0.6	dw	[82]
		Eupatorium sp.	0.4–0.5		
B_3 (mg/100 g)	Brazil	Arecaceae	8.9–13.96	dw	[77]
		Myrcia sp.	13.38		
		Piper sp.	14.43		
		Cestrum sp.	7.27–13.94		
	Thailand	*Z. mays*	7.03	dw	[79]
	U.S. (California)	Rosaceae (*Prunus dulcis*)	7.10–8.05	fw	[83]
	Saudi Arabia	Asteracea (*Helianthus annuus*)	0.77	fw	[76]
	Brazil	*Eucalyptus sp.*	1.5–2.6	dw	[82]
		Eupatorium sp.	1.3		
		Eucalyptus sp.	3.2–3.7		
		Eupatorium sp.	3.0–3.8		
B_5	Turkey	*R. ponticum*	1.94	fw	[75]
	U.S. (California)	*P. dulcis*	0.76–3.38	fw	[83]

(continued)

Table 8.3 (continued)

Vitamins	Bee product origin	Botanical source	Range quantification	Fresh or dry weight	Reference
B_6	Brazil	Arecaceae	0.08–0.71	dw	[77]
		Myrcia sp.	0.09–0.65		
		Piper sp.	0.18–0.44		
		Cestrum sp.	0.09–0.77		
	U.S. (California)	*P. dulcis*	0.61–0.91	fw	[83]
	Turkey	*R. ponticum*	0.45	fw	[75]
	Saudi Arabia	*H. annuus*	0.77	fw	[76]
		C. pepo	0.44		
		M. sativa	0.65		
	Brazil	*Eucalyptus sp.* (mg/100 g)	3.6–3.8	dw	[82]
		Eupatorium sp. (mg/100 g)	1.9–3.4		
Vitamin C (μg/g)	Brazil	Rubiaceae	158.3	dw	[7]
		Brassica sp.	129.6–171.9		
		Mimosa sp.	347.5		
		Machaerium sp.	94.2–456.0		
		Piptocarpha sp.	190.8		
		Eupatorium sp.	145.9		
		Eucalyptus sp.	253.2–796.9		
		Caesalpineaceae	430.3		
		Andira sp.	217.9		
		M. caesapiniaefolia	114–273.9	dw	[17, 71]
		M. scabrella	127	dw	[71]
		Eucalyptus sp.	364.7–537.3	dw	[17]
		Macroptilium sp.	334.7		
		Myrcia sp.	560.3		
	U.S. (California)	*P. dulcis*	20.6 (mg/100 g)	fw	[83]
	Turkey	*R. ponticum*	16.24 (μg/100 g)	fw	[75]
B_9	Saudi Arabia	*P. dactylifera*	1.5	fw	[76]
		B. napus	1.9		
		M. sativa	2.33		
B_{12}	Saudi Arabia	*C. pepo*	2.5		
		B. napus	0.75		

According to the International Honey Commission (IHC) the vitamin composition of BP (mg/100 g) was suggested to be: 0.6–1.3 for B_1, 0.6–2.0 for B_2, and 4.0–11.0 for B_3. Few reports have been conducted on BP vitamin composition, mainly from Brazil, and fewer from Saudi Arabia, Turkey and Thailand. The BP from Brazil was shown to have high B-complex vitamin concentrations [72, 77]. Riboflavin plays an important role in cellular respiration and is found in large quantities in *Zea mays* BP, making it a promising supplement in patients with vitamin B deficiency (pellagra) [78, 79]. According to other studies, the quantity of liposoluble vitamins and vitamin C is insufficient. Conversely, Brazilian BP presents a high level of ascorbic acid (6.0–79.7 mg/100 g) and vitamin E strongly correlates to b* value suggesting that yellow-colored BP is linked to the content of α-tocopherol [7].

β-carotene or pro-vitamin A, contains one-sixth of the biological activity of vitamin A and highly antioxidant. In terms of β-carotene, vitamin A was identified *Zea mays* BP from Thailand in lower amounts (1.53 mg/100 g) [79] and in higher amounts in Brazilian BP (5.63–19.89 mg/100 g), with the lowest content in *Eucalyptus sp.* and the highest in polyfloral BP, followed by *Macroptilium sp.* BP [17]. According to World Health Oranization (WHO) [80, 81] a solid food product is deemed a source of a particular vitamin or mineral if the respective portion contains 15% of the daily recommended intake (RDI). Accordingly, out of the liposoluble vitamins, pro-vitamin A intake should be 500–600 μg RE/day for adults. Therefore, 15% of this amount represents 90 μg of vitamin A for adults. As the RDI portion for dry BP is up to 25 g, Sattler et al. (2015) demonstrated that BP samples predominant in *Machaerium sp.* (Fabaceae) could be considered as a pro-vitamin A source, providing 63.7–137.4% DRI for men and 81.9–176.6% DRI for women [7].

In the case of vitamin E, the RDI is between 7.5 and 10 mg α-TE/day for adults and 15% is equivalent to 1.5 mg/day in the portion.

The RDI for vitamin C is between 40–45 mg/day for adults and 15% corresponds to 6.7 mg/day. According to Sattler et al. (2015), consumption of BP predominant in *Eucalyptus sp.* (Myrtaceae) provides 7.0–22.1% DRI for men and 8.4–26.6% DRI for women, followed by BP samples predominant in *Machaerium sp* [7]. The DRI for B_2 vitamin is 1.3 mg/day for individuals, and 15% of this quantity is comparable to 0.20 of B_2 vitamin. Considering the recommend intake of 25 g of BP daily, the highest content in vitamin B_2 might be provided by BP predominat in *Myrcia sp.* (Myrtaceae) that provides 0.62 mg/25 g BP, followed by *Piper sp.* (Piperaceae) that provides 0.51 mg/25 g BP of vitamin B_2 [77]. A lower DRI of 0.18 mg/day is for vitamin B_1 and the considered BP sources are the same as mentioned above for vitamin B_2 [77].

8.3.2 Health Effects of Vitamins Found in BP

Vitamins are critical elements for human metabolism, acting as co-enzymes or enzymes in many crucial processes for the body's regular performance. Several researches have revealed that vitamins are essential in human health, but also to prevent and treat different ailments.

Vitamin B complex sustain and boost metabolic rate, maintain muscular tone, ensure healthy skin condition, improve neurological and immune system activities, and support cell division and development [84, 85]. B complex vitamins such as B_2, B_6, and B_9 are essential in nutritional support of the immune system, demonstrated in both animal and human studies [86, 87]. Conversely, several studies have shown that over-supplementation of certain B complex vitamins is associated with cancer incidence. Long-term-supplementation of B_{12} was associated with 40% increased risk of lung cancer in men demonstrating that this vitamin proves to be detrimental in the treatment of lung cancer [88] and a three-fold risk of prostate cancer [89]. Conflicting findings highlight the importance of dose administration and response are crucial factors to investigate. In this aspect, it was shown that B_2, B_6 and B_9 inhibit U937 cell proliferation and migration in a dose-dependent manner [90].

Riboflavin is a key-role in cell growth and development, effective in migraine prevention, known to protect the body against oxidative stress [91] and reduce hepatocellular injury following liver I/R [92]. Vitamin C along with other agents (i.e. vitamin E) protects the liver and act in the management against oxidative stress [93–95]. In Table 8.4 are presented *in vivo* preclinical studies of different vitamins present in BP.

8.4 Conclusions and Prospects

Researchers have focused on the consumers demand for a healthier and balanced diet. BP is regarded as a functional product due to its nutritional composition and health promoting effects. Briefly, BP contains a variety of dietary phytochemicals such as carotenoids and vitamins with functional properties. BP mostly contains vitamins from the B complex, C, D, E, and carotenoids (α- and β-carotene, α-tocopherol, β-cryptoxanthin, lutein and zeaxanthin), although there have been little investigations on the relationship between botanical origin and quantification of carotenoids and vitamins. Recent studies focused on the interaction mechanisms of BP with the human body linked with its nutritional and biological activities. Among the carotenoids present in BP, β-carotene and α-tocopherol fight against degenerative diseases such as cataract and osteoporosis, as well as against oxidative stress diabetes, and spinal cord ischemia/reperfusion injury. Lutein is extensively used as a preventive agent against macular degeneration and osteoporosis, whereas zeaxanthin is used against cognitive decline diseases, such as Alzheimer and against lung and liver diseases. Vitamins from the B complex are known to ameliorate hepatotoxicity, diabetes and tuberculosis infection, whereas combined administration of vitamins C and E is effective against oxidative stress and ensure healthy skin condition, improve neurological and immune system activities. Therefore, BP proves to be significant in the production and use as a functional ingredient with multiple health effects.

Table 8.4 *In vivo* preclinical studies of different vitamins present in BP

Vitamins	Disorders	Study model	Treatment scheme	Treatment duration	Biological activity	Reference
Vitamin E	Heavy metals (HM)-induced renal and testicular injuries	MFI male albino mice	G1: control; G2: HM in water; G3: HM + Vit. E (50 IU/kg bw); G4: Vit. E	7 weeks	↓plasma creatinine, urea and uric acid in G3 and G4 compared to G2 ↑ kidney GSH and SOD in G3 and G4 ↑ testis GSH and SOD in G3 and G4 compared to G2	[96]
	I/R induced-oxidative stress	Male SD rats	G1: sham (no ischemia); G2: control (ischemia); G3–5: crocin (10, 20 and 40 mg/kg); G6: Vit. E (100 mg/kg); G7: crocin (40 mg/kg) + vit. E (100 mg/kg)	30 days	↓ TBARS in rat hearts in G6 and G7 compared to groups G3–5 ↑ CAT, SOD and total antioxidant activity levels in heart of rats in G6 and G7 compared to G3–5	[97]
	Mild traumatic brain injury	Male SD rats	G1: sham; G2: mild fluid percussion injury (FPI); G3: FPI + Vit. E (500 IU/kg)	4 weeks	↑BDNF and CaMKII levels in hippocampus of rats in G3 compared to G2 ↑ synapsin I, CREB and SOD levels in G3 compared to G2	[98]
B_1	Diethyl nitrosamine (DEN) induced hepatocellular carcinoma	Male Wistar strain albino rats	G1: control (normal saline, 0.9%); G2: DEN (200 mg/kg bw, single i.p.) + phenobarbital; G3: DEN + phenobarbital (0.05%) + MTX (5 mg/kg bw, i.p.); G4: DEN + phenobarbital + thiamin; G5:DEN + phenobarbital + MTX + thiamin	24 weeks	↑ AFP and VEGF in thiamin-administered groups (G4 and G5) compared with G2 and G3; ↑ liver weight, liver index (%) and LPO in G4 and G5; ↓ body weight and SOD, GPx, CAT levels in G4 and G5	[99]

(continued)

Table 8.4 (continued)

Vitamins	Disorders	Study model	Treatment scheme	Treatment duration	Biological activity	Reference
	Meiotic maturation of mice oocytes	Female 5-week-old ICR mice	G1: control 20% casein diet + thiamin (12 g/kg diet); G2: thuamin–free diet for 20 days and after as in G1 until the end of the experiment	62 days	↓ food intake and anorexia in G2 prior to thiamin administration ↑ significant recovery of body weight after thiamin diet in G2 ↓ abnormal oocytes in G2 after thiamin administration ↑ B1 concentration in urine, liver, and uterus	[100]
	Mycobacterium tuberculosis infection	6 weeks old female and male mice	G1: control; G2: 20 μg VB1/100 g bw	4 weeks	↓ mycobacterial growth in lungs and spleen of mice in G2 ↑ TNF-α, IL-6, and nitrate levels in G2 compared with control	[101]
B_2	Liver ischaemia/ reperfusion (I/R) injury	Male swiss mice	G1: control; G2: I/R; G3: I/R + VB2 (30 μmoles/kg, bw)	2 hours	↓ ALT, AST, parenchymal damage MPO levels in liver of mice from G3 compared with G2 ↑ GSH and SOD levels in G3 compaerd with G2	[92]
	Potassium bromate(PB)-induced hepatotoxicity	8-week-old Swiss albino rats	G1: control; G2: PB (150 mg/kg): G3: VB2 (2 mg/kg); G4: PB+ 2 mg/kg VB2; G5: PB + 4 mg/kg VB2	1 month	↓ liver function markers (ALP, ALT, ASP) in VB2-treatments groups compared with G2 ↓ GST and TR in liver in VB2-administered groups compared with G2 ↑ antioxidative enzymes in VB2 groups	[102]

	Lipopolysaccharide (LPS)-induced lung injury	10–12 weeks old male Wistar albino rats	G1: phosphate buffer saline (PBS) intranasally; G2: LPS (20 mg/rat in 50 mlof PBS); G3: VB2 (30 mg/kg, p.o.) + LPS; G4: VB2 (100 mg/kg, p.o.) + LPS; G5: LPS + dexamethasone (1 mg/kg, p.o.)	7 days	↓ MDA level, MPO activity in lung tissue in VB2-treatment groups ↑ GSH, GPx, GR levels in G3 and G4 compared with G2 ↑ iNOS and CAT gene expressions in G3 and G4 compared with G2	[103]
B_3	Alloxan-induced diabetes	6 months old wistar rats	G1: control; G2: VB3 (15 mg/kg bw); G3: diabetic control; G4: diabetic+VB3 10 mg/kg bw; G5: diabetic+VB3 15 mg/kg bw	30 days	↓ fasting blood glucose levels in G4 and G5 compared with G3 ↓ glucose metabolic enzymes in niacin groups compared with G3 ↑ antioxidant enzyme levels in G4 and G5 ↓ cholesterol and triglyceride levels in G4 and G5	[104]
	Diabetes mellitus-indued brain dysfunctions	tw-month-old Wistar rats	G1: control; G2: diabetes (single dose of STZ of 60 mg/kg); G3: diabetes+VB3 (100 mg/kg, i.p.); G4: diabetes+N-GABA (55 mg/kg, i.p.)	6 weeks	↓ glucose and bw in G3 after diabetes compared to G1 and in G4 compared to G1 and G2 ↓ NF-κB and BAX in G3 compared to G1 and in G4 compared to G1 and G2 ↑ VEGF in brains of rats in G3 compared to G2 ↓ GFAP in G4 compared to G2	[105]
	I/R injury–associated acute and chronic kidney injury	Male SD rats	G1: control; G2: I/R; G3: I/R + VB3 (100 mg/kg/day)	7 days	↓ cTnT, creatinine, BUN, MDA levels in cardiac tissue in G3 compared to G2	[106]

(continued)

Table 8.4 (continued)

Vitamins	Disorders	Study model	Treatment scheme	Treatment duration	Biological activity	Reference
Vitamin C	Renal I/R injury	Wistar strain male rats	G1: control; G2: I/R; G3: I/R + Vit. C (50 mg/kg, i.v.); G4: I/R+ Vit. E (20 mg/kg intramuscularly); G5: I/R + hydrocortisone (50 mg/kg i.v.); G6: combination of vit. C + E + hydrocortisone	1 h	↓ severity of tubular injury and tubular degeneration in G3 and G6 compared to G2 ↓ total histological scores in G3-G6 compared to G2	[107]
	Imidacloprid-induced oxidative stress	Male Swiss albino mice	G1: control (corn oil, 5 ml/kg); G2: Vit. C (200 mg/kg bw); G3: single dose of imidacloprid (1/10 LD50); G4: Vit. C+ 30 min before imidacloprid administration; G5: Vit. C + 30 min after imidacloprid administration	24 h	↓ MDA, CAT, SOD levels and GPx and GST activities in hepatic tissues of mice in G2 and G4 compared to G3	[108]
	Smoke–induced pulmonary emphysema	4-month-old SMP30-KO mice	**Preventive study** G1: 0.0375 g/L Vit. C + air; G2: 0.0375 g/L Vit. C+ smoke; G3: 1.5 g/L Vit. C + air; G4: 1.5 g/L Vit. C+ smoke **Treatment study** TG1: water+0.0375 g/L Vit. C + air; TG2: water+1.5 g/L Vit. C + air; TG3: water+0.0375 g/L Vit. C + smoke; TG4: water+1.5 g/L Vit. C + smoke	2 months	↓ TNFα in TG4 compared to TG3 ↑ total Vit. C in plasma and in the lungs in TG2 and TG4 compared with TG1 and TG3 ↓ VEGF in the lungs and BALF of mice in TG3 compared to the other groups ↑ collagen synthesis in TG2 and TG4 compared to the other groups	[109]

	Mycobacterium tuberculosis infection	6–8 weeks old CBA/J female mice	G1: control; G2: Vit. C (3 g/kg, i.p.); G3: INH (100 mg/L) + RIF (40 mg/L): G4: INH + RIF + Vit. C	4 weeks	↓ CFU in the lungs and spleens of infected mice in G2 and G4 compared to G3	[110]

Note: bronchoalveolar lavage fluid *BALF*, brain-derived neurotrophic factor *BDNF*, blood urea nitrogen *BUN*, cAMP-response element-binding protein *CREB*, serum troponin T *cTnT*, diethyl nitrosamine *DEN*, glutathione-S-transferase *GST*, injected intravenously *i.v.*, isoniazid *INH*, methotrexate *MTX*, nicotinic acid with gamma-aminobutyric acid *N-GABA*, Poly (ADP-ribose) polymerase *PARP-1*, rifampin *RIF*, thioredoxin reductase *TR*, vascular endothelial growth factor *VEGF*

References

1. Kostić AŽ, Milinčić DD, Barać MB, Ali Shariati M, Tešić ŽL, Pešić MB (2020) The application of pollen as a functional food and feed ingredient—the present and perspectives. Biomolecules 10(1):84. https://doi.org/10.3390/biom10010084
2. Čeksterytė V, Kurtinaitienė B, Venskutonis PR, Pukalskas A, Kazernavičiūtė R, Balžekas J (2016) Evaluation of antioxidant activity and flavonoid composition in differently preserved bee products. Czech J Food Sci 34(2):133–142
3. De-Melo AAM, Estevinho LM, Moreira MM, Delerue-Matos C, de Freitas AS, Barth OM et al (2018) A multivariate approach based on physicochemical parameters and biological potential for the botanical and geographical discrimination of Brazilian bee pollen. Food Biosci 25:91–110
4. Campos MGR, Bogdanov S, de Almeida-Muradian LB, Szczesna T, Mancebo Y, Frigerio C et al (2008) Pollen composition and standardisation of analytical methods. J Apicult Res 47(2):154–161. https://doi.org/10.1080/00218839.2008.11101443
5. Campos MGR, Frigerio C, Lopes J, Bogdanov S (2010) What is the future of bee-pollen. J ApiProduct ApiMedi Sci 2(4):131–144
6. Carpes ST, De Alencar SM, Cabral ISR, Oldoni TLC, Mourão GB, Haminiuk CWI et al (2013) Polyphenols and palynological origin of bee pollen of Apis mellifera L. from Brazil. Characterization of polyphenols of bee pollen. CyTA-J Food 11(2):150–161
7. Sattler JAG, de Melo ILP, Granato D, Araújo E, de Freitas AS, Barth OM et al (2015) Impact of origin on bioactive compounds and nutritional composition of bee pollen from southern Brazil: a screening study. Food Res Int 77:82–91
8. Kritchevsky SB (1999) β-Carotene, carotenoids and the prevention of coronary heart disease. J Nutr 129(1):5–8
9. Mărgăoan R, Mărghitaş LA, Dezmirean DS, Dulf FV, Bunea A, Socaci SA et al (2014) Predominant and secondary pollen botanical origins influence the carotenoid and fatty acid profile in fresh honeybee-collected pollen. J Agric Food Chem 62(27):6306–6316. https://doi.org/10.1021/jf5020318
10. Ivanov I, Petkova N, Tumbarski J, Dincheva I, Badjakov I, Denev P et al (2018) GC-MS characterization of n-hexane soluble fraction from dandelion (Taraxacum officinale Weber ex FH Wigg.) aerial parts and its antioxidant and antimicrobial properties. Zeitschrift für Naturforschung C 73(1–2):41–47
11. Stewart-Wade SM, Neumann S, Collins LL, Boland GJ (2002) The biology of Canadian weeds. 117. Taraxacum officinale GH Weber ex Wiggers. Can J Plant Sci 82(4):825–853
12. Barros L, Cabrita L, Boas MV, Carvalho AM, Ferreira ICFR (2011) Chemical, biochemical and electrochemical assays to evaluate phytochemicals and antioxidant activity of wild plants. Food Chem 127(4):1600–1608
13. Gey KF (1998) Vitamins E plus C and interacting conutrients required for optimal health. Biofactors 7(1–2):113–174
14. Maqbool MA, Aslam M, Akbar W, Iqbal Z (2018) Biological importance of vitamins for human health: a review. J Agric Basic Sci 2(3):50–58
15. Rizvi S, Raza ST, Ahmed F, Ahmad A, Abbas S, Mahdi F (2014) The role of vitamin E in human health and some diseases. Sultan Qaboos Univ Med J 14(2):e157
16. Melo ILP, Almeida-Muradian LB (2010) Stability of antioxidants vitamins in bee pollen samples. Química Nova 33:514–518
17. Oliveira KCLS, Moriya M, Azedo RAB, Almeida-Muradian LB, Teixeira EW, Alves ML et al (2009) Relationship between botanical origin and antioxidants vitamins of bee-collected pollen. Química Nova 32:1099–1102
18. Almeida-Muradian LB, Pamplona LC, Coimbra SI, Barth OM (2005) Chemical composition and botanical evaluation of dried bee pollen pellets. J Food Comp Anal 18(1):105–111. https://doi.org/10.1016/j.jfca.2003.10.008

19. Abd Alla AE, Salem RA (2020) Impact of storage period on different types of bee pollen pigments. J Plant Prot Pathol 11(1):9–13
20. Bohoyo-Gil D, Dominguez-Valhondo D, Garcia-Parra JJ, González-Gómez D (2012) UHPLC as a suitable methodology for the analysis of carotenoids in food matrix. Eur Food Res Technol 235(6):1055–1061
21. Thakur M, Nanda V (2020) Composition and functionality of bee pollen: a review. Trends Food Sci Technol 98:82–106
22. Salazar-González CY, Stinco CM, Rodríguez-Pulido FJ, Díaz-Moreno C, Fuenmayor C, Heredia FJ et al (2022) Characterization of carotenoid profile and α-tocopherol content in Andean bee pollen influenced by harvest time and particle size. LWT 170:114065. https://doi.org/10.1016/j.lwt.2022.114065
23. Gardana C, Del Bo' C, Quicazán MC, Corrrea AR, Simonetti P (2018) Nutrients, phytochemicals and botanical origin of commercial bee pollen from different geographical areas. J Food Compos Anal 73:29–38. https://doi.org/10.1016/j.jfca.2018.07.009
24. Conte G, Benelli G, Serra A, Signorini F, Bientinesi M, Nicolella C et al (2017) Lipid characterization of chestnut and willow honeybee-collected pollen: Impact of freeze-drying and microwave-assisted drying. J Food Compos Anal 55:12–19. https://doi.org/10.1016/j.jfca.2016.11.001
25. Karkar B, Şahin S, Güneş ME (2021) Evaluation of antioxidant properties and determination of phenolic and carotenoid profiles of chestnut bee pollen collected from Turkey. J Apic Res 60(5):765–774. https://doi.org/10.1080/00218839.2020.1844462
26. Pacheco S, Peixoto FM, Borguini RG, Nascimento LSM, Bobeda CRR, Santiago MCPA et al (2014) Microscale extraction method for HPLC carotenoid analysis in vegetable matrices. Sci Agric 71:416–419
27. Domínguez-Valhondo D, Bohoyo Gil D, Hernández MT, González-Gómez D (2011) Influence of the commercial processing and floral origin on bioactive and nutritional properties of honeybee-collected pollen. Int J Food Sci Technol 46(10):2204–2211. https://doi.org/10.1111/j.1365-2621.2011.02738.x
28. Schulte F, Mäder J, Kroh LW, Panne U, Kneipp J (2009) Characterization of pollen carotenoids with in situ and high-performance thin-layer chromatography supported resonant Raman spectroscopy. Anal Chem 81(20):8426–8433. https://doi.org/10.1021/ac901389p
29. Şahin S, Karkar B (2019) The antioxidant properties of the chestnut bee pollen extract and its preventive action against oxidatively induced damage in DNA bases. J Food Biochem 43(7):e12888. https://doi.org/10.1111/jfbc.12888
30. Ross AC, Stephensen CB (1996) Vitamin A and retinoids in antiviral responses. FASEB J 10(9):979–985
31. Zhang Y, Zhu X, Huang T, Chen L, Liu Y, Li Q et al (2016) β-Carotene synergistically enhances the anti-tumor effect of 5-fluorouracil on esophageal squamous cell carcinoma in vivo and in vitro. Toxicol Lett 261:49–58. https://doi.org/10.1016/j.toxlet.2016.08.010
32. Darvin ME, Sterry W, Lademann J, Vergou T (2011) The role of carotenoids in human skin. Molecules 16(12):10491–10506
33. Césarini JP, Michel L, Maurette JM, Adhoute H, Béjot M (2003) Immediate effects of UV radiation on the skin: modification by an antioxidant complex containing carotenoids. Photodermatol Photoimmunol Photomed 19(4):182–189. https://doi.org/10.1034/j.1600-0781.2003.00044.x
34. Grether-Beck S, Marini A, Jaenicke T, Stahl W, Krutmann J (2017) Molecular evidence that oral supplementation with lycopene or lutein protects human skin against ultraviolet radiation: results from a double-blinded, placebo-controlled, crossover study. Br J Dermatol 176(5):1231–1240
35. Krinsky NI, Landrum JT, Bone RA (2003) Biologic mechanisms of the protective role of lutein and zeaxanthin in the eye. Annu Rev Nutr 23(1):171–201

36. Cho S, Lee DH, Won C-H, Kim SM, Lee S, Lee M-J et al (2010) Differential effects of low-dose and high-dose beta-carotene supplementation on the signs of photoaging and type I procollagen gene expression in human skin in vivo. Dermatology 221(2):160–171
37. Age-Related Eye Disease Study Research G (2001) A randomized, placebo-controlled, clinical trial of high-dose supplementation with vitamins C and E, beta carotene, and zinc for age-related macular degeneration and vision loss: AREDS report no. 8. Arch Ophthalmol 119(10):1417–1436
38. Lindbergh CA, Renzi-Hammond LM, Hammond BR, Terry DP, Mewborn CM, Puente AN et al (2018) Lutein and zeaxanthin influence brain function in older adults: a randomized controlled trial. J Int Neuropsychol Soc 24(1):77–90
39. Min J-Y, Min K-B (2014) Serum lycopene, lutein and zeaxanthin, and the risk of Alzheimer's disease mortality in older adults. Dement Geriatr Cogn Disord 37(3–4):246–256
40. Ademowo OS, Dias HKI, Milic I, Devitt A, Moran R, Mulcahy R et al (2017) Phospholipid oxidation and carotenoid supplementation in Alzheimer's disease patients. Free Radic Biol Med 108:77–85
41. Karppi J, Laukkanen JA, Mäkikallio TH, Ronkainen K, Kurl S (2013) Serum β-carotene and the risk of sudden cardiac death in men: a population-based follow-up study. Atherosclerosis 226(1):172–177
42. Mužáková V, Kand'ár R, Meloun M, Skalický J, Královec K, Žáková P et al (2010) Inverse correlation between plasma β-carotene and Interleukin-6 in patients with advanced coronary artery. Int J Vitam Nutr Res 80(6):369–377
43. Maria AG, Graziano R, Nicolantonio DO (2015) Carotenoids: potential allies of cardiovascular health? Food Nutr Res 59(1):26762
44. El-Demerdash FM, Yousef MI, Kedwany FS, Baghdadi HH (2004) Cadmium-induced changes in lipid peroxidation, blood hematology, biochemical parameters and semen quality of male rats: protective role of vitamin E and β-carotene. Food Chem Toxicol 42(10):1563–1571. https://doi.org/10.1016/j.fct.2004.05.001
45. Maritim A, Dene BA, Sanders RA, Watkins Iii JB (2002) Effects of β-carotene on oxidative stress in normal and diabetic rats. J Biochem Mol Toxicol 16(4):203–208
46. El-Din SHS, El-Lakkany NM, El-Naggar AA, Hammam OA, Abd El-Latif HA, Ain-Shoka AA et al (2015) Effects of rosuvastatin and/or β-carotene on non-alcoholic fatty liver in rats. Res Pharma Sci 10(4):275
47. Vardi N, Parlakpinar H, Ates B, Cetin A, Otlu A (2009) Antiapoptotic and antioxidant effects of β-carotene against methotrexate-induced testicular injury. Fertil Steril 92(6):2028–2033. https://doi.org/10.1016/j.fertnstert.2008.09.015
48. Zhu L, Song Y, Liu H, Wu M, Gong H, Lan H et al (2021) Gut microbiota regulation and anti-inflammatory effect of β-carotene in dextran sulfate sodium-stimulated ulcerative colitis in rats. J Food Sci 86(5):2118–2130. https://doi.org/10.1111/1750-3841.15684
49. Zhou L, Ouyang L, Lin S, Chen S, Liu Y, Zhou W et al (2018) Protective role of β-carotene against oxidative stress and neuroinflammation in a rat model of spinal cord injury. Int Immunopharmacol 61:92–99
50. Arnal E, Miranda M, Almansa I, Muriach M, Barcia JM, Romero FJ et al (2009) Lutein prevents cataract development and progression in diabetic rats. Graefes Arch Clin Exp Ophthalmol 247(1):115–120. https://doi.org/10.1007/s00417-008-0935-z
51. Qiu X, Gao D-H, Xiang X, Xiong Y-F, Zhu T-S, Liu L-G et al (2015) Ameliorative effects of lutein on non-alcoholic fatty liver disease in rats. World J Gastroenterol: WJG 21(26):8061
52. Li H, Huang C, Zhu J, Gao K, Fang J, Li H (2018) Lutein suppresses oxidative stress and inflammation by Nrf2 activation in an osteoporosis rat model. Med Sci Monit 24:5071
53. Erhan E, Salcan I, Bayram R, Suleyman B, Dilber M, Yazici GN et al (2021) Protective effect of lutein against acrolein-induced ototoxicity in rats. Biomed Pharmacother 137:111281. https://doi.org/10.1016/j.biopha.2021.111281
54. Zhang Y, Ding H, Xu L, Zhao S, Hu S, Ma A et al (2022) Lutein can alleviate oxidative stress, inflammation, and apoptosis induced by excessive alcohol to ameliorate reproductive damage

in male rats. Nutrients 14(12) Available from: https://mdpi-res.com/d_attachment/nutrients/nutrients-14-02385/article_deploy/nutrients-14-02385-v2.pdf?version=1654822307
55. Pour MM, Farjah GH, Karimipour M, Pourheidar B, Ansari MHK (2019) Protective effect of lutein on spinal cord ischemia-reperfusion injury in rats. Iran J Basic Med Sci 22(4):412
56. Yang Y, Tan X, Xu J, Wang T, Liang T, Xu X et al (2020) Luteolin alleviates neuroinflammation via downregulating the TLR4/TRAF6/NF-κB pathway after intracerebral hemorrhage. Biomed Pharmacother 126:110044. https://doi.org/10.1016/j.biopha.2020.110044
57. Ichsan AM, Bukhari A, Lallo S, Miskad UA, Dzuhry AA, Islam IC et al (2022) Effect of retinol and α-tocopherol supplementation on photoreceptor and retinal ganglion cell apoptosis in diabetic rats model. Int J Retina Vitreous 8(1):1–12
58. Wallert M, Ziegler M, Wang X, Maluenda A, Xu X, Yap ML et al (2019) α-Tocopherol preserves cardiac function by reducing oxidative stress and inflammation in ischemia/reperfusion injury. Redox Biol 26:101292. https://doi.org/10.1016/j.redox.2019.101292
59. Tokoro M, Gotoh K, Kudo Y, Hirashita Y, Iwao M, Arakawa M et al (2021) α-Tocopherol suppresses hepatic steatosis by increasing CPT-1 expression in a mouse model of diet-induced nonalcoholic fatty liver disease. Obes Sci Pract 7(1):91–99
60. Elfakhri KH, Abdallah IM, Brannen AD, Kaddoumi A (2019) Multi-faceted therapeutic strategy for treatment of Alzheimer's disease by concurrent administration of etodolac and α-tocopherol. Neurobiol Dis 125:123–134. https://doi.org/10.1016/j.nbd.2019.01.020
61. Biswal MR, Justis BD, Han P, Li H, Gierhart D, Dorey CK et al (2018) Daily zeaxanthin supplementation prevents atrophy of the retinal pigment epithelium (RPE) in a mouse model of mitochondrial oxidative stress. PLoS One 13(9):e0203816. https://doi.org/10.1371/journal.pone.0203816
62. Xu XL, Hu D-N, Iacob C, Jordan A, Gandhi S, Gierhart DL et al (2015) Effects of Zeaxanthin on growth and invasion of human Uveal Melanoma in Nude Mouse Model. J Ophthalmol 2015:392305. https://doi.org/10.1155/2015/392305
63. Kou L, Du M, Zhang C, Dai Z, Li X, Zhang B (2017) The hypoglycemic, Hypolipidemic, and anti-diabetic nephritic activities of Zeaxanthin in diet-Streptozotocin-induced diabetic Sprague Dawley rats. Appl Biochem Biotechnol 182(3):944–955. https://doi.org/10.1007/s12010-016-2372-5
64. Li X, Zhang P, Li H, Yu H, Xi Y (2022) The protective effects of Zeaxanthin on Amyloid-β Peptide 1–42-induced impairment of learning and memory ability in rats. Front Behav Neurosci 16:912896
65. Jin X, Jin W, Li G, Zheng J (2022) Zeaxanthin attenuates OVA-induced allergic asthma in mice by regulating the p38 MAPK/β-catenin signaling pathway. Allergol Immunopathol 50(5):75–83
66. Ozaki K, Okamoto M, Fukasawa K, Iezaki T, Onishi Y, Yoneda Y et al (2015) Daily intake of β-cryptoxanthin prevents bone loss by preferential disturbance of osteoclastic activation in ovariectomized mice. J Pharmacol Sci 129(1):72–77. https://doi.org/10.1016/j.jphs.2015.08.003
67. Liu X-R, Wang Y-Y, Fan H-R, Wu C-J, Kumar A, Yang L-G (2016) Preventive effects of β-cryptoxanthin against cadmium-induced oxidative stress in the rat testis. Asian J Androl 18(6):920
68. Sahin K, Orhan C, Akdemir F, Tuzcu M, Sahin N, Yılmaz I et al (2017) β-Cryptoxanthin ameliorates metabolic risk factors by regulating NF-κB and Nrf2 pathways in insulin resistance induced by high-fat diet in rodents. Food Chem Toxicol 107:270–279. https://doi.org/10.1016/j.fct.2017.07.008
69. Badawi MS (2022) The protective effect of β-cryptoxanthin against cyclophosphamide-induced lung injury in adult male albino rats. Bull Natl Res Cent 46(1):106. https://doi.org/10.1186/s42269-022-00792-2
70. Ball GFM (2012) Water-soluble vitamin assays in human nutrition. Springer
71. Melo ILPd, Freitas ASd, Barth OM, Almeida-Muradian LBd (2009) Relação entre a composição nutricional e a origem floral de pólen apícola desidratado

72. de Arruda VAS, Pereira AAS, Estevinho LM, de Almeida-Muradian LB (2013) Presence and stability of B complex vitamins in bee pollen using different storage conditions. Food Chem Toxicol 51:143–148
73. Rigotti A (2007) Absorption, transport, and tissue delivery of vitamin E. Mol Asp Med 28(5–6):423–436
74. Stahl PH, Wermuth CG (2002) Handbook of pharmaceutical salts: properties, selection and use. Chem Int 24:21
75. Ecem BN (2021) Vitamin, mineral, polyphenol, amino acid profile of bee pollen from Rhododendron ponticum (source of "mad honey"): nutritional and palynological approach. J Food Meas Charact 15(3):2659–2666. https://doi.org/10.1007/s11694-021-00854-5
76. Al-Kahtani SN (2017) Fatty acids and B vitamins contents in honey bee collected pollen in relation to botanical origin. Sci J King Faisal University (Basic Appl Sci) 18(2):41–48
77. de Arruda VAS, Pereira AAS, de Freitas AS, Barth OM, de Almeida-Muradian LB (2013) Dried bee pollen: B complex vitamins, physicochemical and botanical composition. J Food Compos Anal 29(2):100–105. https://doi.org/10.1016/j.jfca.2012.11.004
78. Pinto JT, Zempleni J (2016) Riboflavin. Adv Nutr 7(5):973–975
79. Chantarudee A, Phuwapraisirisan P, Kimura K, Okuyama M, Mori H, Kimura A et al (2012) Chemical constituents and free radical scavenging activity of corn pollen collected from Apis mellifera hives compared to floral corn pollen at Nan. Thailand BMC Complement Altern Med 12(1):45. https://doi.org/10.1186/1472-6882-12-45
80. World Health O (1983) Measuring change in nutritional status: guidelines for assessing the nutritional impact of supplementary feeding programmes for vulnerable groups. World Health Organization, Geneva
81. World Health Organization (2004) Vitamin and mineral requirements in human nutrition. World Health Organization
82. De-Melo AAM, Estevinho MLMF, Sattler JAG, Souza BR, Freitas AS, Barth OM et al (2016) Effect of processing conditions on characteristics of dehydrated bee-pollen and correlation between quality parameters. LWT – Food Sci Technol 65:808–815. https://doi.org/10.1016/j.lwt.2015.09.014
83. Loper GM, Standifer LN, Thompson MJ, Gilliam M (1980) Biochemistry and microbiology of bee-collected almond (Prunus dulcis) pollen and bee bread. I-Fatty acids, sterols, vitamins and minerals. Apidologie 11(1):63–73
84. Mikkelsen K, Apostolopoulos V (2018) B vitamins and ageing. Biochem Cell Biol Ageing: Part I Biomed Sci 90:451–470
85. Mikkelsen K, Stojanovska L, Prakash M, Apostolopoulos V (2017) The effects of vitamin B on the immune/cytokine network and their involvement in depression. Maturitas 96:58–71
86. Adhikari PM, Chowta MN, Ramapuram JT, Rao SB, Udupa K, Acharya SD (2016) Effect of vitamin B12 and folic acid supplementation on neuropsychiatric symptoms and immune response in HIV-positive patients. J Neurosci Rural Pract 7(03):362–367
87. Fukuda S, Koyama H, Kondo K, Fujii H, Hirayama Y, Tabata T et al (2015) Effects of nutritional supplementation on fatigue, and autonomic and immune dysfunction in patients with end-stage renal disease: a randomized, double-blind, placebo-controlled, multicenter trial. PLoS One 10(3):e0119578
88. Brasky TM, White E, Chen C-L (2017) Long-term, supplemental, one-carbon metabolism–related vitamin B use in relation to lung cancer risk in the vitamins and lifestyle (VITAL) cohort. J Clin Oncol 35(30):3440
89. Hultdin J, Van Guelpen B, Bergh A, Hallmans G, Stattin P (2005) Plasma folate, vitamin B12, and homocysteine and prostate cancer risk: a prospective study. Int J Cancer 113(5):819–824
90. Mikkelsen K, Prakash MD, Kuol N, Nurgali K, Stojanovska L, Apostolopoulos V (2019) Anti-tumor effects of vitamin B2, B6 and B9 in promonocytic lymphoma cells. Int J Mol Sci 20(15):3763
91. Hassan I, Chibber S, Naseem I (2010) Ameliorative effect of riboflavin on the cisplatin induced nephrotoxicity and hepatotoxicity under photoillumination. Food Chem Toxicol 48(8–9):2052–2058

92. Sanches SC, Ramalho LNZ, Mendes-Braz M, Terra VA, Cecchini R, Augusto MJ et al (2014) Riboflavin (vitamin B-2) reduces hepatocellular injury following liver ischaemia and reperfusion in mice. Food Chem Toxicol 67:65–71
93. Layachi N, Kechrid Z (2012) Combined protective effect of vitamins C and E on cadmium induced oxidative liver injury in rats. Afr J Biotechnol 11(93):16013–16020
94. Oyinbo CA, Dare WN, Okogun GRA, Anyanwu LC, Ibeabuchi NM, Noronha CC et al (2006) The hepatoprotective effect of vitamin C and E on hepatotoxicity induced by ethanol in Sprague Dawley rats. Pak J Nutr 5(6):507–511
95. Yanardag R, Ozsoy-Sacan O, Ozdil S, Bolkent S (2007) Combined effects of vitamin C, vitamin E, and sodium selenate supplementation on absolute ethanol-induced injury in various organs of rats. Int J Toxicol 26(6):513–523
96. Al-Attar AM (2011) Antioxidant effect of vitamin E treatment on some heavy metals-induced renal and testicular injuries in male mice. Saudi J Biol Sci 18(1):63–72
97. Dianat M, Esmaeilizadeh M, Badavi M, Samarbaf-Zadeh AR, Naghizadeh B (2014) Protective effects of crocin on ischemia-reperfusion induced oxidative stress in comparison with vitamin E in isolated rat hearts. Jundishapur J Nat Pharm Prod 9(2):e17187
98. Wu A, Ying Z, Gomez-Pinilla F (2010) Vitamin E protects against oxidative damage and learning disability after mild traumatic brain injury in rats. Neurorehabil Neural Repair 24(3):290–298
99. Saleem S, Kazmi I, Ahmad A, Abuzinadah MF, Samkari A, Alkrathy HM et al (2020) Thiamin regresses the anticancer efficacy of methotrexate in the amelioration of diethyl nitrosamine-induced hepatocellular carcinoma in Wistar strain rats. Nutr Cancer 72(1):170–181. https://doi.org/10.1080/01635581.2019.1614199
100. Tsuji A, Nakamura T, Shibata K (2017) Effects of mild and severe vitamin B1 deficiencies on the meiotic maturation of mice oocytes. Nutr Metab Insights 10:1178638817693824
101. Hu S, He W, Du X, Huang Y, Fu Y, Yang Y et al (2018) Vitamin B1 helps to limit mycobacterium tuberculosis growth via regulating innate immunity in a peroxisome proliferator-activated receptor-γ-dependent manner. Front Immunol 9:1778
102. Hassan I, Ebaid H, Alhazza IM, Al-Tamimi J (2020) The alleviative effect of vitamin B2 on potassium bromate-induced hepatotoxicity in male rats. Biomed Res Int 2020:8274261
103. Al-Harbi NO, Imam F, Nadeem A, Al-Harbi MM, Korashy HM, Sayed-Ahmed MM et al (2015) Riboflavin attenuates lipopolysaccharide-induced lung injury in rats. Toxicol Mech Methods 25(5):417–423
104. Abdullah KM, Alam MM, Iqbal Z, Naseem I (2018) Therapeutic effect of vitamin B3 on hyperglycemia, oxidative stress and DNA damage in alloxan induced diabetic rat model. Biomed Pharmacother 105:1223–1231. https://doi.org/10.1016/j.biopha.2018.06.085
105. Tykhonenko T, Guzyk M, Tykhomyrov A, Korsa V, Yanitska L, Kuchmerovska T (2022) Modulatory effects of vitamin B3 and its derivative on the levels of apoptotic and vascular regulators and cytoskeletal proteins in diabetic rat brain as signs of neuroprotection. Biochim Biophys Acta Gen Subj 1866(11):130207. https://doi.org/10.1016/j.bbagen.2022.130207
106. Tai ST, Fu YH, Yang YC, Wang JJ (2015) Niacin ameliorates kidney warm ischemia and reperfusion injury–induced ventricular dysfunction and oxidative stress and disturbance in mitochondrial metabolism in rats. Transplant Proc 47(4):1079–1082. https://doi.org/10.1016/j.transproceed.2014.11.057
107. Azari O, Kheirandish R, Azizi S, Abbasi MF, Chaman SGG, Bidi M (2015) Protective effects of hydrocortisone, vitamin C and E alone or in combination against renal ischemia-reperfusion injury in rat. Iran J Pathol 10(4):272
108. El-Gendy KS, Aly NM, Mahmoud FH, Kenawy A, El-Sebae AKH (2010) The role of vitamin C as antioxidant in protection of oxidative stress induced by imidacloprid. Food Chem Toxicol 48(1):215–221
109. Koike K, Ishigami A, Sato Y, Hirai T, Yuan Y, Kobayashi E et al (2014) Vitamin C prevents cigarette smoke–induced pulmonary emphysema in mice and provides pulmonary restoration. Am J Respir Cell Mol Biol 50(2):347–357
110. Vilchèze C, Kim J, Jacobs WR Jr (2018) Vitamin C potentiates the killing of mycobacterium tuberculosis by the first-line tuberculosis drugs isoniazid and rifampin in mice. Antimicrob Agents Chemother 62(3):e02165–e02117

Chapter 9
Important Contaminants (Mycotoxins, Pesticide Residues, Pirolizidine Alkaloids) in Pollen

Miroslava Kačániová, Natália Čmiková, and Vladimíra Kňazovická

9.1 Introduction

From early spring to late summer, honeybees (*Apis mellifera* L.) are collecting pollen from various plants. Being rich in protein, fatty acids, and vitamins, bee pollen is the primary nutritional source for bees [1–3]. Due to these valuable ingredients, bee pollen is also an attractive food supplement in human nutrition [4]. For this purpose, bee pollen can be collected by the installation of pollen traps at the hive entrance from early spring to late summer. However, a thorough control of samples is important because bee pollen may be contaminated with both residues of pesticides applied in agriculture mainly during spring [5–8], and also harmful natural contaminants produced by plants during summer [9]. According to international publications, the most important food safety risks associated with bee pollen are mycotoxins, pesticides, and pyrrolizidine alkaloids [10, 11].

M. Kačániová (✉)
Institute of Horticulture, Faculty of Horticulture and Landscape Engineering,
Slovak University of Agriculture, Nitra, Slovakia

School of Medical and Health Sciences, University of Economics and Human Sciences in Warsaw, Warsaw, Poland
e-mail: miroslava.kacaniova@uniag.sk

N. Čmiková
Institute of Horticulture, Faculty of Horticulture and Landscape Engineering,
Slovak University of Agriculture, Nitra, Slovakia
e-mail: xcmikova@uniag.sk

V. Kňazovická
Institute of Apiculture Liptovský Hrádok, Research Institute for Animal Production Nitra,
National Agricultural and Food Centre, Liptovský Hrádok, Slovakia
e-mail: vladimira.knazovicka@nppc.sk

N. Ecem Bayram et al. (eds.), *Pollen Chemistry & Biotechnology*,
https://doi.org/10.1007/978-3-031-47563-4_9

9.1.1 Mycotoxins

One of the most important criteria for a bee pollen quality standard is microbiological contamination, which is highly influenced by environmental factors that influence the development and reproduction of microscopic fungi and bacteria [12]. The fresh bee pollen is very hygroscopic and therefore very moist. Therefore, pollen is a rich medium for microorganisms to grow and multiply [13]. Microscopic fungi are very difficult to control under natural conditions [14]. They are widespread in nature, and have spores that are resistant to various environmental factors, which not only enter to the bee pollen but can also enter in the honey [15]. Mycotoxicosis caused by fungal mycotoxins can be caused by the harmful effects of mycotoxins, which are secondary metabolites of fungi [16]. Mycotoxin-contaminated food may result in cytotoxic, neurotoxic, immunosuppressive, teratogenic, mutagenic, and carcinogenic effects [17]. Fungi of *Fusarium, Aspergillus, Penicillium, Alternaria, Cladosporium,* and *Mucor* genera were identified as the etiologic factors of the above-mentioned diseases. Special attention should be given to toxins produced by *Fusarium* spp. (fusariotoxins). Deoxynivalenol (DON) and zearalenone (ZEA) produced by *F. culmorum* and *F. graminearum* are mycotoxins that cause a variety of health problems [18]. The recovery of storage fungi (*Aspergillus* and *Penicillium*) in fresh pollen presents a potential risk for human health and must look responsibly of the post-harvest process [19]. Some of the factors that can influence bee pollen storage is water activity (a_w). It plays an important role. It is the amount of water needed for the growth and development of microorganisms. In order to grow microorganisms require a minimum a_w value. Yeasts and fungi are growing to a_w value over 0.61 [20]. In the freshly collected bee pollen, the water content is about 25% [21], which represents a suitable substrate for the growth of microorganisms especially yeasts and fungi [22]. Bee pollen should be dried after harvest to avoid adverse effects on microorganisms. The moisture content of dried bee pollen should not exceed 8 % [23]. The quality of the bee pollen may result in non-compliance with proper handling of the bee pollen after removal from the hives [24]. A very important moment is pollen collection from the traps. The quality of the bee pollen also depends on the time of storage [25]. Improper storage of the pollen (increases umidity) can cause the development of molds and bacteria, which leads to the production of mycotoxins, causing poisoning in humans [26, 27].

The quality of bee pollen is strongly dependent on its preservation. Pollen has specific characteristics linked to floral species or cultivars it comes from [28]. After being collected from hives, bee pollen is carried to packaging centers, where the product is cleaned, dried, and packaged. The techniques vary depending on the available machinery. The quality of the product also depends on these cleaning, drying, and packaging processes currently applied by beekeepers or apicultural traders to achieve longer storage life [29]. Considering its nutrient content, a variety of microorganisms could grow in bee pollen. If collection, storage, and marketing practices are not appropriate fungi might develop in it as it happens in cereal grains [30, 31]. Fungi colonize the terrestrial environment successfully and utilize solid substrates efficiently by growing over their surfaces and penetrating their matrices. Many fungi produce mycotoxins that can cause acute or chronic intoxication and damage to humans and animals after ingestion of contaminated food and feed [32, 33]. Among the mycotoxins, aflatoxins and ochratoxin A (OTA) occupy special places due to their high occurrence and toxicity. The European Commission has established maximum allowable limits for these toxins in some food products [34–36].

9.1.1.1 Aflatoxins

Aflatoxins are the product of the metabolism of different fungi species which belong to *Aspergillus* genus with *A. flavus* and *A. parasiticus* strains as the main producers [37]. They can be synthetized in fungi's spores and mycelium or secreted as exotoxins [38]. The most toxic and dangerous aflatoxins are aflatoxin B_1 and B_2 [39]. Both aflatoxin B_1 and B_2 are carcinogenic for humans and animals and are listed in Group 1 of carcinogenic substances according to International Agency for Research on Cancer (IARC) [40]. The liver is the organ that suffers most from the effects of aflatoxins [41]. Ingestion of these toxins can lead to aflatoxicosis, an acute form of poisoning, or, in the case of long-term exposure, to the development of liver cancer [41]. Hydroxylated AFB-forms presented in milk are aflatoxin M_1 and M_2 [37] which are possibly carcinogenic for humans (IARC Group 2A of carcinogenic substances) [39, 40]. Furthermore, two other forms of AF exist- Aflatoxin G_1 and G_2 [42]. As aflatoxins show detrimental effects on bee health, the incidence of these compounds in hives is undesirable. It is for this reason that the occurrence and production of propolis in hives is an effective way for bees to deal with AFs toxicity [43, 44] which could indicate that this source of pollen contamination with aflatoxins is at least probable. In the past, aflatoxin occurrence in feed and food was a characteristic of tropic or sub-tropic regions due to favourable climatic conditions.

Recently, with climatic changes, which extensively influence weather conditions in temperate areas (such as the majority of Europe), the presence of aflatoxins in these areas is becoming more frequent [42]. Markaki [45] has shown, on Greek bee-collected pollen samples, that there is no detectable AFB1 contamination of batch pollen samples collected directly from beehives during the whole incubation period. But, if samples were inoculated with *Aspergillus parasiticus*, relatively high AFB1contamination was detected. In that sense, it can be assumed that the most likely mode of AFB1 contamination of bee collected samples investigated in this study is after collection and caused by beekeepers themselves or by merchants since all of the samples were commercially provided. The sample with the highest concentration of AFB1far exceeded the upper limit of 5 μg/kg, and even exceeded the limit of 12 μg/kg set by Serbian legislation. The other analysed bee pollen samples had concentrations of AFB1 above the limit of 2 μg/kg, among which only five had values between 2 and 5 μg/kg. Although the aforementioned types of food do not include bee pollen (unless it is classified as dietary food for special medical purposes, for which MPC of AFB1 is 0.1 μg/kg) there is no doubt that the presence of aflatoxins are extremely undesirable. For example, consumption of 100 g of the investigated bee-collected pollen samples as a food supplement could provide an intake of 0.315–1.732 μg of AFB1 per day [42].

The mean concentrations of total aflatoxins were relatively low in products from rape (*Brassica napus* L.), sunflower (*Helianthus annuus* L.) and silver birch (*Betula verrucosa*), but it reached 16 μg/kg in poppy (*Papaver somniferum*) pollen. The amount of aflatoxins and OTA were below the limit of quantification in dried bee pollen samples from Spain [46]. The authors emphasized that results may be related to rapid drying and optimal moisture content. The mean concentration of aflatoxins can exceed 20 μg/kg in both fresh and dried pollen samples [47]. Rodríguez-Carrasco et al. [48] reported very low mycotoxin contamination in bee pollen, although, they tested only *Fusarium* toxins. Of the 15 samples examined, only 2 contained mycotoxins, namely neosolaniol and nivalenol, above the limit of detection. The analysed samples were randomly purchased in supermarkets, so it can be assumed, that their moisture content was reduced to the optimal level.

9.1.1.2 Ochratoxin (OTA)

Ochratoxin (OTA) is nephrotoxic, hepatotoxic, teratogenic, and immunotoxic to animals [49] and has been associated to fatal endemic human nephropathies [50]. Due to its carcinogenicity to mice and rats [51], OTA has been classified as possible carcinogen to humans being included in group 2B [52]. This mycotoxin is produced by *Aspergillus ochraceus, A. alliaceus* [53], *A. carbonarius* [54], *A. niger* [55], and *Penicillium verrucosum* [56]. OTA is widely distributed, and its occurrence has been reported mainly in cereals and coffee [57, 58], milk [59], wine [45] and beer [60]. Significant contamination of bee pollen was determined in a case of Slovakian samples [61]. In total, 45 samples were divided into three groups of 15 samples originating from poppy, rape, and sunflower plants. Determined OTA concentration ranges in poppy, rape, and sunflower pollen samples were 6.12–10.98 μg/kg, 3.24–9.87 μg/kg, and 0.23–6.93 μg/kg, respectively. In Spain, by analysing the toxigenic potential of *A. ochraceus* in various substrates (bee pollen, maize, wheat, and rice) Medina et al. [62] found that OTA production in bee pollen was statistically significantly higher than that found in the production of tested cereals, regardless of the incubation time (7, 14, 21, 28 days). When commercialized, pollen may still contain *A. flavus* spores as reported by several studies [24, 25, 63], highlighting the potential risk for human health in bee pollen consumption due to the high contamination level by these molds and their mycotoxins. As a matter of fact, it has been demonstrated that bee pollen is a substrate stimulating the production of ochratoxin A by *A. ochraceus* [62]; this mycotoxin is highly cytotoxic and is reported for insecticidal effects [64]. A few studies showed contamination of honey. In Northern Italy, *A. flavus* and *A. japonicus* have been identified in a shotgun sequencing of DNA contained in honey [65].

9.1.1.3 Fumonisins

Fumonisins, besides aflatoxin, ochratoxin, zearalenone, and trichothecene mycotoxins are of greatest concern due to their high occurrence and potential toxic effects on human and animal health. Since the first discovery, at least 30 fumonisin analogs have been characterized and classified in four main groups (A, B, C, and P series). Additional groups are also suggested (i.e., D and X series, etc.). These secondary metabolites of *Fusarium verticillioides* (formerly *F. moniliforme*), *F. proliferatum, F. fujikuroi* and some other *Fusarium* species, *Alternaria alternata* f. sp. *Lycopersici, Aspergillus awamori* and *Aspergillus niger* are polyketides. Predominantly, they are produced by *F. verticillioides* and *F. proliferatum* [66–70]. *Fusarium* species often cause diseases of crops in hot climates where favourable conditions such as temperature stress, insect damage and high-water activities. Most commonly they contaminate maize and maize-derived products, although other crops can also be affected. It has been estimated that B-type fumonisins (FB1, FB2, FB3 and FB4) are the most common natural FBs in feed (more than 95 % of the all detected FBs). They are considered heat stable (in common food processing), but a number of different forms have been described as a result of hydrolysis (hydrolysed fumonisins B and partially hydrolysed fumonisins B), formation of Maillard-type modified forms (NCM- and NDF-fumonisins B), and interaction with the matrix (hidden and bound forms [71–73]. The presence of both, *F. proliferatum* and *F. verticillioides* was confirmed in thirty i.e., forty-five bee pollen samples, respectively but FBs were quantified only in the samples originating from sunflower (15 samples) [61].

9.1.1.4 Zearalenone (ZEN)

Zearalenone (ZEN) is mycoestrogen with limited toxicity that is produced by several *Fusarium* species: *F. graminearum*, *F. culmorum*, *F. crookwellense*, and *F. equiseti*. It is regularly present in crops and crop products [37]. According to IARC this macrocyclic lactone is classified in group 3 which means that it is not classifiable as to its carcinogenicity to humans [40] In the case of pollen, the significant contamination with ZEN was recorded in Slovakian bee samples [61].

9.1.1.5 Other Toxins

In a study from Slovakia [61], the authors also reported the contamination of all examined bee pollen samples with T-2 toxin and deoxynivalenol. Both toxins belong to trichothecene compounds, the sesquiterpenoid metabolites obtained after the microbiological activity of several fungi from the following genera: *Fusarium* (primary source), *Trichoderma*, *Myrothecium*, *Phomopsis*, etc., [37]. Together with ZEN, they were the most dominant quantified mycotoxins in the pollen samples. Additionally, the presence of DON and T-2 toxins were checked in 15 pollen samples from Spain.

9.1.2 Pesticide Residues

Sörös et al. [74] defines plant protection products as preparations of natural, synthetic, or chemical origin that are capable of reducing, attracting, or alerting pests or regulating the life processes of pests and plants. The use of pesticides is essential for the food supply of the Earth's population, so residues are generally present in plant foods. Alternatives of these products are grown by organic farming [75]. The main principles of organic farming are sustainability, rural development, animal welfare, and food safety. Due to the benefits of organic agriculture, global interest is constantly growing in these products [76]. Given that organic farming systems tend to promote greater diversity in microbial communities ongoing competition among fungal species, and between fungi and bacteria, should help assure that no one fungal species reaches dangerous levels. Inappropriate post-harvest handling and storage of conventional and organic food greatly increases the risk of mycotoxin contamination and has little or nothing to do with how the food was produced in the field. Unusual and largely uncontrollable environmental conditions play an important role in triggering a significant portion of the serious mycotoxin contamination episodes that have occurred, regardless of how food was grown [77]. The production and labelling of organic foods are regulated at the European level [78]. This document summarizes the requirements for organic beekeeping practices on bee nutrition, veterinary drug use, animal welfare, and hive placement. Under the regulation, apiaries should be located in areas that ensure the availability of pesticide-free nectar and pollen sources. Under favourable environmental conditions, bees typically forage in a radius of 2 km from the hive [79], therefore a distance of at least 3 km must be provided from bee plants treated with conventional pesticides [78]. However, since honeybees are capable to forage at distances of up to 10 km [79], the pesticide contamination of organic beekeeping products cannot be completely excluded.

Pesticide occurrence in honeybee-collected pollen increases by increasing the number of screened pesticides and analysed samples, and by decreasing the limit of detection. In agricultural and urban environments, bees can be exposed to pesticides in several ways, including contact via air particles and consumption of contaminated food and water. Pesticide residues in pollen and nectar represent a relevant route of exposure for bees as they rely on these food sources to grow and survive during the larval stage and to perform their tasks when they become adults [80, 81]. Several studies [82, 83] have shown that pollen and nectar are often contaminated with different combinations of pesticides, which can elicit synergistic effects [81, 84–86]. In ecotoxicology, synergism occurs when the interaction between two or more substances produces an effect greater than the sum of their individual effects. Several pesticides, mainly fungicides, that alone are considered not or slightly toxic for bees, in combination with low residues of other compounds, can produce adverse effects [87]. Synergism can also occur when bees are exposed to very low (not lethal) concentrations of two toxic compounds [88]. Yet, for most pesticide combinations, there is no information on their potential synergistic interactions because most ecotoxicological studies have tested single compounds and multiple-pesticide exposures are not considered in the current bee risk assessment scheme [81]. At present, ca. 1000 different pesticides (e.g. active ingredients) are commercially available [89] and it is obviously impossible to test all potential combinations. However, the most likely pesticide combinations that bees are exposed in the field should be assessed [81]. To achieve this goal robust and large-scale information on the actual pesticide exposure on bees is required, but information on the time space distributions of pesticides at the field scale is not readily available, and therefore exposure and co-exposure levels cannot be quantified [81].

Up to now, several monitoring activities have been conducted at national and regional scales to evaluate the field-realistic exposure of bees to pesticides after their release on the market [86, 90–93]. The percentage of positive samples and the number of detected pesticides have proven to be very variable among countries and years. Pesticide contamination in the pollen can be associated with the number of permitted active ingredients allowed in different countries [94]. However, data gathered from these monitoring initiatives are difficult to compare and harmonize due to differences in the analytical sensitivity, the list of screened pesticides, in the number of analysed samples, in the bee matrices considered, and in the sampling protocols.

Only a few studies from different parts of the world, aimed to determine pesticide residue contents of bee pollen. Relatively little data are available from Asia, Africa, and South America [95]. In the selected studies, several active substances were detected from pollen samples simultaneously using multiresidue methods. One of the most comprehensive studies on the topic was conducted by Johnston et al. [96], who examined the pesticide residue content of 107 bee pollen samples and 25 bee bread samples from 12 European countries. They determined the concentration of 300 active substances [96]. Two-thirds of the products were contaminated with at least one active agent and a total of 53 pesticides were identified including insecticides, acaricides, fungicides, and herbicides. The most common substances were chlorpyrifos-ethyl, boscalid, and thiacloprid. In a few cases, concentrations of some fungicides exceeded 0.001‰, for example boscalid (Sweden), dimethomorph (Italy), fenhexamide (Germany), and cyprodinil (Switzerland). Böhme et al. [5] conducted studies on the pesticide residue content of bee pollen over 5 years in Germany, involving 281 samples. Approximately four-fifths of the samples were contaminated with at least one substance. A total of 73 active substances were detected, mostly fungicides. Most pesticides were revealed in samples from hives located near orchards, followed by products from cereal plantations and meadows. The most common active substance in bee pollen was thiacloprid with the highest concentration of 470 μg/kg. Relatively few studies were conducted on the pesticide contamination of bee pollen in countries outside Europe. Nai et al. [97] screened 155 pollens for 232 pesticides in Taiwan. It was concluded that at least one pesticide was detectable in three-quarter of the samples [97]. A total of 56 active substances were identified, of which fluvalinate and chlorpyrifos showed the highest detection frequency and high residue levels. Tong et al. [98] examined the pesticide contamination of 189 pollen samples originated from major beekeeping areas of China. There were insecticides (imidacloprid, thiametoxam, fenpropathrin, bifenthrin, chlorpyrifos), acaricides (coumaphos, fluvalinate), and fungicides (carbendazim, triadimefon) among the most common active substances in these samples. Mullin et al. [90] conducted a broad survey of pesticide contamination of apicultural products across 23 states of the USA and 1 Canadian province. Most pollen samples contained at least one systemic pesticide and nearly half of the samples contained both in-hive acaricides (fluvalinate, coumaphos) and fungicides (chlorothalonil). Authors emphasized that the scientific knowledge is incomplete on the biological effects of pesticide combinations. Bees from hives deployed near mass flowering crops tend to collect pollen from other species if it is possible. It is generally assumed, that the pesticide residue content of bee pollen is determined by the pesticide treatment of crops. However, Botías et al. [82] suggested that the chronic exposure of bees to neonicotinoids came from residues in non-focal crop pollen. Wildflowers, for example, can become contaminated due to the fact that some pesticides are persistent in the soil [82]. Pesticide residues in bee pollen pose a risk not only to bee health but also raise safety concerns in human nutrition [5]. Considering this fact, a human risk assessment was performed for the most common pesticide residues (chlorpyrifos, fluvalinate, carbendazim, and thiacloprid) and pesticides detected in exceptionally high concentrations in bee pollen.

Of course, the detection of pesticides in bee matrices depends on the location of the apiaries and the time of the season when the samples were gathered. Other factors have been identified to affect the level of pesticide residues in bee matrices such as crop species/variety, dose and mode of application, physicochemical properties of the pesticides, soil type, weather, and sampling time of day [80, 99, 100]. However, we argue that the analytical methodologies and the sampling protocols (e.g. limit of detections, number of screened pesticides, and number of analysed samples), may have limited our ability to detect real pesticide exposure in bee matrices. In fact, we expect a high number of detected pesticides in monitoring studies where a higher number of pesticides and samples have been analysed. We also expect a high frequency of pesticide occurrences in monitoring studies where lower detection limits (LODs) for each screened active ingredient have been used. On the contrary, studies using high LODs might have reported low pesticide occurrences and false-negative samples, underestimating pesticide exposure on bees.

Here, we assess the impact of the analytical methodologies on the occurrence of pesticides in bee matrices by means of a systematic literature review. Specifically, we want to know if the methodologies could have affected the results of pesticide occurrences obtained from the monitoring studies. We have limited our systematic review to the honey bee-collected pollen (beebread – pollen stored in the hive – and corbicular pollen from returning foragers) because it has the highest frequency of pesticide occurrences and it is considered the best matrix for assessing pesticide contamination in the environment [91, 93].

9.1.3 Pyrrolizidine Alkaloids

One class of potential natural contaminants of bee pollen is pyrrolizidine alkaloids (PAs) [73]. Pyrrolizidine alkaloids (PAs) are secondary metabolites that contribute to the survival of certain plant species as a part of their natural defence mechanism against herbivores [101]. To date, more than 660 PAs compounds have been identified that have a structure characteristic of the source plant. Their common feature is the necine base composed of two fused five-membered rings with a nitrogen atom at the bridgehead. Necine base is esterified with one or two nucleic acids. These are mono- or dicarboxylic acids with branched carbon chains containing hydroxy, alkoxy-, epoxy-, and carboxyester groups. Assumedly, around 3 % of all flowering plants may produce PAs [102]. Most of these plants belong to the families of Asteraceae (e.g., Senecioneae (*Senecio* spp.) and Eupatorieae (*Eupatorium* spp.), Boraginaceae (e.g., *Borago* spp. and *Echium* spp.), and Fabaceae (e.g., *Crotalaria* spp.) [103, 104]. Harmful PAs are hepatotoxic to both animals and humans and can cause acute poisoning and chronic effects [105–107]. Chronic exposure to low PAs amounts can result in diseases such as liver cirrhosis or possibly cause cancer as metabolic activation produces genotoxic and cancerogenic metabolites [108]. As a consequence, the US Food and Drug Administration (FDA) decreed to ban PA-containing products from the market. By contrast, maximum levels of PAs are currently not in force for food in the European Union (EU). A novel EU regulation defining maximum PAs levels for pollen and pollen-based food supplements among other foods such as (herbal) teas, herbs, and spices will be effective by July 2022. Then, for pollen and related food supplements a maximum PAs level of 500 μg/kg must not be exceeded. Furthermore, a maximum intake of 1 μg PAs per day was established in pharmaceutical products [109]. Likewise, the European Food Safety Authority (EFSA) has introduced a benchmark dose lower confidence limit 10 % ($BMDL_{10}$) amount of 237 μg riddelliine per kg body weight (BW) in rats as a reference point for the assessment of carcinogenic risks, assuming similar carcinogenic potency for different PAs [73]. Considering a tolerable margin of exposure (MOE) of 10,000 for humans, this corresponds with a maximum intake of 0.024 μg PAs per kg BW per day [110]. This in turn corresponds with a maximum daily intake of 1.8 μg PAs per day for adults (75 kg BW) and ~ 1 μg for juveniles (40 kg BW, 10 years) [111].

Next to foods with a direct botanical background such as teas and herbals, PAs were also exemplarily studied in honey and bee pollen (91,111–117). In one study, 17 of 55 commercial pollen products, mainly from Spain, Romania, Italy, and France, were contaminated with toxic PAs of up to 16,400 ng/g pollen [112]. Similarly, detected PAs in 60 % of 119 bee pollen samples from various countries with ΣPA contents of up to 37,900 ng/g pollen [113]. Recently, Kast et al. [114] presented two complementary methods: a high-performance liquid chromatography coupled to tandem mass spectrometry (LC–MS/MS) method which allowed studying 18 PAs and PANO in commercial bee pollen products as well as daily collected bee pollen samples, and a LC-HRMS method comparing daily collected pollen

Table 9.1 Contaminants in bee pollen [10]

Organochlorine pesticides	< MRL[a]
Organophosphate pesticides	< MRL
Pyrethroids	< MRL
Alfatoxin B_1	Max. 2 μg/kg
Alfatoxin $B_1 + B_2 + G_1 + G_2$	Max. 4 μg/kg
Cloramphenicol (CAP)	Absent
Nitrofuran metabolites	Absent
Sulfonamides	Absent
Heavy metal Pb	Max 0,5 mg/kg
Heavy metal Hg	Max 0,01 mg /kg
Heavy metal Cd	Max 0,03 mg/kg
Radioactivity (Cs-134 and Cs-137)	<600 Bq/kg

[a] Should be smaller than the values established for honey

samples to flower heads, by detecting of all PA types, including saturated, non-cancerogenic PAs [115]. To achieve higher confidence for the identification of the plant source, nearly all PA-containing plants occurring in Switzerland were also analysed by LC-HRMS. Bee-collected pollen indicated the presence of the highest PAs concentrations in "*Echium*-type PA" samples collected in June and "*Eupatorium*-type PA" samples (assigned as intermedine and lycopsamine (N-oxides) mainly from mid-July and August [115]. However, no reliable information existed about PA levels in bee-collected pollen from Germany.

Given the geographic neighborhood of Switzerland (the observation site Basel of Kast et al. [114] is directly at the border to Germany) and Southern Germany, we aimed to carry out a thorough study by collecting bee pollen by means of traps in 57 locations in Baden-Wuerttemberg (Southern Germany). Knowing the fact that pollen samples in spring (from April to June) could be highly contaminated with pesticides in Southern Germany [7], the aim was to determine whether the pollen samples were otherwise contaminated in a later period (Table 9.1).

References

1. Avni D, Hendriksma HP, Dag A, Uni Z, Shafir S (2014) Nutritional aspects of honey bee-collected pollen and constraints on colony development in the eastern Mediterranean. J Insect Physiol 69:65–73. https://doi.org/10.1016/j.jinsphys.2014.07.001
2. Mărgăoan R, Mărghitaş LA, Dezmirean DS, Dulf FV, Bunea A, Socaci SA, Bobiş O (2014) Predominant and secondary pollen botanical origins influence the carotenoid and fatty acid profile in fresh honeybee-collected pollen. J Agric Food Chem 62:6306–6316. https://doi.org/10.1021/jf5020318
3. Taha E-KA, Al-Kahtani S, Taha R (2019) Protein content and amino acids composition of bee-pollens from major floral sources in Al-Ahsa, eastern Saudi Arabia. Saudi J Biol Sci 26:232–237. https://doi.org/10.1016/j.sjbs.2017.06.003

4. Feás X, Vázquez-Tato MP, Estevinho L, Seijas JA, Iglesias A (2012) Organic bee pollen: botanical origin, nutritional value, bioactive compounds. Antioxid Act Microbiol Qual Mol 17:8359–8377. https://doi.org/10.3390/molecules17078359
5. Böhme F, Bischoff G, Zebitz CPW, Rosenkranz P, Wallner K (2018) Pesticide residue survey of pollen loads collected by honeybees (Apis mellifera) in daily intervals at three agricultural sites in South Germany. PLoS One 13:e0199995. https://doi.org/10.1371/journal.pone.0199995
6. Drummond FA, Ballman ES, Eitzer BD, Du Clos B, Dill J (2018) Exposure of honey bee (Apis mellifera L.) colonies to pesticides in pollen, A statewide assessment in Maine. Environ Entomol 47:378–387. https://doi.org/10.1093/ee/nvy023
7. Friedle C, Wallner K, Rosenkranz P, Martens D, Vetter W (2021) Pesticide residues in daily bee pollen samples (April–July) from an intensive agricultural region in southern Germany. Environ Sci Pollut Res 28:22789–22803. https://doi.org/10.1007/s11356-020-12318-2
8. Traynor KS, Pettis JS, Tarpy DR, Mullin CA, Frazier JL, Frazier M, van Engelsdorp D (2016) In-hive pesticide Exposome: assessing risks to migratory honey bees from in-hive pesticide contamination in the Eastern United States. Sci Rep 6:33207. https://doi.org/10.1038/srep33207
9. Pyrrolizidine alkaloids in herbal teas and teas. 29
10. Campos MGR, Bogdanov S, de Almeida-Muradian LB, Szczesna T, Mancebo Y, Frigerio C, Ferreira F (2008) Pollen composition and standardisation of analytical methods. J Apic Res 47:154–161. https://doi.org/10.1080/00218839.2008.11101443
11. Thakur M, Nanda V (2020) Composition and functionality of bee pollen: a review. Trends Food Sci Technol 98:82–106. https://doi.org/10.1016/j.tifs.2020.02.001
12. Xue X, Selvaraj JN, Zhao L, Dong H, Liu F, Liu Y, Li Y (2014) Simultaneous determination of aflatoxins and Ochratoxin A in bee pollen by low-temperature fat precipitation and Immunoaffinity column cleanup coupled with LC-MS/MS. Food Anal Methods 7:690–696. https://doi.org/10.1007/s12161-013-9723-4
13. DeGrandi-Hoffman G, Chen Y, Simonds R (2013) The effects of pesticides on queen rearing and virus titers in honey bees (Apis mellifera L.). Insects. 4:71–89. https://doi.org/10.3390/insects4010071
14. Finola MS, Lasagno MC, Marioli JM (2007) Microbiological and chemical characterization of honeys from Central Argentina. Food Chem 100:1649–1653. https://doi.org/10.1016/j.foodchem.2005.12.046
15. Maria P, Mihaela V, Roxana A (2009) Study concerning the honey qualities in Transylvania region. JASO 2:1034–1040. https://doi.org/10.29302/oeconomica.2009.11.2.49
16. Hani B, Dalila B, Dahamna S, Harzallah D, Mouloud G, Khennouf S (2012) Microbiological sanitary aspects of pollen. Adv Environ Biol 6:1415–1420
17. Krnjaja V, Lević J, Stanković S, Petrović T, Stojanović L, Radović Č, Gogić M (2012) Distribution of moulds and mycotoxins in maize grain silage in the trench silo. Biotechnol Anim Husb 28:845–854. https://doi.org/10.2298/BAH1204845K
18. Smith JE, Lewis CW, Anderson JG, Solomon GW (1994) Mycotoxins in human nutrition and health. In: Studies of the European Commission, Directorate General XII, Brussels, pp 104–123
19. Nardoni S, D'Ascenzi C, Rocchigiani G, Moretti V, Mancianti F (2016) Occurrence of moulds from bee pollen in Central Italy – a preliminary study. Ann Agric Environ Med 23. https://doi.org/10.5604/12321966.1196862
20. Jay JM (2007) Modern food microbiology. Aspen Publication, Maryland
21. Bogdanov S (2012) Pollen: production, nutrition and health a review. Bee Prod Sci
22. Brindza J, Gróf J, Bacigálová K, Ferianc P, Tóth D (2010) Pollen microbial colonization and food safety. Acta Chimica Slovaca 8
23. Sinkevičienė J, Marcinkevičienė A, Baliukonienė V, Jovaišienė J (2019) Fungi and mycotoxins in fresh bee pollen. Proceedings of the international scientific conference "Rural development". 69–72

24. Deveza MV, Keller KM, Lorenzon MCA, Nunes LMT, Sales ÉO, Barth OM (2015) Mycotoxicological and palynological profiles of commercial brands of dried bee pollen. Braz J Microbiol 46:1171–1176. https://doi.org/10.1590/S1517-838246420140316
25. González G, Hinojo MJ, Mateo R, Medina A, Jiménez M (2005) Occurrence of mycotoxin producing fungi in bee pollen. Int J Food Microbiol 105:1–9. https://doi.org/10.1016/j.ijfoodmicro.2005.05.001
26. Estevinho ML, Afonso SE, Feás X (2011) Antifungal effect of lavender honey against Candida albicans, Candida krusei and Cryptococcus neoformans. J Food Sci Technol 48:640–643. https://doi.org/10.1007/s13197-011-0243-1
27. de Arruda VAS, Pereira AAS, de Freitas AS, Barth OM, de Almeida-Muradian LB (2013) Dried bee pollen: B complex vitamins, physicochemical and botanical composition. J Food Compos Anal 29:100–105. https://doi.org/10.1016/j.jfca.2012.11.004
28. Serra BJ (1988) Plant origin of honey collected pollen produced in Spain. Anales de la Asociación de Palinólogos de Lengua Española 73–78
29. Serra BJ, Lopez A (1986) Études microbiologiques du pollen d'abeilles. Revue Française d'Apiculture:259–266
30. Magan N, Lacey J (1988) Ecological determinants of mould growth in stored grain. Int J Food Microbiol 7:245–256. https://doi.org/10.1016/0168-1605(88)90043-8
31. Lacey J, Magan N (1991) Fungi colonising cereal grain: their occurrence and water and temperature relationships. In: Cereal grain—mycotoxins, fungi and quality in drying and storage. Elsevier Science, Amsterdam, pp 77–118
32. Marasas WFO, Nelson PE (1987) Mycotoxicology. The Pennsylvania State University Press, University Park
33. Moss MO (1996) Mycotoxins. Mycol Res 100:513–523. https://doi.org/10.1016/S0953-7562(96)80001-3
34. Commission Regulation (EC) No 466/2001 of 8 March 2001 setting maximum levels for certain contaminants in foodstuffs (Text with EEA relevance.). (2001)
35. Commission Regulation (EC) No 472/2002 of 12 March 2002 amending Regulation (EC) No 466/2001 setting maximum levels for certain contaminants in foodstuffs (Text with EEA relevance) – Publications Office of the EU, https://op.europa.eu/en/publication-detail/-/publication/997b6e24-351e-4cce-81a9-0e9cb0ffc244
36. Commission Regulation (EC) No 257/2002 of 12 February 2002 amending Regulation (EC) No 194/97 setting maximum levels for certain contaminants in foodstuffs and Regulation (EC) No 466/2001 setting maximum levels for certain contaminants in foodstuffs (Text with EEA relevance), https://www.legislation.gov.uk/eur/2002/257/adopted
37. Bennett JW, Klich M (2003) Mycotoxins. Clin Microbiol Rev 16:497–516. https://doi.org/10.1128/CMR.16.3.497-516.2003
38. Hanssen E, Jung M (1973) Control of aflatoxins in the food industry. In: Krogh P (ed) Control of mycotoxins. Butterworth-Heinemann, pp 239–250
39. Van EHP (2013) Mycotoxins: risks, regulations and European co-operation. Zbornik Matice srpske za prirodne nauke 125:7–20
40. Mycotoxin Biomarkers of Exposure: A comprehensive review – Vidal – 2018 – Comprehensive reviews in food science and food safety – Wiley Online Library, https://ift.onlinelibrary.wiley.com/doi/abs/10.1111/1541-4337.12367
41. Neal GE (1995) Genetic implications in the metabolism and toxicity of mycotoxins. Toxicol Lett 82–83:861–867. https://doi.org/10.1016/0378-4274(95)03600-8
42. Kostić AŽ, Milinčić DD, Petrović TS, Krnjaja VS, Stanojević SP, Barać MB, Tešić ŽL, Pešić MB (2019) Mycotoxins and mycotoxin producing fungi in pollen: review. Toxins (Basel) 11:64. https://doi.org/10.3390/toxins11020064
43. Niu G, Johnson RM, Berenbaum MR (2011) Toxicity of mycotoxins to honeybees and its amelioration by propolis. Apidologie 42:79. https://doi.org/10.1051/apido/2010039

44. Temiz A, Mumcu AŞ, Tüylü AÖ, Sorkun K, Salih B (2013) Antifungal activity of propolis samples collected from different geographical regions of Turkey against two food-related molds, Aspergillus versicolor and Penicillium aurantiogriseum. Gida 38:135–142
45. Markaki P, Delpont-Binet C, Grosso F, Dragacci S (2001) Determination of Ochratoxin A in red wine and vinegar by Immunoaffinity high-pressure liquid chromatography. J Food Prot 64:533–537. https://doi.org/10.4315/0362-028X-64.4.533
46. Garcia-Villanova RJ, Cordón C, González Paramás AM, Aparicio P, Garcia Rosales ME (2004) Simultaneous immunoaffinity column cleanup and HPLC analysis of aflatoxins and Ochratoxin A in Spanish bee pollen. J Agric Food Chem 52:7235–7239. https://doi.org/10.1021/jf048882z
47. Nuvoloni R, Meucci V, Turchi B, Sagona S, Fratini F, Felicioli A, Cerri D, Pedonese F (2021) Bee-pollen retailed in Tuscany (Italy): labelling, palynological, microbiological, and mycotoxicological profile. LWT 140:110712. https://doi.org/10.1016/j.lwt.2020.110712
48. Rodríguez-Carrasco Y, Font G, Mañes J, Berrada H (2013) Determination of mycotoxins in bee pollen by gas chromatography–tandem mass spectrometry. J Agric Food Chem 61:1999–2005. https://doi.org/10.1021/jf400256f
49. Dirheimer G (1998) Recent advances in the genotoxicity of mycotoxins. In: Revue de Medecine Veterinaire (France)
50. Pleština R (1992) Some features of Balkan endemic nephropathy. Food Chem Toxicol 30:177–181. https://doi.org/10.1016/0278-6915(92)90030-O
51. Pohland AE, Nesheim S, Friedman L (1992) Ochratoxin A: a review (technical report). Pure Appl Chem 64:1029–1046. https://doi.org/10.1351/pac199264071029
52. IARC: Some naturally occurring substances: food items and constituents, heterocyclic aromatic amines and mycotoxins
53. Bayman P, Baker JL, Doster MA, Michailides TJ, Mahoney NE (2002) Ochratoxin production by the aspergillus ochraceus group and aspergillus alliaceus. Appl Environ Microbiol 68:2326–2329. https://doi.org/10.1128/AEM.68.5.2326-2329.2002
54. Joosten HMLJ, Goetz J, Pittet A, Schellenberg M, Bucheli P (2001) Production of ochratoxin A by aspergillus carbonarius on coffee cherries. Int J Food Microbiol 65:39–44. https://doi.org/10.1016/S0168-1605(00)00506-7
55. Abarca ML, Bragulat MR, Castellá G, Cabañes FJ (1994) Ochratoxin A production by strains of aspergillus Niger var. Niger. Appl Environ Microbiol 60:2650–2652. https://doi.org/10.1128/aem.60.7.2650-2652.1994
56. Frisvad JC, Filtenborg O (1989) Terverticillate Penicillia: chemotaxonomy and mycotoxin production. Mycologia 81:837–861. https://doi.org/10.1080/00275514.1989.12025674
57. Trucksess MW, Giler J, Young K, White KD, Page SW (1999) Determination and survey of Ochratoxin A in wheat, barley, and coffee—1997. J AOAC Int 82:85–89. https://doi.org/10.1093/jaoac/82.1.85
58. Vrabcheva T, Usleber E, Dietrich R, Märtlbauer E (2000) Co-occurrence of Ochratoxin A and Citrinin in cereals from Bulgarian villages with a history of Balkan endemic nephropathy. J Agric Food Chem 48:2483–2488. https://doi.org/10.1021/jf990891y
59. Breitholtz-Emanuelsson A, Olsen M, Oskarsson A, Palminger I, Hult K (1993) Ochratoxin A in cow's milk and in human milk with corresponding human blood samples. J AOAC Int 76:842–846. https://doi.org/10.1093/jaoac/76.4.842
60. Visconti A, Pascale M, Centonze G (2000) Determination of ochratoxin A in domestic and imported beers in Italy by immunoaffinity clean-up and liquid chromatography. J Chromatogr A 888:321–326. https://doi.org/10.1016/S0021-9673(00)00549-5
61. Kačániová M, Juráček M, Chlebo R, Kňazovická V, Kadasi-Horáková M, Kunová S, Lejková J, Haščík P, Mareček J, Šimko M (2011) Mycobiota and mycotoxins in bee pollen collected from different areas of Slovakia. J Environ Sci Health B 46:623–629. https://doi.org/10.1080/03601234.2011.589322

62. Medina Á, González G, Sáez JM, Mateo R, Jiménez M (2004) Bee pollen, a substrate that stimulates Ochratoxin A production by aspergillus ochraceus Wilh. Syst Appl Microbiol 27:261–267. https://doi.org/10.1078/072320204322881880
63. Bucio Villalobos CM, López Preciado G, Martínez Jaime OA, Torres Morales JJ (2010) Mycoflora associated to bee pollen collected by domesticated bees (Apis mellifera L). Nova Scientia 2:93–103
64. Boguś MI, Wrońska AK, Kaczmarek A, Boguś-Sobocińska M (2021) In vitro screening of 65 mycotoxins for insecticidal potential. PLoS One 16:e0248772. https://doi.org/10.1371/journal.pone.0248772
65. Bovo S, Utzeri VJ, Ribani A, Cabbri R, Fontanesi L (2020) Shotgun sequencing of honey DNA can describe honey bee derived environmental signatures and the honey bee hologenome complexity. Sci Rep 10:9279. https://doi.org/10.1038/s41598-020-66127-1
66. Marasas WF (2001) Discovery and occurrence of the fumonisins: a historical perspective. Environ Health Perspect 109(Suppl 2):239–243
67. Huffman J, Gerber R, Du L (2010) Recent advancements in the biosynthetic mechanisms for polyketide-derived mycotoxins. Biopolymers 93:764–776. https://doi.org/10.1002/bip.21483
68. Perrone G, Stea G, Epifani F, Varga J, Frisvad JC, Samson RA (2011) Aspergillus Niger contains the cryptic phylogenetic species A. Awamori. Fungal Biol 115:1138–1150. https://doi.org/10.1016/j.funbio.2011.07.008
69. Mbundi L, Gallar-Ayala H, Khan MR, Barber JL, Losada S, Busquets R (2014) Chapter two – Advances in the analysis of challenging food contaminants: nanoparticles, bisphenols, mycotoxins, and brominated flame retardants. In: Fishbein JC, Heilman JM (eds) Advances in molecular toxicology. Elsevier, pp 35–105
70. Navale V, Vamkudoth KR, Ajmera S, Dhuri V (2021) Aspergillus derived mycotoxins in food and the environment: prevalence, detection, and toxicity. Toxicol Rep 8:1008–1030. https://doi.org/10.1016/j.toxrep.2021.04.013
71. Bullerman LB, Ryu D, Jackson LS (2002) Stability of Fumonisins in food processing. In: DeVries JW, Trucksess MW, Jackson LS (eds) Mycotoxins and food safety. Springer, Boston, pp 195–204
72. Rheeder JP, Marasas WFO, Vismer HF (2002) Production of Fumonisin analogs by fusarium species. Appl Environ Microbiol 68:2101–2105. https://doi.org/10.1128/AEM.68.5.2101-2105.2002
73. Knutsen HK, Alexander J, Barregård L, Bignami M, Brüschweiler B, Ceccatelli S, Cottrill B, Dinovi M, Edler L, Grasl-Kraupp B, Hogstrand C, Hoogenboom LR, Nebbia CS, Oswald IP, Petersen A, Rose M, Roudot A-C, Schwerdtle T, Vleminckx C, Vollmer G, Wallace H, Gomes JAR, Binaglia M, EFSA Panel on Contaminants in the Food Chain (CONTAM) (2017) Risks for human health related to the presence of pyrrolizidine alkaloids in honey, tea, herbal infusions and food supplements. EFSA J 15:e04908. https://doi.org/10.2903/j.efsa.2017.4908
74. Sörös C, Csóka M, Végh R, Sipos L (2021) Food safety hazards of bee pollen – a review. Trends Food Sci Technol 114:490–509. https://doi.org/10.1016/j.tifs.2021.06.016
75. Farkas Z, Országh E, Engelhardt T, Csorba S, Kerekes K, Zentai A, Süth M, Nagy A, Miklós G, Molnár K, Rácz C, Dövényi-Nagy T, Ambrus Á, Győri Z, Dobos AC, Pusztahelyi T, Pócsi I, Jóźwiak Á (2022) A systematic review of the efficacy of interventions to control aflatoxins in the dairy production Chain—feed production and animal feeding interventions. Toxins 14:115. https://doi.org/10.3390/toxins14020115
76. Bostan I, Onofrei M, Gavriluţă Vatamanu AF, Toderaşcu C, Lazăr CM (2019) An integrated approach to current trends in organic food in the EU. Foods 8:E144. https://doi.org/10.3390/foods8050144
77. Zhang H, Zheng X, Bai N, Li S, Zhang J, Lv W (2019) Responses of soil bacterial and fungal communities to organic and conventional farming systems in East China. J Microbiol Biotechnol 29:441–453. https://doi.org/10.4014/jmb.1809.09007

78. Regulation (EU) 2018/848 of the European Parliament and of the Council of 30 May 2018 on organic production and labelling of organic products and repealing Council Regulation (EC) No 834/2007, https://www.legislation.gov.uk/eur/2018/848/contents
79. Garbuzov M, Couvillon MJ, Schürch R, Ratnieks FLW (2015) Honey bee dance decoding and pollen-load analysis show limited foraging on spring-flowering oilseed rape, a potential source of neonicotinoid contamination. Agric Ecosyst Environ 203:62–68. https://doi.org/10.1016/j.agee.2014.12.009
80. Rortais A, Arnold G, Halm M-P, Touffet-Briens F (2005) Modes of honeybees exposure to systemic insecticides: estimated amounts of contaminated pollen and nectar consumed by different categories of bees. Apidologie 36:71–83. https://doi.org/10.1051/apido:2004071
81. Sgolastra F, Hinarejos S, Pitts-Singer TL, Boyle NK, Joseph T, Lūckmann J, Raine NE, Singh R, Williams NM, Bosch J (2019) Pesticide exposure assessment paradigm for solitary bees. Environ Entomol 48:22–35. https://doi.org/10.1093/ee/nvy105
82. Botías C, David A, Horwood J, Abdul-Sada A, Nicholls E, Hill E, Goulson D (2015) Neonicotinoid residues in wildflowers, a potential route of chronic exposure for bees. Environ Sci Technol 49:12731–12740. https://doi.org/10.1021/acs.est.5b03459
83. David A, Botías C, Abdul-Sada A, Nicholls E, Rotheray EL, Hill EM, Goulson D (2016) Widespread contamination of wildflower and bee-collected pollen with complex mixtures of neonicotinoids and fungicides commonly applied to crops. Environ Int 88:169–178. https://doi.org/10.1016/j.envint.2015.12.011
84. Thompson HM, Fryday SL, Harkin S, Milner S (2014) Potential impacts of synergism in honeybees (Apis mellifera) of exposure to neonicotinoids and sprayed fungicides in crops. Apidologie 45(5):545–553
85. Sgolastra F, Medrzycki P, Bortolotti L, Renzi MT, Tosi S, Bogo G, Teper D, Porrini C, Molowny-Horas R, Bosch J (2017) Synergistic mortality between a neonicotinoid insecticide and an ergosterol-biosynthesis-inhibiting fungicide in three bee species. Pest Manag Sci 73:1236–1243. https://doi.org/10.1002/ps.4449
86. Xavier JC, Hordijk W, Kauffman S, Steel M, Martin WF (2020) Autocatalytic chemical networks at the origin of metabolism. Proc R Soc B Biol Sci 287(20192377). https://doi.org/10.1098/rspb.2019.2377
87. Wade A, Lin C-H, Kurkul C, Regan ER, Johnson RM (2019) Combined toxicity of insecticides and fungicides applied to California almond orchards to honey bee larvae and adults. Insects 10:20. https://doi.org/10.3390/insects10010020
88. Berenbaum MR, Johnson RM (2015) Xenobiotic detoxification pathways in honey bees. Curr Opin Insect Sci 10:51–58. https://doi.org/10.1016/j.cois.2015.03.005
89. Lewis KA, Tzilivakis J, Warner DJ, Green A (2016) An international database for pesticide risk assessments and management. Hum Ecol Risk Assess Int J 22:1050–1064. https://doi.org/10.1080/10807039.2015.1133242
90. Mullin CA, Frazier M, Frazier JL, Ashcraft S, Simonds R, van Engelsdorp D, Pettis JS (2010) High levels of Miticides and agrochemicals in north American apiaries: implications for honey bee health. PLoS One 5:e9754. https://doi.org/10.1371/journal.pone.0009754
91. Chauzat M-P, Martel A-C, Cougoule N, Porta P, Lachaize J, Zeggane S, Aubert M, Carpentier P, Faucon J-P (2011) An assessment of honeybee colony matrices, Apis mellifera (Hymenoptera: Apidae) to monitor pesticide presence in continental France. Environ Toxicol Chem 30:103–111. https://doi.org/10.1002/etc.361
92. Orantes-Bermejo FJ, Pajuelo AG, Megías MM, Fernández-Píñar CT (2010) Pesticide residues in beeswax and beebread samples collected from honey bee colonies (Apis mellifera L.) in Spain. Possible implications for bee losses. J Apic Res 49:243–250. https://doi.org/10.3896/IBRA.1.49.3.03
93. Porrini C, Mutinelli F, Bortolotti L, Granato A, Laurenson L, Roberts K, Gallina A, Silvester N, Medrzycki P, Renzi T, Sgolastra F, Lodesani M (2016) The status of honey bee health in Italy: results from the Nationwide bee monitoring network. PLoS One 11:e0155411. https://doi.org/10.1371/journal.pone.0155411

94. Donley N (2019) The USA lags behind other agricultural nations in banning harmful pesticides. Environ Health 18:44. https://doi.org/10.1186/s12940-019-0488-0
95. Toselli G, Sgolastra F (2020) Seek and you shall find: an assessment of the influence of the analytical methodologies on pesticide occurrences in honey bee-collected pollen with a systematic review. Chemosphere 258:127358. https://doi.org/10.1016/j.chemosphere.2020.127358
96. Johnstone P, Huxdorff H, Simon G, Santillo D (2014) The bee's burden: an analysis of pesticide residues in comb pollen (beebread) and trapped pollen from honey bees (Apis mellifera) in 12 European countries – Greenpeace Research Laboratories, https://www.greenpeace.to/greenpeace/?p=1776
97. Nai Y-S, Chen T-Y, Chen Y-C, Chen C-T, Chen B-Y, Chen Y-W (2017) Revealing pesticide residues under high pesticide stress in Taiwan's agricultural environment probed by fresh honey bee (hymenoptera: Apidae) pollen. J Econ Entomol 110:1947–1958. https://doi.org/10.1093/jee/tox195
98. Tong Z, Duan J, Wu Y, Liu Q, He Q, Shi Y, Yu L, Cao H (2018) A survey of multiple pesticide residues in pollen and beebread collected in China. Sci Total Environ 640–641:1578–1586. https://doi.org/10.1016/j.scitotenv.2018.04.424
99. Bonmatin J-M, Giorio C, Girolami V, Goulson D, Kreutzweiser DP, Krupke C, Liess M, Long E, Marzaro M, Mitchell EAD, Noome DA, Simon-Delso N, Tapparo A (2015) Environmental fate and exposure; neonicotinoids and fipronil. Environ Sci Pollut Res 22:35–67. https://doi.org/10.1007/s11356-014-3332-7
100. Gierer F, Vaughan S, Slater M, Thompson HM, Elmore JS, Girling RD (2019) A review of the factors that influence pesticide residues in pollen and nectar: future research requirements for optimising the estimation of pollinator exposure. Environ Pollut 249:236–247. https://doi.org/10.1016/j.envpol.2019.03.025
101. Mädge I, Gehling M, Schöne C, Winterhalter P, These A (2020) Pyrrolizidine alkaloid profiling of four Boraginaceae species from northern Germany and implications for the analytical scope proposed for monitoring of maximum levels. Food Addit Contam Part A Chem Anal Control Expo Risk Assess 37:1339–1358. https://doi.org/10.1080/19440049.2020.1757166
102. Smith LW, Culvenor CCJ. Plant sources of Hepatotoxic Pyrrolizidine Alkaloids, https://pubs.acs.org/doi/pdf/10.1021/np50014a001
103. Knutsen HK, Alexander J, Barregård L, Bignami M, Brüschweiler B, Ceccatelli S, Cottrill B, Dinovi M, Edler L, Grasl-Kraupp B, Hogstrand C, Hoogenboom LR, Nebbia CS, Oswald IP, Petersen A, Rose M, Vleminckx C, Vollmer G, Wallace H, Bodin L, Cravedi J-P, Halldorsson TI, Haug LS, Johansson N, Loveren H, van Gergelova P, Mackay K, Levorato S, Manen M, van Schwerdtle T, EFSA Panel on Contaminants in the Food Chain (CONTAM) (2018) Risk to human health related to the presence of perfluorooctane sulfonic acid and perfluorooctanoic acid in food. EFSA J 16, e05194. https://doi.org/10.2903/j.efsa.2018.5194
104. Hartmann T, Toppel G (1987) Senecionine n-oxide, the primary product of pyrrolizidine alkaloid biosynthesis in root cultures of Senecio vulgaris. Phytochemistry 26:1639–1643. https://doi.org/10.1016/S0031-9422(00)82261-X
105. Colegate SM, Stegelmeier BL, Edgar JA (2012) 14 – Dietary exposure of livestock and humans to hepatotoxic natural products. In: Fink-Gremmels J (ed) Animal feed contamination. Woodhead Publishing, pp 352–382
106. Beuerle T, Benford D, Brimer L, Cottrill B, Doerge D, Dusemund B, Farmer P, Fürst P, Humpf H, Mulder PPJ (2011) Scientific opinion on pyrrolizidine alkaloids in food and feed. EFSA J 9:1–134. https://doi.org/10.2903/j.efsa.2011.2406
107. Kakar F, Akbarian Z, Leslie T, Mustafa ML, Watson J, van Egmond HP, Omar MF, Mofleh J (2010) An outbreak of hepatic Veno-occlusive disease in Western Afghanistan associated with exposure to wheat flour contaminated with pyrrolizidine alkaloids. J Toxicol 2010:e313280. https://doi.org/10.1155/2010/313280
108. Edgar JA, Colegate SM, Boppré M, Molyneux RJ (2011) Pyrrolizidine alkaloids in food: a spectrum of potential health consequences. Food Addit Contam: Part A. 28:308–324. https://doi.org/10.1080/19440049.2010.547520

109. Federal Institute for Drugs and Medical Devices (BfArM) – Bundesgesundheitsministerium, https://www.bundesgesundheitsministerium.de/en/en/ministry/authorities-within-the-remit/federal-institute-for-drugs-and-medical-devices-b.html
110. Bundesinstitut Für Risikobewertung: Updated risk assessment on levels of 1,2-unsaturated pyrrolizidine alkaloids (PAs) in foods: BfR Opinion No 026/2020 of 17 June 2020. BfR-Stellungnahmen. 2020, no. 026 (2020). https://doi.org/10.17590/20200805-100055
111. Fakayode OA, Aboagarib EAA, Zhou C, Ma H (2020) Co-pyrolysis of lignocellulosic and macroalgae biomasses for the production of biochar – a review. Bioresour Technol 297:122408. https://doi.org/10.1016/j.biortech.2019.122408
112. Kempf M, Wittig M, Schönfeld K, Cramer L, Schreier P, Beuerle T (2011) Pyrrolizidine alkaloids in food: downstream contamination in the food chain caused by honey and pollen. Food Addit Contam: Part A. 28:325–331. https://doi.org/10.1080/19440049.2010.521771
113. Dübecke A, Beckh G, Lüllmann C (2011) Pyrrolizidine alkaloids in honey and bee pollen. Food Addit Contam: Part A 28:348–358. https://doi.org/10.1080/19440049.2010.541594
114. Kast C, Kilchenmann V, Reinhard H, Droz B, Lucchetti MA, Dübecke A, Beckh G, Zoller O (2018) Chemical fingerprinting identifies Echium vulgare, Eupatorium cannabinum and Senecio spp. as plant species mainly responsible for pyrrolizidine alkaloids in bee-collected pollen. Food Addit Contam: Part A. 35:316–327. https://doi.org/10.1080/19440049.2017.1378443
115. Kast C, Kilchenmann V, Reinhard H, Bieri K, Zoller O (2019) Pyrrolizidine alkaloids: the botanical origin of pollen collected during the flowering period of Echium vulgare and the stability of pyrrolizidine alkaloids in bee bread. Molecules 24:2214. https://doi.org/10.3390/molecules24122214

Chapter 10
Other Bioactive Constituents of Pollen

José Bernal, Silvia Valverde, Adrián Fuente-Ballesteros, Beatriz Martín-Gómez, and Ana M. Ares

Abbreviations

BRCs	betaines and related compounds
CDA	canonical discriminant analysis
DAD	diode array detector
FID	flame ionization detector
FT-IR	fourier transform infrared spectroscopy
GC	gas chromatography
GRA	glucoraphanin
GSLs	glucosinolates
HCCAs	hydroxycinnamic acid amides
HILIC	hydrophilic interaction liquid chromatography
HPLC	high performance liquid chromatography
IT	irradiation
MS	mass spectrometry
MS/MS	tandem mass spectrometry
NMR	nuclear magnetic resonance
PDA	photodiode array
SE	solvent extraction
SFN	sulforaphane
SPE	solid phase extraction
UHPLC	ultra-high performance liquid chromatography
UHT	ultra-high temperature instantaneous sterilization

J. Bernal · S. Valverde · A. Fuente-Ballesteros · B. Martín-Gómez · A. M. Ares (✉)
Analytical Chemistry Group (TESEA group), IU CINQUIMA, University of Valladolid, Valladolid, Spain
e-mail: ana.maria.ares@uva.es; https://tesea.uva.es/

N. Ecem Bayram et al. (eds.), *Pollen Chemistry & Biotechnology*,
https://doi.org/10.1007/978-3-031-47563-4_10

10.1 Introduction

The consumption of pollen has increased considerably in recent years due to its beneficial effects on health along with its nutritional properties, which are directly related to its constituents [1, 2]. They comprise a large number of different substances, including a high protein content, carbohydrates, lipids like fatty acids, phenolic compounds, essential amino acids and their esters, vitamins, minerals such as sodium, potassium and magnesium, as well as other bioactive compounds like organic acids or glucosinolates [3, 4]. It should be mentioned that the composition and quality of pollen strongly depend on the plant of origin and the geographical origin, along with other factors such as climatic conditions, type of soil, the moment in which the pollen grows, and the age and nutritional condition of the plant [5–7]. Therefore, it is not surprising that in recent years many papers dedicated to investigating the composition of pollen have been published in order to valorize this product, highlighting its nutritional value and its possible health-promoting effects, and consequently several articles dealing with this topic have also been reported [4, 5, 8–12]. It should be highlighted that most works have focused mainly on the overall content of proteins, amino acids, lipids, and phenolic compounds in pollen. However, not much attention has been paid to the individual determination of these components [13–16]. In any case, this trend is changing, since more and more importance is being given to distinguishing the composition of pollen in detail. This helps establish the compounds responsible for nutritional activity and benefits, enabling the selection of the most beneficial pollen varieties or extraction of these compounds to use in food supplements or nutraceuticals [5]. Consequently, families of compounds that were not previously investigated in pollen, or to which little attention had been paid to their determination, are currently being tackled for different purposes, like phenolamides, glucosinolates, betaines, or biopolymers (sporopollenin, among other compounds) [17–20], but to our knowledge, no specific study has been published devoted specifically to summarize the publications related to their determination.

Thus, the main goal of this work is to present an overview of the publications that focus their attention on determining other pollen constituents that are not so often investigated in this matrix. This chapter is laid out according to the different group of compounds (glucosinolates, betaines, phenolamides, stilbenes, organic acids, and other compounds). A brief introduction about the main characteristics of each family, including some comments about their bioactive effects, has been included in each section, although most of the attention was focused on discussing some of the studies related to their determination in pollen, paying special attention to the publications of the last 10 years (2012–2022). It should be also noted that the analytical methodologies employed for their determination have been also briefly discussed. However, those readers who are interested in more specific aspects relating to extraction/separation methods can check the specific publications included in the reference list.

10.2 Phenolamides

Phenolamides, which are also known as hydroxycinnamic acid amides (HCAAs) or phenylamides, represent a major group of secondary metabolites commonly distributed in the plant kingdom predominating in flowers and pollen grains [21]. They are the result of the conjugation of polyamines or arylmonoamines and a phenolic moiety, mostly phenylpropanoids like the hydroxycinnamates [22]. Phenolamides have been proved to manifest different biological activities such as antioxidant, anti-inflammatory, antimicrobial, anti-cancer, and antimicrobial properties, as well as protective effects for the cardiovascular system and against atherosclerosis, metabolic syndrome or neurodegenerative diseases [21]. In addition, their potential could be established in cosmetic industry due to their tyrosinase inhibitory effects that could be used as active principle in skin whitening products [21, 23].

Ferulic, *p*-coumaric, caffeic and sinapic acids, or their glycosylated forms are the most common hydroxycinnamic acid derivatives found in nature; while aromatic phenolic (dopamine, octopamine, tyramine, noradrenaline, anthranilic acid), indolic (serotonin, tryptamine) or aliphatic (agmatine, putrescine, cadaverine, spermidine, and spermine) are the main amine moieties. The diversity of the phenolamides is due to all the possible combinations to form different structures from one to four N-substitutions, giving rise to more than 800 possible theoretical structures [21, 24]. Experimental data support that acyltransferase enzyme families are involved in the synthesis of the phenylpropanoids [17]. HCCAs are involved in growth and development phenomena of plants, and their structurally diverse presence in almost all plant organs supports the hypotheses about their involvement in floral initiation and fertility, as well as plant defense [22, 25]. Considering their presence in pollen, HCAAs are found on pollen surface of both monocots and dicots of all higher plants, but the concentration found specifically in pollen grains remains an unsettled matter mystery [17].

The bibliographic research indicates the determination of these compounds, either individually or jointly with other compounds from the same family or from other families of bioactive compounds, studied for different purposes, as can be seen in Table 10.1. Both types of pollen, directly taken from plants [26–32] and bee pollen [7, 33–45] have been examined, being bee pollen predominantly investigated in the last decade. Analytical methodologies employed for its identification comprise mostly the use of high-performance or ultra-high-performance liquid chromatography (HPLC or UHPLC) coupled to spectrophotometric (UV, photodiode array PDA, or diode array detector, DAD) or mass spectrometry (MS) detectors, including high resolution mass spectrometers and tandem mass spectrometry (MS/MS), which are preferred for identifying, confirm and even quantifying low values of concentration [7, 27–36, 39, 42–47]. In some studies, focused on separating fraction rich in phenolamides or even to isolate them [26, 37, 38, 40, 41], other complementary techniques like matrix assisted laser desorption/ionization (MALDI) together with the use of MS detectors or nuclear magnetic resonance spectroscopy (NMR) were employed to confirm their structures and molecular weight [26, 38, 40, 41]. Taking into consideration the sample

Table 10.1 Applications in the analysis of phenolamides in pollen

Matrix	Botanical origin (country)	Number of compounds	Reported bioactivity	Sample treatment	Determination technique	Amount (average values)	Ref.
P	*Quercus alba* L. (NS)	3	Antifungal	SE (MeOH)	HPLC-UV purification, HPLC-MS/MS, NMR	NS	[26]
P	*Helianthus annuus* L. (China)	1	Antiparasitic	SE (MeOH)	HPLC-MS	12 mg/g	[27]
P	*Helianthus* L. (Belgium)	5	Antiparasitic	SE (MeOH:water)	HPLC-MS/MS	10–25 mg/g	[28]
P	*Ambrosia artemisiifolia* L. (Serbia)	2	Antioxidant	SE (water)	UHPLC-ESI-LTQ-MS/MS	NS	[29]
P	*Fragaria* × *ananassa*, *Camellia sinensis* (Japan)	7	Antioxidant	SE (MeOH)	HPLC-UV and UHPLC-ESI-qTOF	37 mg/g	[30]
P	*Malus domestica*, *Arabidopsis thaliana* (France)	3	Biosynthesis	SE (MeOH)	HPLC-DAD, UPLC-ESI-MS	10–150 μg/g	[31]
P	*Arabidopsis thaliana* (UK)	33	NS	SE (MeOH:water)	UPLC-PDA-ESI-qTOF-MS and UPLC-PDA-ESI-QTrap-MS	Relative quantification	[32]
BP	*Helianthus* L. (Slovakia)	5	Antifungal	SE (EtOH)	NMR, FT-IR, FT-Raman and LC–Qorbitrap (ESI/APCI)	NS	[33]
BP	*Mimosa pudica* L. (Brazil)	12	Antioxidant	SE (MeOH; EtOH; MeOH:water)	UHPLC-DAD-ESI-qTOF-MS	NS	[34]
BP	Fabaceae (Brazil)	9	Antioxidant	SE (EtOH)	UPLC-DAD-ESI-qTOF-MS/MS	NS	[35]
BP	Multifloral (Brazil)	2	Antioxidant	SE (MeOH:water)	HPLC-PDA-ESI-MS/MS	NS	[36]

BP	*Brassica napus* (China)	12	Antioxidant	(i) SE (Hexane); (ii) SE (EtOH); (iii) Column fractionation	HPLC-ESI-qTOF-MS/MS	NS	[37]
BP	*Camellia sinensis* L., *Mimosa diplotricha, Helianthus annuus* L., *Nelumbo nucifera, Xyris complanata, Ageratum conyzoide* (Thailand)	6	Antioxidant and antilipoxygenase	(i) SE (MeOH); (ii) Partition (DCM, Hexane, MeOH); (iii) Column purification; (iv) HPLC preparation	HPLC-UV, MALDI-TOF and NMR	0.03–57 mg/g	[38]
BP	Multifloral: *Plantago* sp., *Crepis capillaris* and *Cytisus striatus* (Portugal)	14	Antioxidant and bioaccessibility	SE (EtOH:water)	HPLC-DAD-ESI-MS/MS	Raw BP: total 8–38 mg/g BP after digestion: total 1–15 mg/g	[39]
BP	*Quercus mongolica* (Japan)	18 (9 new)	Antityrosinase	(i) SE (MeOH); (ii) Partition (DCM, Hexane, EtOAc); (iii) Column purification; (iv) HPLC preparation	HRESI-qTOF-MS and NMR	1–114 mg from 15 kg of BP	[40]
BP	*Helianthus annuus* L. (Thailand)	2	Antityrosinase	(i) SE (MeOH); (ii) Partition (DCM); (iii) silica gel purification; (iv) HPLC preparation	*micrOTOF* and NMR	0.04 mg/g of each compound	[41]
BP	*Silybum marianum, Prunus* sect. *Armeniaca, Fagopyrum, Nelumbu nucifera, Rosa* L., *Camellia* L., *Nigela sativa* L., *Helianthus* L. (China)	26	Antityrosinase	(i) Lipid removation (ether); (ii) SE (EtOH:water); (iii) Concentration and dilution (water); (iv) SE (EtOAc); (v) Concentration and dilution (MeOH)	UPLC-ESI-qTOF-MS	NS	[42]

(continued)

Table 10.1 (continued)

Matrix	Botanical origin (country)	Number of compounds	Reported bioactivity	Sample treatment	Determination technique	Amount (average values)	Ref.
BP	NS fermented (China)	22	Compound stability against microbes	SE (EtOH:water)	UPLC-ESI-MS/MS	NS	[43]
BP	*Attalea funifera* (Brazil)	18	NS	(i) SE (EtOH); (ii) Dissolve (MeOH); (iii) SE (EtOAc)	UPLC-DAD-ESI-qTOF-MS/MS	NS	[44]
BP	NS	18	NS	(i) SE (hexane); (ii) Concentration and dilution (MeOH:THF)	UHPLC-DAD-ESI-Orbitrap	NS	[7]
BP	*Brassica napus* L., *Vicia faba* L., *Papaver rhoeas*, *Phellodendron amurense*, *Chrysanthemum* L., *Fagopyrum esculentum*, oil-tea, *Helianthus* L., *Rose* L., *Cucumis melo*, *Pyrus* L., *Crataegeus*, *Prunus* sect. *Prunus*, *Prunus* sect. *Armeniaca*, *Actinidia deliciosa*, *Rhus chinensis*, *Zea mays*, *Citrullus lanatus*, *Nelumbo nucifera* (China)	31 (19 new)	NS	SE (EtOH:water)	HPLC-PDA and HPLC-ESI-QTOF	2–39 mg/g	[45]

APCI atmospheric pressure chemical ionization, *BP* bee pollen, *DAD* diode array detector, *DCM* dichloromethane, *EtOH* ethanol, *ESI* electrospray ionization, *EtOAc* ethylacetate, *FT-IR* fourier transform-infrared spectroscopy, *FT-Raman* fourier transform-Raman spectroscopy, *HPLC* high performance liquid chromatography, *HS* high resolution, *LTQ* linear ion trap quadrupole, *MALDI* matrix-assisted laser desorption ionization, *MeOH* methanol, *MS* mass spectrometry, *MS/MS* tandem mass spectrometry, *NMR* nuclear magnetic resonance, *NS* not specified, *P* plant pollen, *PDA* photodiode array, *QTrap* triple quadrupole/linear ion trap, *qTOF* quadrupole-time-of-flight, *SE* solvent extraction, *THF* tetrahydrofuran, *TOF* time of flight, *UHPLC* ultra-high performance liquid chromatography, *UV* ultraviolet treatment, solvent extraction (SE) with pure organic solvents or their aqueous mixtures have been the predominant procedure to extract the compounds previous to analysis.

Considering the values of the amounts reported for the analysis of HCAAs, ignoring the number of compounds identified and the data for preparative purposes, it has been seen to be comparable in pollen or bee pollen [30, 39, 45] with values up to 35 mg/g. Significant outcomes have been found in 11 types of bee pollen out of 20 analyzed in which 5 phenolamides conjugated putrescine, 10 phenolamide conjugated spermidines and 16 phenolamide conjugated with spermine constituted over than 1% of the total weight, particularly with a high value in rose pollen followed to pear pollen [45]. Phenolamide conjugated with spermidine were the ones which presented the highest values among the other two types [45].

Although similar content has been described in both types of pollen, a greater number of compounds in the bee matrix were usually reported [7, 34, 37, 39, 40, 42–45]. Recently, Qiao et al. [45] have examined phenolamides and flavonoid glycosides in 20 types of monofloral bee pollen, reporting 19 new structures in this matrix for the first time. Only in one case [32], more than 30 different bis- and tris-substituted spermidine conjugates have been determined, in the methanol-soluble fraction of the *Arabidopsis thaliana* pollen exine. Moreover, the presence of spermidine derivates in both sub-pollen particles and pollen from *Ambrosia artemisiifolia* L. were reported for the first time.

Focusing on research goals from studies of HCAAs, one work [31] showed that three spermidine-based phenolamides accumulated specifically in pollen grain of the apple tree, revealing the probable evolution of the enzyme dedicated to the synthesis of trihydroxycinnamoyl spermidines in pollen coat. Regarding the examination of their biological activity, most investigations were devoted to evaluate the potential of the individual compounds, or the extracts that contain them, as antioxidants in both types of pollen: bee pollen [34–39] and plant pollen [29, 30]. The most obvious function of phenolamides in general is UV protection [17] which has been exposed in strawberry pollen due to their spermidine derivates by the examination of oxygen radical absorbance capacity of the extracts [30]. Other methods to determine the antioxidant activity and protective effects against oxidative stress are based on their determination by 2,2-diphenyl-1-picrylhydrazyl radical, 2,2′-azino-bis-3-ethylbenzothiazoline-6-sulphonic acid, and ferric reducing antioxidant power assays. All or some of those methods have been examined to conclude that extracts that contained phenolamides together with other phenolic compounds in Fabaceae [35] or multifloral [36] bee pollen showed antiradical activity. However, in one case, the evaluation was carried out with the individual and separate groups of compounds, in which the activity of phenolamides group (12 compounds) was found to be significantly higher than that of flavonoids in rape bee pollen [37]. Relations with inflammatory and immune response and lipoxygenase inhibition properties have been reported in different monofloral bee pollen, being the highest value for *Mimosa diplotricha*, which suggested that the major active compounds could be the 6 HCAAs identified [38]. In addition, bioaccessibility and antioxidant activity have been also assessed in multifloral bee pollen [39]. In this case, the findings indicated that of the 35 phytochemicals identified (total of 12 spermine and spermidine derivates), the most affected by gastrointestinal digestion were phenylamides, with a complete digestibility at the end of the intestinal phase.

Additionally, some studies have been conducted in which three spermidine derivates were isolated from *Quercus alba*. In them, the mycelial growth of the oat leaf stripe pathogen and powdery mildew was reduced [26]. Moreover, this antifungal activity was also observed in sunflower bee pollen against three types of widespread spoilage fungi, exhibiting different resistance each one of them [33]. On the other hand, the same class of bee pollen (*Helianthus* L.) has been tested for evaluating its effect on bumble bees gut parasite (*Crithidia bombi)* in *Bombi impatients* [27] or *Bombi terrestris* [28]. However, the results did not turn out to be satisfactory because even though sunflower pollen consistently reduced the *Crithidia* parasites relative to control pollen (buckwheat [27] or willow pollen [28]), none of the spermidine derivates tested had significant effects. Furthermore, they had surprising diverging effects increasing parasite load, what led to think that caution is necessary when working with plant–bee–parasite studies, control diets and biological models [28].

Furthermore, bee pollen matrix has exhibited antyrosinase activity, which could be a therapeutic target in melanin-related disorders and active ingredients for functional cosmetic whitening. The oxidative reaction of tyrosine to dopaquinone is regulated by tyrosinase, which is linked to melanin synthesis. Two safflospermidines from sunflower bee pollen have been found to be potential natural tyrosinase inhibitors [41], while 21 spermidine derivates and 5 spermine derivates were first reported as main contributors in eight types of monofloral bee pollen [42]. Besides, 18 constituents including new compounds were isolated, and their antityrosinase evaluation was verified in *Quercus mongolica* bee pollen, revealing that the methoxy group in HCAAs reduced the effect [40].

To summarize, although the investigation of phenolamides has increased in the last years, there are few references reporting quantitative data, possibly because of the challenge in individually synthesizing the standards. The major focus of the study of phenolamides has been to verify their potential antioxidant activity. Nevertheless, taking into consideration the importance of the floral origin, as it affects both the nutritional value and bioactivity, and the lack of evidence from studies based on its determination using HCAAs, this chapter could emphasize the need for a starting point to evaluate these compounds as markers of botanical and/or geographic origin of the bee pollen.

10.3 Glucosinolates and Related Compounds

Glucosinolates (GSLs) are S-glycosides in which the glycone is β-D-thioglucose and the aglycone is a sulfated oxime. They are natural compounds of the secondary metabolism of plants mainly present in dicotyledonous plants and especially abundant in the Brassicaceae (cruciferous) family [18]. They were also originally called mustard oil glycosides or thioglycosides, as they were discovered to be derivatives of mustard oils. There are different types of GSLs depending on what amino acid they come from, and they can be aliphatic, indole or aromatic. The evolutionary appearance of GSLs in certain groups of plants occurs thanks to the competitive advantage that they confer as anti-nutritional compounds, nematicides, fungicides and herbicides, protecting the plant from numerous possible enemy attacks [46]. Under normal conditions, GSLs are chemically stable and truly useful in response to plant injuries, as well as in defense against pests and pathogens. They require "activation" by myrosinase enzymes, which will break them down (hydrolyze) and release the GSLs real functional compounds, whether they are isothiocyanates (sulforaphane), nitriles, thiocyanates, epithionitriles or oxazolidines [47], which are responsible for most of their biological activities and the reason why they produce the bioactive effect [48–50]. Although the compounds derived from glucosinolates are toxic to many organisms, since they are used by plants to defend themselves, the presence of any of them in our diet can have a very beneficial impact on our health. For example, one of the most common degradation products, isothiocyanates, can act as powerful weapons against cancer, due to the activation of proteins that activity can prevent the appearance of cancer cells or act against those already present [50]. In addition, they protect against cardiovascular, neurodegenerative, diabetes-related diseases, etc. However, if we want to avoid any type of effect that this type of enzyme can cause us, it would be enough to bring the pollen samples to a temperature close to 70 °C, since, like many other enzymes, it denatures at high temperatures, losing its activity [48]. GSLs have been investigated in pollen in some publications ([18, 46, 51–55]; see Table 10.2).

It should be highlighted that different methods for analyzing GSLs depending on the presence (intact or non-intact desulfo-derivatives) of a sulfate group have been reported in the literature. Although the desulfation stage decreases the polarity of GSLs and improves their chromatographic resolution in reversed-phase liquid chromatography, it is a time-consuming procedure, and faster methods have been proposed for the direct analysis of intact- GSLs [18]. The first work was published more than 30 years ago [51], in which authors determined the GSLs composition of pollen from plants of rapeseed (*Brassica napus* cv Midas) and Indian mustard (*B. juncea.* cv Early Yellow) by using HPLC and UV detection. Sample treatment consisted of a SE, a partition of the resulting extracts, a precipitation of the protein material and an enzymatic desulfation; meanwhile, gas chromatography (GC) with MS detection (GC-MS) was used to identify and confirm the presence of certain GSLs. This study addressed glucosinolate content as an important chemical parameter of seed meal quality in cruciferous oilseed crops, like those investigated in this work. The authors provided the normalized GSL composition of these samples, being concluded that both types of pollen appear to contain predominantly GSLs with an aliphatic side chain. However, authors concluded that further quantitative analysis was necessary to establish whether relative amounts of GSLs in pollen can be correlated with the relative levels of GSLs in seed and thereby provide a basis for pollen selection.

Many years later, GSLs were investigated in bee pollen by the same research group [18, 46, 52, 53, 55] for different reasons: (i) to check for the first time their

Table 10.2 Applications in the analysis of glucosinolates and related compounds in pollen

Matrix	Botanical origin (country)	Number of compounds	Reported bioactivity	Sample treatment	Determination technique	Average values (µg/kg, DW)	Ref.
P	*Brassica napus* cv Midas and *B. juncea*. cv Early Yellow (Australia)	12	NS	(i) SE (MeOH and CF); (ii) PT (Hexane and Water); (iii) EV	HPLC-UV	NS	[51]
BP	Multifloral (Spain)	12	NS	(i) SE (water); (ii) SPE (NH_2); (iii) EV	HPLC-ESI-MS/MS	4–2226	[18]
BP	*Brassica* t., *Prunus* t., *Cistus* t., Leguminosae, Brassicaceae, *Quercus* sp., *Diplotaxis* t., and *Papaver*, Multifloral, *Retama* t [53]. *Rosa* t [53]., *Vicia* t [53]. (Spain)	15	NS	(i) SE (water); (ii) SPE (NH_2); (iii) EV	UHPLC-ESI-qTOF-MS/MS	22–9806 [46] 40–581786 [52] 19–9936 [53]	[46, 52, 53]
BP	*Cocos nucifera* L, *Corianrum sativum* L, *Brassica napus* L., Multifloral (India)	1	Antioxidant	(i) SE (MeOH and Water); (ii) SE (MeOH); (iii) EV	UHPLC-DAD-ESI-qTOF-MS/MS	NS	[54]
BP	Multifloral (Spain)	1	NS	(i) SE (MeOH); (ii) EV	HPLC-ESI-MS/MS	13–22	[55]

BP bee pollen, *CF* chloroform, *DAD* diode array detector, *ESI* electrospray ionization, *DW* dry weight, *EV* evaporation, *HPLC* high-performance liquid chromatography, *MeOH* methanol, *MS/MS* tandem mass spectrometry, *NS* not specified, *P* plant pollen, *PT* partition, *qTOF* quadrupole time-of-flight, *SE* solvent extraction, *SPE* solid-phase extraction, *UHPLC* ultra-high performance liquid chromatography

presence in this bee product; (ii) because if they are present in a product that would be consumed by humans, they would be positively affected by their beneficial health properties; also; (iii) because GSLs could be used as a parameter that helps to identify the origin of pollen. This is a quite important issue, since the consumption of bee products has become increasingly significant, and as their production cannot increase rapidly in the short term, this can result in substantial economically motivated adulteration [56]. Therefore, authentication of bee products is essential in order to protect consumer health and to avoid competition that could create a destabilized market [57]. The first two studies [18, 46] were based on developing analytical methodologies that allowed a fast and reliable determination of GSLs in bee pollen. In a first instance, an efficient sample treatment based on a SE with heated water followed by a solid-phase extraction (SPE) with NH_2-based cartridges was proposed; while the objective of the second work was to reduce the chromatographic run and the solvent consumption of the mobile phase, as well as improving the selectivity of the method. This was done by employing UHPLC coupled to a MS/MS detector. In both works, several bee pollen samples from experimental apiaries were analyzed, and as it was preliminary thought, GSLs were found over a wide concentration range (4–9806 μg/kg) in most of the analyzed samples, being sinigrin and gluconapin the GSLs detected at the highest concentrations. Nevertheless, significant differences in GSL content were observed between the samples. This latter finding induced the authors to continue investigating the potential of GSLs as markers of the origin of bee pollen, and consequently two new works were recently published [52, 53]. In the first work [52], the GSL content of 49 bee pollen samples, which were separated by color from four different experimental apiaries located in the same area (Marchamalo, Guadalajara, Spain) were determined by using the previously developed UHPLC-MS/MS method. Results showed that GSL residues were detected in most of the samples (40–581,786 μg/kg), and these differed as expected in number and concentration. It was possible to directly differentiate all the apiaries from the other three by comparing GSL content and employing a canonical discriminant analysis (CDA). However, the number of samples in some of the apiaries was not large, and they were selected according to their different colors. Subsequently, in the most recent study [53], authors tried to confirm their previous findings by determining the GSL content in a larger number of bee pollen samples from the same four apiaries than in the previous work, but in this case, there was no prior separation by color, and harvesting took place in three consecutive foraging periods. In addition, the GSL content of commercial bee pollen samples was also determined. GSLs were found in most of the samples analyzed, although these differed in terms of number and concentration (19–9936 μg/kg). Only one of the GSLs under study (glucoiberin) was not detected in any of the 83 samples. It was possible to differentiate the commercial samples from the experimental apiaries by means of a CDA based on their GSL content. In addition, it was also demonstrated for the first time that bee pollen samples could be classified according to their harvesting period by studying the GSL content. Thus, these results have confirmed the potential of GSLs as markers of the geographical origin (apiary) and the harvesting period, which may help to increase the traceability of the bee pollen samples and facilitate their authentication.

Finally, it should be mentioned that a recent study was devoted to performing a screening of Indian bee pollen by using UHPLC-DAD-MS/MS based on antioxidant properties and its composition, paying special attention to the phenolic compounds [54]. Sample treatment involved a double SE. Sixty compounds were identified, most of them belonging to the flavonoid family, and only one GSL was detected, sinigrin, although its concentration was not provided. Results showed that a significant botanical differentiation existed between the Indian bee pollen composition and its antioxidant properties. Authors also concluded that the variation in studied parameters could be useful for beekeepers to produce bee pollen of high quality.

Sulforaphane (SFN) is an isothiocyanate with antimicrobial and anticancer properties, which is obtained from the hydrolysis of glucoraphanin (GRA), and its formation depends on several factors such as pH, hydrolysis time and temperature [58]. It has also been suggested in several *in vivo* and *in vitro* studies that SFN has the potential to reduce the risk of various types of cancer, diabetes, atherosclerosis, respiratory diseases, neurodegenerative and ocular disorders, in addition to some cardiovascular diseases [59, 60]. As GRA was found in bee pollen in previous studies [18, 46, 52, 53, 55], the same authors decided to investigate the potential presence of its isothiocyanate compound [55]. Therefore, authors developed a new HPLC-MS/MS method for determining SFN in bee pollen, which required a prior SE of SFN from bee pollen, not only for checking its potential presence in this product for the first time, but also as it might contribute to the nutritional and health promoting properties of bee pollen. SFN residues were detected in most of the bee pollen samples at low concentration levels (13–22 μg/kg). The presence of SFN in bee pollen was an interesting finding hitherto unreported, and although the observed concentrations were low, this study could be used as a starting point for further investigation.

To recapitulate, GSLs have demonstrated their potential as bee pollen markers, which could have a direct influence in authenticating the origin of this product, and consequently combat the fraud that it is currently affecting its commercialization. Moreover, GSLs have been also used as quality indicators of the plant origins. Finally, it should be mentioned that the detection of SFN in bee pollen, a compound with a great bioactivity, confirms pollen as one of the products with the largest number of bioactive compounds that can be found in nature.

10.4 Organic Acids

Organic acids are a family of chemical compounds that contain carboxylic acid group(s) in their structures and are classified based on the number of these groups. In general, they could be catalogued as mono-carboxylic, dicarboxylic, α-hydroxyl, and sugar acids. They usually behave as weak acids and dissolve completely in aqueous solutions and most of them exist in dimeric form [61]. In recent years, organic acids from natural products have gained interest in different industrial areas as important phytochemicals due to their special characteristics such as bactericidal, antimicrobial, and antioxidant activity. In addition, they are involved in several metabolic pathways [62]. They have a wide range of applications such as animal feeds, as a new growth promoter as an alternative of antibiotics or to improve plant quality. In addition, the understanding of the chemical composition of bee pollen, particularly the organic acid content, is a key parameter to provide information about its quality and could be used as criteria against adulteration. This profile will be strongly influenced by the botanical origin and potentially contribute to the aroma and flavor of pollen. It should be mentioned that few works were reported about the investigation of the organic acid profile in pollen, and most of them were published in the last years ([63–68]; see Table 10.3).

According to the examined literature, HPLC has been the analytical technique selected in most works to analyze organic acids in pollen, and particularly in bee pollen [63–68], coupled to several detectors such as refractive index [64, 65], DAD [68] or MS/MS [63, 66, 67]. MS/MS detectors could be the most interesting option in recent years thanks to metabolomics tools that allow to detect a multitude of metabolites that cannot be detected by traditional technologies. This kind of analysis allows to deepen the knowledge of the nutritional value and biological activity, as well as to discover some potential characteristic metabolites. A cheaper alternative methodology based on capillary electrophoresis coupled to DAD with previous derivatization step was proposed [63]. The sample treatment was quite similar in most cases based on a SE with water, acidified water, or methanol [63, 64, 67, 68]. Exceptions were made in other publications [63, 65] that used ultrasound assisted extraction with ethanol as solvent. Malic acid and citric acid have been reported in large quantities as the main compounds in a wide concentration range comprised between 0.13 and 5.62 g/kg [63, 64, 66, 68]. The highest concentrations were reported in flower pollen instead of bee pollen. Both compounds are involved in Krebs cycle and the properties of citric acid make it suitable for use as a food supplement, flavoring and as a preventer of kidneys stones formation [68]. However, other organic acids have been reported as relevant acids in bee pollen samples such as gluconic acid, lactic acid, acetic acid, succinic acid, γ-aminobutyric acid, 2-hydroxyisobutyric acid (GABA) and suberic acid [63–65, 67, 68], because of fermentation process in the bees' stomach. It should be noted that GABA acts as a neurotransmitter in the mammalian central nervous system and has been demonstrated to have multiple pharmacological properties, including antihypertensive and antidepressant activities. Formic acid and butyric acid have been used as quality markers due to the fact that their absence represents that the pollen is not contaminated with microorganisms [64]. Furthermore, different organic acid types such as tartaric acid, fumaric acid, ketoglutaric acid, pyruvic acid and metabolites were detected in some bee pollen samples [63, 67].

As can be deduced from the related literature, organic acids have been scarcely investigated in bee pollen, which may be due to the difficulty of their analysis. However, in the last few years the number of published papers has increased, and the trend has focus on the study of derivates from organic acids thanks to advanced analytical techniques. Nevertheless, more research would be needed in order to be able to draw firm conclusions about the organic acid profile based on different botanical origins or geographical areas.

Table 10.3 Applications in the analysis of organic acids in pollen

Matrix	Botanical origin (country)	Number of compounds	Reported bioactivity	Sample treatment	Determination technique	Average values (g/kg)	Ref.
BP	Multifloral, *Castanea*, *Fagopyrum esculentum*, *Quercus* L. (Turkey)	8	Antibiotic	SE (water)	CE-DAD	0.07–32	[63]
P	*Quercus* L. (Korea)	3	Flavor, antibiotic	SE (water)	HPLC-RID	3–9	[64]
BP	Asteraceae, Fabaceae, Plantaginaceae, Rosaceae (Turkey)	28	Antibiotic	UAE (EtOH)	HPLC-ESI-MS/MS	0.8–1.5	[63]
BP	*Brassica napus* (Brazil)	3	NS	NS	HPLC-RID	0.9–2.1	[65]
BP	Multifloral (Germany)	52	Flavor, aroma, biological activities	UAE (EtOH)	HPLC-ESI-MS/MS	5	[66]
BP	NS (China)	75	NS	SE (MeOH)	UPLC-ESI-MS/MS	NS	[67]
BP	*Helianthus annuus*, *Robinia*, *Quercus*, Pinaceae, *Prunus*, *Tillia* (Romania)	6	Antihypertensive and antidepressant	SE (HPO_3)	HPLC-DAD	NS	[68]

BP bee pollen, *CE* capillary electrophoresis, *DAD* diode array detector, *EtOH* ethanol, *ESI* electrospray ionization, *HPLC* high performance liquid chromatography, HPO_3 meta-phosphoric acid, *MeOH* methanol, *MS* mass spectrometry, *MS/MS* tandem mass spectrometry, *NS* not specified, *P* plant pollen, *RID* refractive index detector, *SE* solvent extraction, *UAE* ultrasound assisted extraction

10.5 Stilbenes

Stilbenes are a class of polyphenols that are present in plants and are considered beneficial for health. They are phytoalexins which are synthesized by the enzyme stilbenosynthetase. The best known is resveratrol, a compound synthesized by some plant organisms as a result of their exposure to both biotic and abiotic stresses [69]. It presents two geometric isomers that appear naturally, the *trans* being generally the predominant one, although its glycosides, piceids, also appear with similar properties [70, 71]. Several studies have shown that stilbenes in general, and resveratrol in particular, have beneficial effects on health, so their introduction into both human and animal diets is of great interest. In addition to their predominant role as antioxidants, several studies have shown its efficacy in cancer prevention, inhibiting the initiation and development of tumors, and they have also exhibit anti-inflammatory, antithrombotic and immunomodulatory activities [72, 73]. It is well-known that phenolic compounds are among the predominant compounds in pollen, and subsequently, it is not surprising that several publications have focused their attention in determining stilbenes, and in most cases resveratrol, in pollen (see Table 10.4).

It should be mentioned that only one study investigated the content of the different isomers of stilbenes (*cis* and *trans*), of resveratrol and piceid in bee pollen from experimental apiaries and local markets [74]. The primary objective of this work was to check residues of these compounds that can be found in this bee products, after the stilbenes were supplied to the beehive for controlling *Nosema* infection in honeybees. Authors developed a new method that implied a sample treatment composed of a cleaning step with hexane, a further extraction of the analytes with a mixture of hexane and water, and an evaporation of the extract prior to their analysis by HPLC-MS. *Trans*-resveratrol residues were detected in all the samples from experimental apiaries over a wide concentration range (90–2450 μg/kg), and *cis*-resveratrol was also found in some samples. Moreover, *trans*-resveratrol was detected in most of the commercial bee pollen samples, while *cis*-piceid was found in only one of these samples. This was the first time that the presence of resveratrol and piceid isomers in commercial bee pollen was reported.

Resveratrol has been determined in other works along with several other phenolic compounds [76–78]. In two of these studies [76, 77], authors employed the same analytical method, which consisted of an extraction of the phenolic compounds with a mixture of ethanol and water followed by a UHPLC-DAD analysis for the evaluation of the antidiabetic and/or antioxidant effects of bee pollen due to its bioactive compounds, including resveratrol, which was detected at low concentrations in bee pollen (40 mg/kg). Firstly, authors investigated the preventive effect of propolis, bee pollen and their combination on Type 2 diabetes induced by D-glucose in rats [76]. The results of this study showed for the first time that the coadministration of propolis and bee pollen extracts were able to attenuate the Type 2 diabetes. Meanwhile, in the most recent work [77], the potential of bee pollen as an added-value food ingredient was evaluated by investigating its composition regarding bioactive constituents. Resveratrol was detected in all the bee pollen samples analyzed in a wide

Table 10.4 Applications in the analysis of stilbenes in pollen

Matrix	Botanical origin (country)	Number of compounds	Reported bioactivity	Sample treatment	Determination technique	Average values	Ref.
BP	Multifloral (Spain)	4	NS	(i) CL (Hexane); (ii) SE (EtOH and Water); (iii) EV	HPLC-ESI-MS	70–2450 (μg/kg, DW)	[74]
BP	*Papaver rhoeas* L., *Prunus* L., *Rubus ulmifolius* Schott., *Trifolium alexandrinum* L., *Trifolium pratense* gr., (Italy)	10	Antioxidant	SE (EtOH and Water)	UHPLC-ESI-MS/MS	NS	[75]
BP	Multifloral, *Reseda luteola* L., *Thymus vulgaris*, *Cystus* sp. (Morocco)	1	Antioxidant and Antidiabetic	SE (EtOH and Water)	UHPLC-DAD	44 [76] 37–257 [77] (mg/kg, DW)	[76, 77]
BP	NS (Brazil)	1	Antibacterial and Antioxidant	SE (EtOH and Water)	HPLC-PDA	26–126 (mg/kg, DW)	[78]
BP	*Ranunculus* L., *Brassica* L., *Filipendula* Mill, *Rubus* L., *Vicia* L., *Trifolium repens*, *Trifolium pratense*, *Epilobium* L., *Taraxacum* (Finland)	14	Antioxidant	(i) SE (MeOH); (ii) EV	HPLC-DAD	20–2340 (μg/kg, DW)	[79]

BP bee pollen, *CL* cleaning, *DW* dry weight, *DAD* diode array detector, *EtOH* ethanol, *EV* evaporation, *HPLC* High performance liquid chromatography, *MeOH* methanol, *MS/MS* tandem mass spectrometry, *NS* not specified, *PDA* photodiode array, *SE* solvent extraction, *UHPLC* ultra-high performance liquid chromatography

concentration range (37–257 mg/kg). The overall results of the study showed interesting functional characteristics important for food industry applications, as well as high antioxidant and antidiabetic activities, which were mainly associated to phenolic compounds. The antioxidant and antibacterial effects of bee pollen from *Apis mellifera* and *Trigona spinipes* stored under different storage conditions (lyophilization and drying in stove) were evaluated on the basis of their phenolic composition [78], which was determined by analyzing the extracts obtained after a SE with ethanol and water by HPLC coupled to a PDA detector. Resveratrol was found in both types of samples in a concentration that varied depending on the storage conditions of bee pollen and its origin, but it was always comprised between 26 and 125 mg/kg. Authors concluded that conservation methods influenced in a different way the phenolic composition and the biological properties of bee pollen from the two bee species, and it was also found that the phenolic composition, including resveratrol, was quite different between both bee pollen types. It is important to point out that other stilbenes apart from resveratrol and piceid have been found in bee pollen. For example, ten different stilbenes have been found in Italian bee pollen from different botanical origins by means of UHPLC-DAD-MS/MS [75]. The main aim of this work was to establish the phenolic profile of bee pollen to investigate its relation to the antioxidant capacity and to the differences between botanical origins, and consequently, no specific concentration was provided for the studied compounds. Resveratrol, pinosylvin, pallidol and piceatannol were among the stilbenes detected in the analyzed samples. As could be expected, the botanical origin of pollen had a strong influence on the phenolic profile, and, therefore, the antioxidant activity of the extracts was quite different depending on this factor. A similar study was carried out with bee pollen pellets from Finland [79], in which the protein content was also evaluated. The phenolic compound content, including fourteen stilbenes that were not identified, was determined by HPLC-DAD analysis of the extracts obtained after performing a SE with a mixture of methanol and water. The observed concentrations of stilbenes were within 20–2340 μg/kg, with the highest amount of these compounds being found in *Rubus* species pollen. It was also reported that protein and phenolic content, as well as the antioxidant capacity of bee pollen pellets, presented different values according to their botanical origin, and it was demonstrated that this product could be considered as a source of bioactive compounds.

To conclude this section, it can be said that within the family of phenolic compounds, stilbenes have not received as much attention in recent years as, for example, flavonoids. In any case, the detection of resveratrol in pollen in the middle of the last decade was an important milestone, since this compound is of great interest due to its bioactive properties. This has led to its determination on more occasions over the years, along with other stilbenes and other phenolic compounds in order, in many cases, to relate its content to antioxidant activity or to the nutritional value of pollen.

10.6 Betaines and Related Compounds

Betaines are zwitterionic quaternary ammonium compounds, produced by specific biosynthetic pathways involving amino acid and imino acid nitrogen methylation or enzymatic oxidation of choline [19, 80, 81]. They have physiological functions as osmolytes and as donors of methyl groups. When they act as an osmolyte, they accumulate in the cytoplasm and intracellular fluids, protecting cells and their components (proteins, nucleic acids, cell membranes, and enzymes) from environmental stress, like exposure to drought, high salinity or thermal stress [19]. When acting as a donor of methyl groups, betaines participate in many biochemical pathways such as the methionine cycle in the liver and kidneys [80]. Humans ingest betaines exogenously from the diet, but they can also synthesize them endogenously from its precursor, choline, using the choline dehydrogenase enzyme [82]. There are numerous types of betaines such as glycine betaine (commonly called betaine), proline betaine found in citrus fruits, and trigonelline found in coffee beans, tomatoes, and alfalfa [83, 84]. Betaines protect internal organs, prevent chronic diseases, improve liver function, have anti-nociceptive, antioxidant and anti-inflammatory activities [19, 80, 82]. Some recent studies suggest that betaine consumption is capable of preventing the development of alcohol-associated breast cancer [81]. Other studies reveal that trigonelline has positive biological activities against caries, cancer cells, Alzheimer's disease, the hypocholesterolemic effect and diabetes, and inhibits the degranulation of mast cells responsible for allergic reactions [85]. However, if betaines are consumed at high levels, they can trigger cardiovascular diseases [19]. In plants, betaine levels are characterized according to their species, and can be used as botanical biomarkers of bee pollen. This is an interesting point in order to avoid possible fraud in the origin of the pollen that is marketed, since it is a factor to take into account by the consumer and the beekeeping industry [19], along with its bioactive properties. Indeed, the potential of betaines as bee pollen markers has been recently investigated ([19, 86]; see Table 10.5).

It should be highlighted that these works were the first studies devoted to investigate the presence of betaines and related compounds (BRCs), like choline, in bee pollen. Thus, it was necessary to develop an analytical method that allowed the determination of those compounds. This performed in one of the publications [19], in which an analytical method composed of a SE with a mixture of acetonitrile and water combined with hydrophilic interaction liquid chromatography (HILIC) coupled to MS was proposed. Several bee pollen samples from different botanical origins were analyzed with the proposed methodology, and three betaines (betaine, choline and trigonelline) were detected in all of them in a wide range of concentrations (57–62236 mg/kg). Moreover, significant differences in these concentrations were observed in accordance with the origin of the bee pollen, suggesting the potential of these compounds as botanical biomarkers. Therefore, the second publication [86] was focused on demonstrating the suitability of those compounds and markers not only of the geographical and botanical origin, but also of the harvesting period. Bee pollen samples were collected from four experimental apiaries located in the same geographical area (Guadalajara, Spain), and sampled during three consecutive harvest periods in the same year. They were analyzed by the above-mentioned method, and variable amounts of betaines and related compounds were found in the samples, with four of these being identified in all of them (betonicine, betaine, trigonelline and choline) in a wide concentration range (7 to 52384 mg/kg). Furthermore, it was also found that the BRC content per sample was highest in the driest period,

Table 10.5 Applications in the analysis of betaines and related compounds in pollen

Matrix	Botanical origin (country)	Number of compounds	Reported bioactivity	Sample treatment	Determination technique	Average values (mg/kg, DW)	Ref.
BP	*Zea Mays* L. [19], Multifloral, *Raphanus raphanistrum* subsp. *sativus* [19], *Brassica napus* L. [19], *Cistus* L. [19], *Helanthus* L. [19] (Spain)	9 [19] 8 [86]	NS	(i) SE (Water + ACN); (ii) EV	HILIC-ESI-MS	57–62236 [19] 7–52834 [86]	[19, 86]
P	*Pinus yunnanensis* (China)	1	NS	SE (EtOH)	UHPLC-ESI-MS	NS	[87]
P	*Fraxinus* L., *Celtis occidentalis* L., *Morus* L., *Quercus* L., *Pinus* (USA)	2	NS	(i) SE (Water); (ii) EV	NMR	NS	[88]

ACN acetonitrile, *BP* bee pollen, *DW* dry weight, *ESI* electrospray ionization, *EtOH* ethanol, *EV* evaporation, *HILIC* hydrophilic interaction liquid chromatography, *MS* mass spectrometry, *NMR* nuclear magnetic resonance, *NS* not specified, *P* plant pollen, *SE* solvent extraction, *UHPLC* ultra-high performance liquid chromatography

which confirmed the existing knowledge that the presence of these compounds is favored by lack of water. Results of the CDA demonstrated that these compounds have a potential as markers of the origin (apiary) and the harvest period. This finding is quite a relevant, as previous studies had not focused attention on BRCs. Betaines, among other compounds, have also been investigated in bee pollen with the aim of evaluating the effects of sterilization and storage on the quality of pine pollen [87]. Fresh pine pollen is difficult to store because of its high water content, so it is essential to dry and sterilize it. However, traditional heat sterilization caused changes in food nutrition and flavor substances to some extent, and consequently, methods that do not require the use of heat are being evaluated. Thus, this study was devoted to investigate the effect of different sterilization methods (irradiation-IT and ultra-high temperature instantaneous sterilization-UHT) and storage periods (fresh, one and 2 years) on the quality of pine pollen. Those samples were analyzed from the perspective of metabolomics by using UHPLC-MS/MS. The content of 36 metabolites, including kaempferol, cetyl betaine, and naringenin, which were extracted from pollen with ethanol, was evaluated. Compared with UHT, IT had less effect on pine pollen. In addition, it was seen that the longer the storage time, the greater the effect on pollen content, since it was found that several metabolites were lost during the storage and sterilization. Finally, some BRCs, along with other compounds, which were extracted with ultrapure water, have been determined in airborne pollen particles by NMR. The aim was to evaluate the chemical composition of the airborne pollen for a better understanding of the mechanisms for the observed climate and health-related effects (rhinitis, conjunctivitis, asthma, and wheeze) [88]. Amino acids and carbohydrates were predominant in the samples, and only two BRCs (trigonelline and choline) were found. However, it should be specified that although the compounds were tentatively identified, the potential of NMR combined with statistical tools to evaluate the chemical content of airborne pollen was demonstrated.

It can be concluded that, although BRCs have been scarcely investigated to date in pollen-based matrices, these compounds present a great potential, not only because they could be helpful to assess the quality of pollen along with other constituents, but also for their promising role as bee pollen biomarkers, which could help authenticate the origin of this food supplement.

10.7 Other Compounds

10.7.1 Lignin

Lignin is a phenolic polymer that is currently attracting a lot of attention due to its considerable quantity in plants (15–30%), as it is the second most abundant component of the plant cell wall after cellulose, because of its favorable renewability [89]. It has some significant biological activities, such as antioxidative, antibacterial and anti-UV, and it also regulates enzymatic hydrolysis in biorefineries, promotes plant growth, and improves plant stress resistance in agricultural production [90]. Indeed, the abundant functional groups in lignin impart special structural characteristics and biological activities, as its phenolic hydroxyl groups not only have excellent antioxidative, antibacterial, and anti-UV functions, but also serve as modifiable active sites in lignin [91, 92]. Therefore, it is not surprising that lignin and its derived compounds have been investigated in pollen ([93–98]; see Table 10.6). For example, two Chinese pollens (*Nelumbo nucifera* and *Brassica campestris* L.) were characterized by elemental analysis and advanced solid-state ^{13}C NMR after a Soxhlet extraction and a further hydrolyzation [93].

Their constituents were largely aliphatic components (including sporopollenin), carbohydrates, protein, and lignin. Moreover, lignin-derived compounds can be also investigated in pollen to demonstrate their potential environmental applications. Thus, the extraction of phenols from lignin can be achieved with the aim of using these analytes as biomarkers of sedimentary organic matter sources to obtain information on terrestrial environmental changes [94, 95]. Pollen can be isolated from sedimentary matrices without damaging the pollen tissues, making it an ideal matrix for the analysis of lignin phenols. In one of these studies [95], Japanese pollens from gymnosperms and angiosperms plants of different botanical origins were

Table 10.6 Applications in the analysis of lignin and sporopollenin derivates in pollen

Matrix	Botanical origin (country)	Number of compounds	Reported bioactivity	Sample treatment	Determination method	Average values (mg/g)	Ref.
P	*Nelumbo nucifera* and *Brassica campestris* L. (China)	2 (SPD and LD)	NS	(i) Soxhlet (DCM and MeOH) (ii) Hydrolyzation (TFA)	^{13}C NMR	NS	[93]
P	*Picea glauca* and *Picea mariana* (USA)	4 (LD)	NS	(i) Alkaline oxidation (DEE) (ii) Derivatization (NS)	GC-FID	NS	[94]
P	Gymnosperm and angiosperm plants (Japan)	4 (LD)	NS	(i) Alkaline oxidation (DEE) (ii) Derivatization (BSTFA)	GC-FID	0.8–10.9	[95]
P	Gymnosperm and Angiosperm plants (France)	3 (LD)	NS	(i) SE (DCM) (ii) Thermochemolysis (TMAH)	GC-MS	NS	[96]
P	Gymnosperm and Angiosperm plants (Russia)	3 (LD)	NS	(i) SE (MeOH) (ii) Thermochemolysis (TMAH)	GC-MS	NS	[97]
P	Picea pollen (Germany)	NS (LD)	NS	(i) Alkaline oxidation (DEE) (iii) SPE (NS) (ii) Derivatization (NS)	GC-FID	NS	[98]
P	*Pinus pinaster*, *Betula alba*, *Ambrosia elatior* and *Capsicum annuum* (Spain)	3 (SPD)	NS	(i) Isolation (HF in pyridine; (ii) acidolysis (HCl in dioxane)	SEM, FT-IR, and GC-MS	NS	[99]
P	(*Helianthus annuus*) (Spain and Czech Republic)	NS (SPD)	NS	Acidolysis	TGA, FT-IR and SEM	NS	[101]
P	*Brassica napus* L. (China)	NS (SPD)	NS	Deffating and reflux	FT-IR and XPS	NS	[104]
P	*Pinus ponderosa* (Canada)	15 (SPD)	NS	Enzymatic digestion	FT-IR and XPS	NS	[108]

(continued)

Table 10.6 (continued)

Matrix	Botanical origin (country)	Number of compounds	Reported bioactivity	Sample treatment	Determination method	Average values (mg/g)	Ref.
P	*Pinus rigida* (USA)	3 (SPD)	NS	Thioacidolysis	NMR	NS	[109]
P	*Lilium brownii* and *Cryptomeria fortune* (China)	4 (SPD)	NS	Thioacidolysis	NMR	NS	[110]
P	*Lycopodium clavatum* L. (United Kingdom)	5 (SPD)	NS	Reflux	MALDI-TOF, NMR and XPS	NS	[111]
BP	*Brassica napus, Helianthus annuus,* MF, and *Zea mays* (Hungary and Romania)	NS (SPD)	NS	(i) Soxhlet; (ii) Washing (Water); (iii) Enzymatic and methods	FT-IR and NMR	NS	[105]

BSTFA bis(trimethylsilyl) trifluoroacetamide, *DCM* dichloromethane, *DEE* diethyl ether, *FT-IR* fourier transform infrared spectroscopy, *GC-FID* gas-chromatography flame ionization detector, *GC-MS* gas-chromatography mass spectrometry, *LD* lignin derivates, *MALDI* matrix assisted laser desorption ionization, *MeOH* methanol, *NMR* nuclear magnetic resonance, *NS* not specified, *P* pollen, *SE* solvent extraction, *SEM* scanning electron microscopy, *SPD* sporopollenin derivates, *SPE* solid-phase extraction, *TFA* trifluoroacetic acid, *TGA* thermogravimetric analysis, *TMAH* tetramethylammonium hydroxide, *TOF* time-of-flight, *XPS* X-ray photoelectron spectroscopy analyzed by GC with a flame ionization detector (FID). Prior to analysis, phenols were extracted after an alkaline oxidation with diethyl ether and were derivatized. Results showed that vanillyl, syringyl, cinnamyl and *p*-hydroxyl phenols were mainly present in those samples, although they differed in number and concentration depending on the botanical origin. It should be mentioned that similar findings and methods were reported in a previous publication in which sediment core from an Alaskan lake was analyzed for a suite of phenolic products from oxidation of lignin polymers (vanillyl, syringyl, and cinnamyl phenols), and their composition was compared with pollen data from the same core to assess lignin phenols as sedimentary biomarkers [94]. Results provided strong evidence of a direct linkage between lignin biomarkers and fossil pollen in sediment records. Lignin phenols (vanillyl, syringyl, and cinnamyl) were also identified by thermochemolysis with tetramethylammonium hydroxide combined with GC-MS after solvent extraction with dichloromethane [96] or methanol [97] in lake sediments core sample, revealing features of organic matter composition and vegetational. Indeed, in one of the studies [97], it was suggested that the ratio of cinnamyl phenols to vanillyl phenols could play a role in indicating pollen contribution to the total organic matter in the sediment. Thus, it can be concluded that fossil pollen analysis (lignin phenols) of sediments is a suitable technique for studying vegetation and it provides useful data on the terrestrial paleoenvironment. Finally, the analysis of lignin via lignin-derived phenols makes it possible to distinguish between residues of angiosperm and gymnosperms in soils and sediment of two German lakes [98]. In this study, the sample treatment was different than in previous works, as a SPE prior to their GC-FID determination was performed.

To sum up, biological potential of lignin from pollen has become a hot research topic, but due to the complex nature and undefined chemical structure of lignin, which required the used of advanced techniques like NMR and GC, its applications in the field of biomedicine and the food industry are yet limited. For example, the action mechanism and metabolic pathway of lignin in the human body should be further explored. Nevertheless, the number of studies devoted to investigate lignin is growing day by day, and this, in turn, may be associated with greater attention to pollen, as the main source of this product.

10.7.2 Sporopollenin

Sporopollenin is a biopolymer that constitutes the exine (solid outer covering) of pollen, although it is also present in other matrices including algae, fungi, and ferns [99]. This natural substance is known as one of the most chemically and mechanically stable substances. Therefore, its mechanical properties are of great interest for its use as a material for reinforcement fillers, sensors, and adhesives [100]. Since pollen grains are microcapsules composed of sporopollenin, this property has been used in applications as biomaterials for drug delivery by designing hollow pollen microcapsules [100, 102], sporopollenin-based functional ingredients for gastrointestinal tract targeted delivery [103], or fungicides using sporopollenin plant materials as a controlled release carrier [104]. For those reasons, several publications have investigated the presence of sporopollenin in pollen ([99, 101, 104, 105, 108–111]; see Table 10.6). For example, to obtain sporopollenin exine capsules from various types of bee pollen, different analytical protocols have been tested showing that Soxhlet extractions with solvents of increasing polarity, hot water washing, and enzymatic methods remove most of the interfering material from the matrix [105]. In this latter study, the combined use of NMR and fourier transform infrared spectroscopy (FT-IR) confirmed that the ratio between aromatic and *O*-aliphatic compounds derived from sporopollenin is higher in rapeseed pollen than in sunflower pollen. Sporopollenin is insoluble in most solvents, so solid-state studies through NMR and FT-IR are potentially the most useful techniques for their structural study [106]. These techniques have shown that sporopollenin is mainly composed of long aliphatic chains and aromatic compounds in different proportions, although the understanding of its composition is a real challenge. Several studies have drawn conclusions on the composition of sporopollenin, highlighting an aliphatic polyketide backbone, an aromatic moiety and extensive subunit coupling via ester and ester cross-linking [107]. Other methods have been tested in the scientific literature to study the sporopollenin chemical composition in depth. However, the techniques employed must be able to remove the internal cellular content without affecting the chemical composition of the remaining material. An example of that issue could be one work [99] in which sporopollenin is extracted from pollen grains (100–200 mg) and from different species by treatment with hydrogen fluoride in pyridine. Subsequently, purification was performed by removing the lignin-derived phenolic material by acidolysis (hydrochloric acid in dioxane), while the analysis was performed with FT-IR. Results showed the absence of polysaccharides and phenolic material and the presence of carboxylic acids. Moreover, sporopollenin samples were degraded by ozonolysis, and further GC-MS analysis allowed the elucidation of dicarboxylic acids (hexanedioic and heptanedioic acids) as major components of sporopollenin, and to a lesser extent fatty acids and n-alkanes. The use of FT-IR and X-ray photoelectron spectroscopy (XPS) combined with an enzymatic digestion [108] or a defatting and reflux [104] was also employed to determine sporopollenin and related compounds [104, 108]. Another alternative was the use of a thioacidolysis degradative method together with solid-state NMR techniques to determine the structure of sporopollenin in pine pollen showing that was primarily composed of aliphatic-polyketide-derived polyvinyl alcohol units and 7-*O*-p-coumaroylated C_{16} aliphatic units [109]. A similar approach was followed in a different study [110] in which it was revealed that sporopollenin from pollen contained phenylpropanoid derivatives, including *p*-hydroxybenzoate, *p*-coumarate, ferulate, and lignin guaiacyl. Other complementary techniques such as MALDI-MS have helped to further elucidate the structure by identifying two main blocks of the biopolymer: (i) macrocyclic oligomer and/or polymer composed of polyhydroxylated tetraketide-like monomeric units; (ii) poly(hydroxy acid) network [111]. Finally, it should be highlighted that the sporopollenin composition found by different authors varies slightly with the botanical origin of the pollen sample analyzed, which is in good agreement with the behavior observed for other pollen constituents.

Finally, it can be concluded that sporopollenin isolated from pollen grains has received attention in translational and applied research in diverse areas of drug delivery, vaccines, and environment. It is a structural polymer with remarkable resistance to chemical degradation that can be easily studied by spectroscopic techniques finding monomeric units and inter-unit linkages. A continuous investigation of the biochemical composition and potential of sporopollenin is of the utmost interest to clarify the information available on a daily basis.

10.8 Conclusions

In this chapter we present an overview on the publications dealing with the determination of other bioactive compounds (phenolamides, GSLs, organic acids, stilbenes, BRCs and biopolymers) present in pollen. Scientific interest is demonstrated by the number of research papers (> 60) published on this topic in many different countries around the world during the period reviewed. Special attention has been given to explaining the bioactivity of the studied compounds, along with some of the most relevant publications devoted to their determination. In addition, specific analytical techniques employed for determining each group of bioactive compounds in pollen have been indicated, and a summary of the conditions was included in the corresponding tables. It should be also noted that a brief comment regarding the relevance or potential of studying each family/group of compounds has been included at the end of each section. According to the number of publications, phenolamides seem to be the most promising compounds to be studied, mainly due to their antioxidant activities. However, other compound families such as stilbenes, organic acids, betaines and glucosinolates, although little studied in pollen to date, they have been investigated extensively in other matrices. Moreover, it may be stated that the compounds discussed in this chapter deserve the attention of researchers and food industry due not only to their bioactive effects or nutritional value, but also for their many different applications, like as biomarkers, or in quite diverse areas of drug delivery, vaccines, and environment. In conclusion, all the data and information summarized in this chapter should be useful to increase the knowledge about the presence in pollen of other bioactive constituents and how they could be determined. Furthermore, the chapter aims to attract more attention to pollen from various sectors of the society due to its numerous benefits associated to its consumption. Additionally, it brings attention to the potential use of some of its constituents in areas beyond the food industry.

Acknowledgements Authors thank the Spanish "Ministerio de Economía y Competitividad" and the "Instituto Nacional de Investigación y Tecnología Agraria y Alimentaria" for the funding (RTA 2015-00013-C03-03). A. Fuente-Ballesteros wishes also to thank the University of Valladolid for his PhD grant.

References

1. Gabriele M, Parri E, Felicioli A et al (2015) Phytochemical composition and antioxidant activity of Tuscan bee pollen from different botanical origins. Ital J Food Sci 27:248–259
2. Sattler JAG, de Melo ILP, Granato D et al (2015) Impact of origin on bioactive compounds and nutritional composition of bee pollen from southern Brazil: a screening study. Food Res Int 77:82–91
3. Conte G, Benelli A, Serra A et al (2017) Lipid characterization of chestnut and willow honeybee-collected pollen: impact of freeze-drying and microwave-assisted drying. J Food Compos Anal 55:12–19
4. Ares AM, Valverde S, Bernal JL et al (2018) Extraction and determination of bioactive compounds from bee pollen. J Pharm Biomed Anal 147:110–124
5. Kostić AŽ, Milinčić DD, Nedić N et al (2021) Phytochemical profile and antioxidant properties of bee-collected artichoke (*Cynara scolymus*) pollen. Antioxidants 10:1091
6. Hsu PS, Wu TH, Huang MY et al (2021) Nutritive value of 11 bee pollen samples from major floral sources in Taiwan. Foods 10:2229
7. Gardana C, Del Bo C, Quicazán MC et al (2018) Nutrients, phytochemicals and botanical origin of commercial bee pollen from different geographical areas. J Food Compos Anal 73:29–38
8. Mohammad SM, Mahmud-Ab-Rashid NK, Zawawi N (2021) Stingless bee-collected pollen (bee bread): chemical and microbiology properties and health benefits. Molecules 26:957
9. Khalifa SAM, Elashal MH, Yosri N et al (2021) Bee pollen: current status and therapeutic potential. Nutrients 13:1876
10. Thakur M, Nanda V (2020) Composition and functionality of bee pollen. Trends Food Sci Technol 98:82–106
11. Campos MGR, Bogdanov S, de Almeida-Muradian LB et al (2008) Pollen composition and standardisation of analytical methods. J Apic Res 47:154–161
12. Gonçalves PJS, Estevinho LM, Pereira AP et al (2018) Computational intelligence applied to discriminate bee pollen quality and botanical origin. Food Chem 267:36–42
13. Bayram NE, Gercek YC, Çelik S et al (2021) Phenolic and free amino acid profiles of bee bread and bee pollen with the same botanical origin – similarities and differences. Arab J Chem 14:103004
14. Taha EKA (2015) Chemical composition and amounts of mineral elements in honeybee-collected pollen in relation to botanical origin. J Apic Sci 59:75–81
15. Zhou J, Qi Y, Ritho J et al (2015) Flavonoid glycosides as floral origin markers to discriminate of unifloral bee pollen by LC-MS/MS. Food Control 57:54–61
16. Kaškonienė V, Ruočkuvienė G, Kaškonas P et al (2015) Chemometric analysis of bee pollen based on volatile and phenolic compound compositions and antioxidant properties. Food Anal Methods 8:1150–1163
17. Vogt T (2018) Unusual spermine-conjugated hydroxycinnamic acids on pollen: function and evolutionary advantage. J Exp Bot 69:5311–5315
18. Ares AM, Nozal MJ, Bernal J (2015) Development and validation of a liquid chromatography-tandem mass spectrometry method to determine intact glucosinolates in bee pollen. J Chromatogr B 1000:49–56
19. Ares AM, Toribio L, Nozal MJ et al (2020) Simultaneous determination of betaines and other quaternary ammonium related compounds in bee pollen by hydrophilic interaction liquid chromatography-mass spectrometry. Microchem J 157:105000
20. Maruthi YA, Ramakrishna S (2022) Sporopollenin – invincible biopolymer for sustainable biomedical applications. I J Biol Macromol 222:2957–2965
21. Roumani M, Duval RE, Ropars A et al (2020) Phenolamides: plant specialized metabolites with a wide range of promising pharmacological and health-promoting interests. Biomed Pharmacother 131:110762

22. Edreva AM, Velikova VB, Tsonev TD (2007) Phenylamides in plants. Russ J Plant Physiol 54:287–301
23. Taofiq O, González-Paramás AM, Barreiro MF et al (2017) Hydroxycinnamic acids and their derivatives: cosmeceutical significance, challenges and future perspectives, a review. Molecules 22(2):281
24. Liu S, Jiang J, Ma Z et al (2022) The role of Hydroxycinnamic acid amide pathway in plant immunity. Front Plant Sci 13:922119
25. Roumani M, Besseau S, Gagneul D et al (2020) Phenolamides in plants: an update on their function, regulation, and origin of their biosynthetic enzymes. J Exp Bot 72(7):2334–2355
26. Walters D, Meurer-Grimes B, Rovira I (2001) Antifungal activity of three spermidine conjugates. FEMS Microbiol Lett 201(2):255–258
27. Gekière A, Semay I, Gérard M et al (2022) Poison or potion: effects of sunflower Phenolamides on bumble bees and their gut parasite. Biology (Basel) 11(4):545
28. Adler LS, Fowler AE, Malfi RL et al (2020) Assessing chemical mechanisms underlying the effects of sunflower pollen on a gut pathogen in bumble bees. J Chem Ecol 46:649–658
29. Mihajlovic L, Radosavljevic J, Burazer L et al (2015) Composition of polyphenol and polyamide compounds in common ragweed (*Ambrosia artemisiifolia* L.) pollen and sub-pollen particles. Phytochemistry 109:125–132
30. Yuan L, Mori S, Haruyama N et al (2021) Strawberry pollen as a source of UV-B protection ingredients for the phytoseiid mite *Neoseiulus californicus* (Acari: Phytoseiidae). Pest Manag Sci 77:851–859
31. Elejalde-Palmett C, de Bernonville TD, Glevarec G et al (2015) Characterization of a spermidine hydroxycinnamoyltransferase in *Malus domestica* highlights the evolutionary conservation of trihydroxycinnamoyl spermidines in pollen coat of core Eudicotyledons. J Exp Bot 66(22):7271–7285
32. Handrick V, Vogt T, Frolov A (2010) Profiling of hydroxycinnamic acid amides in *Arabidopsis thaliana* pollen by tandem mass spectrometry. Anal Bioanal Chem 398(7–8):2789–2801
33. Kyselka J, Bleha R, Dragoun M et al (2018) Antifungal polyamides of hydroxycinnamic acids from sunflower bee pollen. J Agric Food Chem 66(42):11018–11026
34. Pereira AN, Amorim C, dos Santos A et al (2022) Chemical composition and free radical-scavenging activities of monofloral bee pollen from *Mimosa pudica* L. J Apic Res:1–8. https://doi.org/10.1080/00218839.2022.2056290
35. Caldas FRL, Filho FA, Facundo HT et al (2019) Cheminal antiradicalar and antimicrobial activity of Facaceae pollen bee. Quim Nova 42(1):49–56
36. Negri G, Teixeira EW, Alves ML et al (2011) Hydroxycinnamic acid amide derivatives, phenolic compounds and antioxidant activities of extracts of pollen samples from Southeast Brazil. J Agric Food Chem 59(10):5516–5522
37. Zhang H, Liu R, Lu Q (2020) Separation and characterization of phenolamines and flavonoids from rape bee pollen, and comparison of their antioxidant activities and protective effects against oxidative stress. Molecules 25(6):1264
38. Khongkarat P, Phuwapraisirisan P, Chanchao C (2022) Phytochemical content, especially spermidine derivatives, presenting antioxidant and antilipoxygenase activities in Thai bee pollens. PeerJ 10:e13506
39. Aylanc V, Tomás A, Russo-Almeida P et al (2021) Assessment of bioactive compounds under simulated gastrointestinal digestion of bee pollen and bee bread: bioaccessibility and antioxidant activity. Antioxidants (Basel) 10(5):651
40. Kim SB, Liu Q, Ahn JH et al (2018) Polyamine derivatives from the bee pollen of *Quercus mongolica* with tyrosinase inhibitory activity. Bioorg Chem 81:127–133
41. Khongkarat P, Ramadhan R, Phuwapraisirisan P et al (2020) Safflospermidines from the bee pollen of *Helianthus annuus* L. exhibit a higher *in vitro* antityrosinase activity than kojic acid. Heliyon 6(3):e03638

42. Zhang X, Yu M, Zhu X et al (2022) Metabolomics reveals that phenolamides are the main chemical components contributing to the anti-tyrosinase activity of bee pollen. Food Chem 389:133071
43. Zhang H, Lu Q, Liu R (2022) Widely targeted metabolomics analysis reveals the effect of fermentation on the chemical composition of bee pollen. Food Chem 375:131908
44. Gomes ANP, Camara CA, dos Santos SA et al (2021) Chemical composition of bee pollen and leishmanicidal activity of rhusflavone. Rev Bras 31:176–183
45. Qiao J, Feng Z, Zhang Y et al (2023) Phenolamide and flavonoid glycoside profiles of 20 types of monofloral bee pollen. Food Chem 405(Pt A):134800
46. Bernal J, González D, Valverde S et al (2019) Improved separation of intact glucosinolates in bee pollen by using ultra-high-performance liquid chromatography coupled to quadrupole time-of-flight mass spectrometry. Food Anal Methods 12:1170–1178
47. Francisco M, Velasco P, Moreno DA et al (2010) Cooking methods of *Brassica rapa* affect the preservation of glucosinolates, phenolics and vitamin C. Food Res Int 43(5):1455–1463
48. Becker TM, Juvik JA (2016) The role of Glucosinolate hydrolysis products from brassica vegetable consumption in inducing antioxidant activity and reducing cancer incidence. Diseases 4(2):22
49. Hanschen FS, Pfitzmann M, Witzel K et al (2018) Differences in the enzymatic hydrolysis of glucosinolates increase the defense metabolite diversity in 19 *Arabidopsis thaliana* accessions. Plant Physiol Biochem 124:126–135
50. Yaday K, Dhankhar J, Kundu P (2022) Isothiocyanates – a review of their health benefits and potential food applications. Cur Res Nutr Food Sci 10(2):476–502
51. Dungey SG, Sang JP, Rothnie NE et al (1988) Glucosinolates in the pollen of rapeseed and indian mustard. Phytochemistry 27(3):815–817
52. Ares AM, Redondo M, Tapia J et al (2020) Differentiation of bee pollen samples according to their intact-glucosinolate content using canonical discriminant analysis. LWT 129:109559
53. Ares AM, Tapia JA, González-Porto AV et al (2022) Glucosinolates as markers of the origin and harvesting period for discrimination of bee pollen by UPLC-MS/MS. Foods 11(10):1446
54. Thakur M, Nanda V (2020) Screening of Indian bee pollen based on antioxidant properties and polyphenolic composition using UHPLC-DAD-MS/MS: a multivariate analysis and ANN based approach. Food Res Int 140:110041
55. Ares AM, Ayuso I, Bernal JL, Nozal MJ, Bernal J (2016) Trace analysis of sulforaphane in bee pollen and royal jelly by liquid chromatography-tandem mass spectrometry. J Chromatogr B Analyt Technol Biomed Life Sci 1012:130–136
56. Wang Z, Ren P, Wu Y, He Q (2021) Recent advances in analytical techniques for the detection of adulteration and authenticity of bee products – a review. Food Addit Contam Part A Chem Anal Control Expo Risk Assess 38(4):533–549
57. Wang R, Zhong M, Hao N et al (2022) Botanical origin authenticity control of pine pollen food products using multiplex species-specific PCR method. Food Anal Methods 15:421–427
58. Alvarez-Jubete L, Smyth TJ, Valverde J et al (2014) Simultaneous determination of sulphoraphane and sulphoraphane nitrile in Brassica vegetables using ultra-performance liquid chromatography with tandem mass spectrometry. Phytochem Anal 25(2):141–146
59. Mutinelli F (2000) European legislation governing the use of veterinary medicinal products with particular reference to varroa control. Bee World 81(4):164–171
60. Campas-Baypoli ON, Sánchez-Machado DI, Bueno-Solano C et al (2010) HPLC method validation for measurement of sulforaphane level in broccoli by-products. Biomed Chromatogr 24(4):387–392
61. Anyasi TA, Jideani AIO, Edokpayi JN, Anokwuru CP (2017) Organic acids, characteristics, properties and synthesis. In: Vargas C (ed) Application of organic acids in food preservation. Nova Publishers, New York, pp 1–46
62. Yin X, Li J, Shin H et al (2015) Metabolic engineering in the biotechnological production of organic acids in the tricarboxylic acid cycle of microorganisms: advances and prospects. Biotechnol Adv 33:830–841

63. Kalaycıoğlu Z, Kaygusuz H, Döker S et al (2017) Characterization of Turkish honeybee pollens by principal component analysis based on their individual organic acids, sugars, minerals, and antioxidant activities. LWT 84:402–408
64. Park Y, Kim JH, Kim SY (2017) Free sugar and organic acid contents of pollens from *Quercus* spp. in Korea. J Apic 32:375–379
65. Kim HJ, Hwang J, Ullah Z et al (2022) Comparison of physicochemical properties of pollen substitute diet for honey bee (*Apis mellifera*). J Asia Pac Entomol 25:101967
66. Çelik S, Kutlu N, Gerçek YC et al (2022) Optimization of ultrasonic extraction of nutraceutical and pharmaceutical compounds from bee pollen with deep eutectic solvents using response surface methodology. Foods 11:3652
67. Zhan H, Lu Q, Liu R (2022) Widely targeted metabolomics analysis reveals the effect of fermentation on the chemical composition of bee pollen. Food Chem 375:131908
68. Oraian M, Dranca F, Ursachi F (2022) Characterization of Romanian bee pollen—an important nutritional source. Foods 11:2633
69. Soto ME, Bernal J, Martín MT et al (2012) Liquid chromatographic determination of resveratrol and piceid isomers in honey. Food Anal Methods 5:162–171
70. Grippi F, Crosta L, Aiello G et al (2008) Determination of stilbenes in Sicilian pistachio by high-performance liquid chromatographic diode array (HPLC-DAD/FLD) and evaluation of eventually mycotoxin contamination. Food Chem 107(1):483–488
71. Montsko G, Pour Nikfardjam MS, Szabo Z et al (2008) Determination of products derived from trans-resveratrol UV photoisomerisation by means of HPLC–APCI-MS. J Photochem Photobiol A 196(1):44–50
72. Gülçin I (2010) Antioxidant properties of resveratrol: a structure–activity insight. Innovative Food Sci Emerg Technol 11(1):210–218
73. Juan ME, Alfaras I, Planas JM (2012) Colorectal cancer chemoprevention by *trans*-resveratrol. Pharmacol Res 65(6):584–591
74. Ares AM, Soto ME, Nozal MJ et al (2015) Determination of resveratrol and piceid isomers in bee pollen by liquid chromatography coupled to electrospray ionization-mass spectrometry. Food Anal Methods 8:1565–1575
75. Rocchetti G, Castiglioni S, Maldarizzi G et al (2019) UHPLC-ESI-QTOF-MS phenolic profiling and antioxidant capacity of bee pollen from different botanical origin. Int J Food Sci Technol 54:335–346
76. Laaroussi H, Bakour M, Ousaaid D et al (2020) Effect of antioxidant-rich propolis and bee pollen extracts against D-glucose induced type 2 diabetes in rats. Food Res Int 138:109802
77. Laaroussi H, Ferreira-Santos P, Genisheva Z et al (2023) Unveiling the techno-functional and bioactive properties of bee pollen as an added-value food ingredient. Food Chem 30:134958
78. Santa Bárbara MF, Moreira MM, Machado CS et al (2021) Storage methods, phenolic composition, and bioactive properties of *Apis mellifera* and *Trigona spinipes* pollen. J Apic Res 60(1):99–107
79. Salonen A, Lavola A, Virjamo V et al (2021) Protein and phenolic content and antioxidant capacity of honey bee-collected unifloral pollen pellets from Finland. J Apic Res 60(5):744–750
80. Craig SA (2004) Betaine in human nutrition. Am J Clin Nutr 80(3):539–549
81. Hong Z, Lin M, Zhang Y et al (2020) Role of betaine in inhibiting the induction of RNA Pol III gene transcription and cell growth caused by alcohol. Chem Biol Interact 325:109129
82. Hassanpour S, Rezaei H, Razavi SM (2020) Anti-nociceptive and antioxidant activity of betaine on formalin- and writhing tests induced pain in mice. Behav Brain Res 390:112699
83. de Zwart FJ, Slow S, Payne RJ et al (2003) Glycine betaine and glycine betaine analogues in common foods. Food Chem 83(2):197–204
84. Servillo L, Giovane A, Casale R et al (2016) Betaines and related ammonium compounds in chestnut (Castanea sativa Mill.). Food Chem 196:1301–1309
85. Nugrahini AD, Ishida M, Nakagawa T et al (2020) Trigonelline: an alkaloid with anti-degranulation properties. Mol Immunol 118:201–209

86. Ares AM, Martín MT, Tapia JA et al (2022) Differentiation of bee pollen samples according to the betaines and other quaternary ammonium related compounds content by using a canonical discriminant analysis. Food Res Int 160:111698
87. Mei S, Yang C, Song X et al (2021) Analysis of the effect of sterilization and storage on the quality of *Pinus yunnanensis* pollen based on untargeted metabonomics. J Food Process Preserv 45:e16033
88. Chalbot MCG, Gamboa da Costa G, Kavouras IG (2013) NMR Analysis of the Water-Soluble Fraction of Airborne Pollen Particles. Appl Magn Reson 44:1347–1358
89. Mariana M, Alfatah T, Khalil HPSA, Bashir Yahya E et al (2021) A current advancement on the role of lignin as sustainable reinforcement material in biopolymeric blends. J Mater Res Technol 15:2287–2316
90. Shu F, Jiang B, Yuan Y et al (2021) Biological activities and emerging roles of lignin and lignin-based products—a review. Biomacromolecules 22(12):4905–4918
91. Kumar Thakur V, Kumari Thakur M, Raghavan P et al (2014) Progress in green polymer composites from lignin for multifunctional applications: a review. ACS Sustain Chem Eng 5:1072–1092
92. Ten E, Vermerris W (2015) Recent developments in polymers derived from industrial lignin. J Appl Polym Sci 132:42069
93. Zhang D, Duan D, Huang Y et al (2016) Novel phenanthrene sorption mechanism by two pollens and their fractions. Environ Sci Technol 50(14):7305–7314
94. Hu FS, Hedges JI, Gordon LS et al (1999) Lignin biomarkers and pollen in postglacial sediments of an Alaskan Lake. Geochim Cosmochim Acta 63(9):1421–1430
95. Tareq SM, Ohta K (2015) Lignin and isotope signatures in pollen: a caveat of lignin phenol biomarker for reconstructing paleovegetation. Asian J Water Environ Pollut 12(1):1–9
96. Bertrand O, Mansuy-Huault L, Montargès E et al (2013) Recent vegetation history from a swampy environment to a pond based on macromolecular organic matter (lignin and fatty acids) and pollen sedimentary records. Org Geochem 64:47–57
97. Ishiwatari R, Yamamoto S, Shinoyama S (2006) Lignin and fatty acid records in Lake Baikal sediments over the last 130 Kyr: a comparison with pollen records. Org Geochem 37(12):1787–1802
98. Kappenberg A, Amelung W, Conze N (2021) Fire–vegetation relationships during the last glacial cycle in a low mountain range (Eifel, Germany). Palaeogeogr Palaeoclimatol Palaeoecol 562:110140
99. Domínguez E, Mercado JA, Quesada MA et al (1999) Pollen sporopollenin: degradation and structural elucidation. Sex Plant Reprod 12(3):171–178
100. Qu Z, Meredith JC (2018) The atypically high modulus of pollen exine. J R Soc Interface 15(146):20180533
101. Ageitos JM, Robla S, Valverde-Fraga L et al (2021) Purification of hollow sporopollenin microcapsules from sunflower and chamomile pollen grains. Polymers (Basel) 13(13):2094
102. Uddin MJ, Gonzalez-Cruz P, Warzywoda J et al (2020) Sporopollenin spikes augment antigen-specific immune response and generate long-lived humoral immunity. Adv Ther 3(10):1–14
103. Schouten Pien JC, Soto-Aguilar D, Aldalbahi A et al (2022) Design of sporopollenin-based functional ingredients for gastrointestinal tract targeted delivery. Curr Opin Food Sci 44:100809
104. Fan TF, Xiang S, Li L et al (2022) Fabrication of ultrafine sporopollenin particles and its application as pesticide carrier. Appl Mater Today 27:101454
105. Hegedüs K, Fehér C, Jalsovszky I et al (2021) Facile isolation and analysis of sporopollenin exine from bee pollen. Sci Rep 11(1):9952
106. Grienenberger E, Quilichini TD (2021) The toughest material in the plant kingdom: an update on sporopollenin. Front Plant Sci 12:703864
107. Quilichini TD, Grienenberger E, Douglas CJ (2015) The biosynthesis, composition and assembly of the outer pollen wall: a tough case to crack. Phytochemistry 113:170–182

108. Lutzke A, Morey KJ, Medford JI et al (2020) An FT-IR and XPS spectroscopy dataset of *Pinus ponderosa* sporopollenin and related samples to elucidate sporopollenin structural features. Data Brief 29:105129
109. Li FS, Phyo P, Jacobowitz J et al (2019) The molecular structure of plant sporopollenin. Nat Plants 5(1):41–46
110. Xue JS, Zhang B, Zhan H et al (2020) Phenylpropanoid derivatives are essential components of sporopollenin in vascular plants. Mol Plant 13(11):1644–1653
111. Mikhael A, Jurcic K, Schneider C et al (2020) Demystifying and unravelling the molecular structure of the biopolymer sporopollenin. Rapid Commun Mass Spectrom 34(10):e8740

Chapter 11
Microbiology of Pollen

Vladimíra Kňazovická and Miroslava Kačániová

Abbreviations

a_w	water activity
BCP	bee-collected pollen (bee/corbicular pollen)
BSP	bee-stored pollen (bee bread, perga, honeycomb pollen)
CFU	colony forming unit
DWV	*Deformed wing virus*
FLAB	fructophilic lactic acid bacteria
FOS	fructooligosaccharides
hbs LAB	honeybee specific lactic acid bacteria
LAB	lactic acid bacteria
MFF	microscopic filamentous fungi (moulds)
TPC	total plate count

11.1 Introduction

Pollen is a food/feed supplement with a great potential because of high quality nutrients and positive effects on health of people and animals [1–8]. But, on the other side, presence of contaminants could be the serious risk. Besides chemical

V. Kňazovická (✉)
Institute of Apiculture Liptovský Hrádok, Research Institute for Animal Production Nitra, National Agricultural and Food Centre, Liptovský Hrádok, Slovakia
e-mail: vladimira.knazovicka@nppc.sk

M. Kačániová
Institute of Horticulture, Faculty of Horticulture and Landscape Engineering, Slovak University of Agriculture, Nitra, Slovakia

Department of Bioenergy, Food Technology and Microbiology, Institute of Food Technology and Nutrition, University of Rzeszow, Rzeszow, Poland
e-mail: miroslava.kacaniova@uniag.sk

N. Ecem Bayram et al. (eds.), *Pollen Chemistry & Biotechnology*,
https://doi.org/10.1007/978-3-031-47563-4_11

contamination (toxic elements, residues of pesticides, herbicides, or antibiotics), bacterial and mycological contamination seems to be a problem [4, 9–12]. However, it cannot be stated clearly, because some microorganisms could have also a useful effect on the organisms. Intake of microorganisms or synbiotics (mixture of prebiotics and probiotics) is important for people, whose beneficial gut microbiota is suppressed [13]. In general, present microbiota of pollen originates in nature, bees, and hive environment. Pollen provides a unique microhabitat [14]. Total plate count (TPC), bacteria from family Enterobacteriaceae (or coliform bacteria), lactic acid bacteria (LAB) and microscopic fungi (MF) belong to target microorganisms in the pollen [15, 16].

The aim of the chapter was to characterize the factors affecting the pollen microbiota, concrete microorganisms, which are typical for pollen and possible general relations between the pollen and microorganisms.

11.2 Factors

Various factors can influence physico-chemical properties and present microbiota in pollen – type of pollen (in term of bee processing and bee species) as well as its processing by beekeeper/processing company. From physico-chemical properties of pollen, water content and activity are the main factors affecting the life of microorganisms. Present microbiota profile is also influenced by the method used to its identification.

11.2.1 Type of Pollen

In term of bee processing level, there are recognized three types of pollen: plant pollen, bee-collected pollen (BCP) and bee-stored pollen (BSP). Plant (or flower) pollen is collected/used without bee work. Human coprolites (desiccated excreta) dating back to 1400–200 B.C. yielded evidence that, plant pollen (mainly from mustard *Sinapis alba*, cabbage *Brassica oleracea*, willow *Salix* spp. and maize *Zea mays*) was common for consumption of American Indians, who obtained it by simple brushes and used to use it to e.g. bread dough [1]. In general, plant pollen is ingested accidentally, e.g., with herbal tea or blossom honey. BCP (or bee/corbicular pollen) is caught on bee hair and joined into the loads/corbiculars by bee saliva during their flight. It is a mixture of flower pollen, nectar and salivary secretions of bees [17]. BCP has a little bit higher water content then plant pollen, because of these aqueous additions. BCP is obtained by pollen traps by beekeeper. BSP (or honeycomb pollen, bee bread, perga) is brought into the wax combs by bees and processed in the hive by honey and bee enzymes addition with subsequent maturation to conserve the main protein feed of bees. BSP is consumed by adult bees and is fed to larvae [18].

What could be the reason for the differences in present microbiota between BCP and BSP? BCP is transformed to BSP through the microbial metabolism and biochemical changes of pollen grains. Bacteria and yeasts are multiplied, lactic acid fermentation is performed, lactic acid increases, pH decreases. This type of lactic acid fermentation is similar to that in yoghurts (and other fermented milk products) and renders the end product more digestible and enriched with new nutrients [19]. At first, LAB and indole-producing bacteria (*Escherichia* spp.), aerobic bacteria and yeasts are developed, then the anaerobic LAB (*Streptococcus* spp.) use the nutrients created by bacteria and yeasts, then streptococci decrease, and lactobacilli increase [20]. During the process, pollen wall is disrupted. Destruction of the pollen coating, exine, intine and bond structures is catalysed by enzymes – protease, cellulase, pectinase and other enzymes, which are produced by fermentation species [21]. Spoilage of BSP is avoided by the low pH, caused by lactic, acetic and alcoholic fermentation, and by the presence of natural products with antibiotic properties produced by resident microbiota of BSP [22].

Plant species and pollination type also significantly influenced structure and diversity of the pollen microbiota [14]. BCP of *Apis mellifera* contains approximately 10,000 times (by 4 log CFU/g) more microorganisms in compare to ripe honey [15]. TPC of BCP was found at level 5 log CFU/g [15, 16, 23]. TPC of BSP was a little bit lower, at about 3.00 log CFU/g [15]. Stingless bee *Melipona* spp. is spread in warm areas of the Americas, including Brazil [24, 25]. Bees of *Melipona* spp. create a special nest and collect the pollen. Local beekeepers obtain this stored pollen and called it "samburá" (pot-pollen). Surprisingly, Oliveira et al. [24] did not detect any aerobic psychrotrophic bacteria, bacilli, sulphite-reducing clostridia, staphylococci, faecal coliforms, moulds, and yeasts in this pot-pollen of *Melipona* spp. from Brazil. However, pot-pollen of stingless bee *Tetragonula biroi* from Philippines had similar TPC comparing to BSP of *Apis mellifera* pollen, i.e., at level 3–4 log CFU/g and bacilli and clostridia were dominant because of low water activity and pH of pot-pollens [26]. Dharampal et al. [27] performed 16S rRNA and ITS gene-based analyses in pollen of bumblebee *Bombus impatiens* to identify the present microbiota and found two ubiquitous bee-associated groups – *Lactocbacillus* and *Wickerhamiella/Starmerella* clade yeasts.

11.2.2 Water Content and Water Activity (a_w)

Water content and water activity are the main factors, which influence the life of microorganisms present in the pollen. High humidity of fresh BCP (about 20–30% w/w) is an ideal culture media for microorganisms like bacteria and yeasts [28–30]. Fresh BCP and BSP with high moisture are easily contaminated by the microbial infection [31]. If the high bacterial counts are found in dried samples, it is an indication of massive contamination of BCP before the drying treatment or incorrect production practises [16]. Important marker in predictive microbiology is a_w. Bacteria grow at a_w higher than 0.91, while yeasts and moulds generally grow when a_w is above 0.61 [32]. Average value of a_w in fresh and dried BCP is 0.67 (with range from 0.58 to 0.77) and 0.33 (with range from 0.19 to 0.47), respectively [16, 33]. According to these statements, the main risk seems to be the growth of microscopic fungi in fresh pollen. In general, positive correlation was found between a_w and some targeted microorganisms (total bacterial count, Enterobacteriaceae, LAB, MF) in the pollen [16, 34].

11.2.3 Processing

Obtaining of BCP from pollen traps is optimal in each day to avoid moulds development. There is a necessity of fast processing of fresh pollen because of high water content [30]. The fresh BCP can be frozen, dried, or lyophilized. Another possible treatment of pollen is UV radiation [35]. Interesting way of pollen processing is its fermentation, which is partial simulation of BSP production. Krell [19] described the process of fermentation as pollen can ("home bee bread") production by addition of pollen to solution of water and honey and possible addition of lactic acid bacteria.

The best way of pollen drying is in an electric oven (at max 30 °C), where humidity can continuously escape and then it is purified by a special machine, similar to seed cleaning machine [29]. Drying of pollen leads to lower counts of microorganisms [16]. Microbial counts of fresh BCP and BCP frozen for 6 months are very similar without significant differences (unpublished data), what indicated a necessity of fast processing of pollen after defrosting.

Fermentation of BCP (during the production of pollen can) leaded to increase of water content by 60%, increase of free acidity by 40%, decrease of pH by 17%, decrease of fat by 1–2% (as fat in dry matter) in pollen can comparing to pollen as a raw material [30]. Low pH has protective function. After fermentation of BCP, there were recorded decrease of bacteria from Enterobacteriaceae family and counts of microscopic filamentous fungi, while counts of sporulating aerobic microorganisms, yeasts and preliminary LAB were stable in comparison to bee pollen as raw material [30, 33]. In general, fermentation of BCP results in decrease of microbial counts and diversity [30, 33, 36]. Zhang et al. [37] confirmed that fermentation improved the antioxidant and anti-inflammatory activities of bee pollen and increased the bioactivity of the material.

After processing of pollen, there is important to use appropriate package material to avoid fast spoilage. The most appropriate packaging material for storing of pollen are glass and hard plastic to protect the product against high humidity [11].

11.2.4 Method of Microorganism Identification

Microorganisms may be identified according to different levels: group (e.g., bacteria, microscopic fungi), family, order, genus, species, or subspecies. The methods used for this purpose can be divided into three groups – methods of cultivation, biochemical methods, and molecular-genetic methods. We can combine and join them to obtain the correct results and according to the reliability of the method used. The basic method is the cultivation of microorganisms. There is a necessity to provide available conditions for target microbial group, including energy and carbon sources, temperature, oxygen, humidity, and possible special nutritional requests. Biochemical methods for identification of microorganisms are based on analyses of their biochemical characteristics (enzyme production or metabolic products) [15]. Microorganisms are also evaluated by microscopic methods. There are two main performances of microscopic methods: observation of morphological signs of cells from cultivated strain or direct observation of examined material processed e.g., by fluorescence in situ hybridization (FISH).

Molecular methodologies offer an alternative laboratory mechanisms for the identification of microorganisms [38]. Microbial DNA can be identified directly from the tested material or from the strain cultivated in Petri plate. Metagenomic analysis, particularly 16S rRNA gene sequencing on high throughput sequencing platform Illumina became the most common and accurate analyses to detect DNA from bacteria [39]. Friedle et al. [40] sequenced amplicons of V3-V4 region in bacterial 16S rRNA gene and amplicons of the ITS1 region in the fungal 18S rRNA gene. Methods based on DNA sequencing (metabarcoding) are also used for determination of plant origin [41] and studies aimed on correlations between plant origin and present microbiota in pollen are also emerging [42, 43]. Another methodology used for identification of bacteria is based on ribosomal protein analysis on an Autoflex speed matrix-assisted laser desorption/ionisation time-of-flight (MALDI-TOF) mass spectrometer [30, 44]. Differences in the results obtained by various types of microbiota identification methods were observed and their explanation are still developed. In general, more microorganisms are identified by cultivation-independent methods, when DNA is isolated directly from the investigated material. Ambika Manirajan et al. [14] found that about 17% of all bacterial families from cultivation-independent approach were also isolated and concluded that it is typical bias related to the culturability of most environmental microbes.

Viruses are not detectable by cultivation methods. However, use of metagenomics, metatranscriptomics and metaviromics was increased in recent time and consequently higher number and diversity of viruses were observed [45].

11.2.5 Proposal of Microbiological Limits for Pollen

Pollen production is developed and there is a need for an international directive for quality control [41]. Proposal of limits for pollen contaminants exists, including limits for microorganisms, as follows: Total (aerobic) plate count max. 5.00 log CFU/g, microscopic fungi max. 4.70 log CFU/g, Enterobacteriaceae max. 2.00 log CFU/g and absent *Salmonella* spp. in 10 g as well as absent *Staphylococcus aureus* and *Escherichia coli*, both in 1 g [28].

11.3 Bacteria

Bacteria associated with pollen are poorly investigated [14]. In the past, mainly bacilli were investigated [46]. In the current research [40], 48% Proteobacteria (*Acinetobacter* spp., *Actibacterium* spp., *Arsenophonus* spp., *Batronella* spp., *Bradyrhizobium* spp., *Carnimonas* spp., *Citrobacter* spp., *Duganella* spp., *Erwinia* spp., *Escherichia* spp., *Frischella* spp., *Gilliamella* spp., *Gluconacetobacter* spp., *Halotalea* spp., *Massilia* spp., *Neokomagataea* spp., *Pantoea* spp., *Pectobacterium* spp., *Phaseolibacter* spp., *Pseudomonas* spp., *Rickettsia* spp., *Rosenbergiella* spp., *Saccharibacter* spp., *Serratia* spp., *Snodgrassella* spp., *Sphingomonas* spp.), 44% Firmicutes (*Fusicatenibacter* spp., *Lactobacillus* spp., *Lactococcus* spp., *Staphylococcus* spp.), 4% Bacteroidetes (*Apibacter* spp., *Bacteroides* spp., *Chryseobacterium* spp., *Epilithonimonas* spp., *Flavobacterium* spp., *Pedobacter* spp.), 3% Actinobacteria (*Arthrobacter* spp.) and 0.1% Cyanobacteria (*Cyanobium* spp.) were found in all samples of bee pollen from Germany. In German plant pollen, Proteobacteria was also found as dominant phylum in all pollen species (birch *Betula pendula*, rape *Brassica napus*, rye *Secale cereal* and autumn crocus *Colchicum autumnale*) [14]. In BCP and BSP from Turkey, the dominant bacterial phylum was Firmicutes, followed by Proteobacteria and from the families, Bacillaceae, Clostridiaceae, Enterococcaceae and Enterobacteriaceae were found as dominant [47]. According to Barta et al. [20], *Pseudomonas* spp. and *Lactobacillus* spp. were the most abundant in BCP and BSP. Bacteria as *Pseudomonas* spp., *Xanthomonas* spp., *Acinetobacter* spp. or *Sphingomonas* spp. are potentially from floral sources and bacteria as *Lactobacillus* spp., *Bifidobacterium* spp., *Gilliamella* spp. or *Snodgrassella* spp. seem to be bee-associated bacteria [48]. Bacterial communities in pollen varied only slightly in term of different locations [40].

11.3.1 Sporulating Bacteria

Sporulating bacteria have G^+ cells and are able to form endospores. Due to the resistance of their endospores to environmental stress, as well as their long-term survival under adverse conditions, most of them are ubiquitous and can be isolated from a wide variety of sources [15]. There are two main classes – bacilli and clostridia. Bacilli are aerobic while clostridia are anaerobic.

In the past, Gilliam [46] found bacilli in the pollen. While plant pollen contained only *Bacillus subtilis*, in the BCP, she found *B. subtilis*, *Priestia megaterium* (formerly *Bacillus megaterium*), *Bacillus licheniformis*, *Niallia circulans* (formerly *Bacillus circulans*) and in BSP *Bacillus subtilis*, *B. pumilus* and *B. licheniformis* and she compared the role of bacilli in pollen to silage, where they have the ability to contribute to the lactic acid and acetic acid content of silage. In Argentina, thirty fresh bee pollen samples were analysed in term of aerobic spore-forming bacteria on cultivation media followed by biochemical and molecular-genetic identification

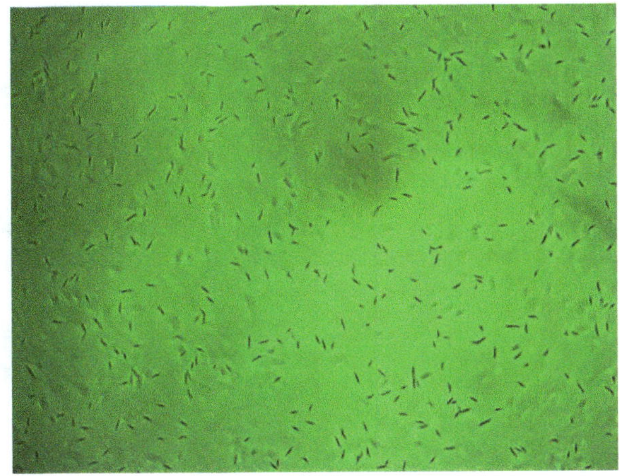

Fig. 11.1 Cells of *Bacillus sonorensis*, cultivated from fermented pollen and shown by phase contrast. (Magnification 40 × 10 × 1.5, photo: V. Kňazovická, 2018)

of cultures, and were found that *Bacillus cereus* sensu stricto was the most predominant species (50%), followed by *Priestia megaterium* (40%), *B. subtilis* (40%), *B. pumilus* (13%), *B. licheniformis* (13%), *B. amyloliquefaciens* (10%), *Weizmannia coagulans* (formerly *Bacillus coagulans*; 7%), *B. thuringiensis* (7%), *Shouchella clausii* (formerly *Bacillus clausii*; 3%), *Peribacillus simplex* (formerly *B. simplex*; 3%), *Paenibacillus polymyxa* (20%), *P. larvae* (17%), *P. alvei* (3%), *P. amylolyticus* (3%), *Lysinibacillus sphaericus* (7%) and *Rummeliibacillus stabekisii* (7%) [17]. From pollen can (home bee bread), *B. licheniformis* (four isolates), *Priestia megaterium* (two isolates) and *B. sonorensis* (one isolate, Fig. 11.1) were isolated and identified by MALDI-TOF [30]. *Bacillus licheniformis* and *Priestia megaterium* are frequently present all over the world and their strains are used in biotechnology [49, 50]. *B. sonorensis* is named after the Sonoran Desert (USA), where the organism was collected [51]. *B. sonorensis* was also isolated from decayed fruit and vegetables [52]. Spores of bacilli and clostridia are normal part of nature. They are found e.g., in the soil, dust, air. With the exception of two species (*B. anthracis* and many *B. cereus* toxin-producer strains), *Bacillus* spp. are considered safe [53].

However, vegetative clostridia of some species are pathogenic and can produce the toxins. In dried Brazilian BCP, sulphite-reducing clostridia were not detected [42, 54]. In pollen can (homemade bee bread), two clostridia species were found: *Clostridium perfringens* (two isolates) and *C. baratii* (three isolates, Fig. 11.2) [30]. *C. perfringens* spores are spread in the soil and intestinal tract of animals and humans and in vegetative form and under specific conditions, the bacteria can produce more than 15 toxins [55]. *C. baratii* can produce botulotoxin in rare cases and result in infant botulism [56]. On the other hand, Stefka et al. [57] demonstrated that the allergy-protective capacity is conferred by a clostridia-containing microbiota, where clostridia regulate innate lymphoid cell function and intestinal epithelial permeability to protect against allergen sensitization. According to Grieger and Vařejka [58], food containing clostridia at dose less than 2.00 log CFU/g cannot cause the illness, because the counts are reduced during the transport by gastrointestinal tract,

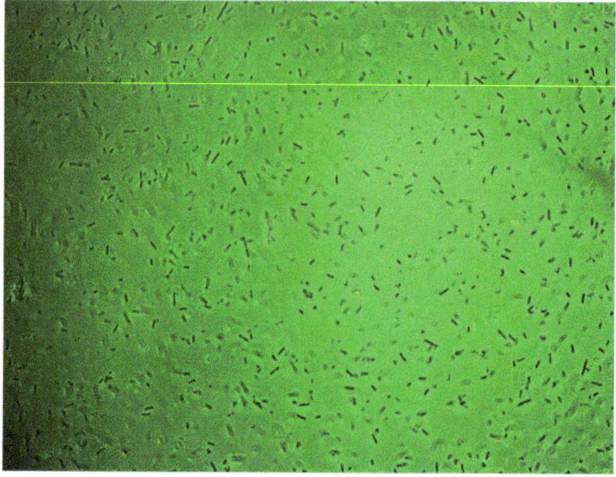

Fig. 11.2 Cells of *Clostridium baratii*, cultivated from fermented pollen and shown by phase contrast. (Magnification 40 × 10 × 1.5, photo: V. Kňazovická, 2018)

but risk groups are children, older people and people with weakened immunity; and food with pH below 4.5 is considered as safe food, because clostridia (especially *C. botulinum*) are not able to multiply and produce botulotoxin under the pH 4.5.

11.3.2 Lactic Acid Bacteria (LAB)

LAB are grouped in one order, six families, thirty three described genera and the main feature is production of lactic acid as a major catabolic end product from glucose [59]. *Lactobacillus* spp. and *Bifidobacterium* spp. are the most important representatives of LAB, finding a widespread use as probiotics for humans and animals [5]. Microorganisms, especially LAB, have an important role in the process of pollen transformation [30]. A few years ago, an important discovery was made in term of bee microbiota related also with bee products. Olofsson and Vásquez [60] detected the indigenous bacterial microbiota in honey crop (also known as honey stomach or honey sack), dominated by *Lactobacillus* and *Bifidobacterium* phylotype related to *Apilactobacillus kunkeei* (formerly *L. kunkeei*), *B. asteroides* and *B. indicum* (formerly *B. coryneforme*). Later, FLAB (fructophilic lactic acid bacteria – LAB, which prefer fructose over glucose) were analysed in the samples from bee hive, including bee pollen and *Apilactobacillus kunkeei* was found as dominant FLAB [61]. The term honeybee specific Lactic Acid Bacteria (hbs LAB) is also known and used for 13 LAB from genus *Lactobacillus* and related genera (*Apilactobacillus kunkeei*, *A. apinorum* (formerly *L. apinorum*), *Bombilactobacillus mellis* (formerly *L. mellis*), *B. mellifer* (formerly *L. mellifer*), *L. apis*, *L. helsingborgensis*, *L. melliventris*, *L. kimbladii* and *L. kullabergensis*) and *Bifidobacterium* (three strains of *B. asteroides* and one strain of *B. indicum*), which are originating in *Apis mellifera* honey crop [62]. These bacteria probably play significant role in pollen fermentation. In Slovak pollen, *Bifidobacterium* spp. were detected in fresh and frozen BCP at level 4–5 log CFU/g and *Lactobacillus* spp. were detected in BSP at level 3 log CFU/g by cultivation method [15]. Di Cagno et al. [36] tested plant pollen, BCP and BSP from ivy (*Hedera helix*) in Italy and found LAB in BCP samples at level 5 log CFU/g, also in BSP, but not in plant pollen. Acid resistant and osmotolerant bacteria *Apilactobacillus kunkeei* was found in greatest abundance in BSP [63]. Plant pollen and fresh BCP is characterized by a large microbial diversity, however, throughout the BSP maturation, many of them disappeared, giving the way to *Apilactobacillus kunkeei* and *Fructobacillus fructosus* (formerly *L. fructosus*) to dominate in BSP [36]. Lactobacilli occur spontaneously in the honeycombs [5]. Anderson et al. [63] stated that BSP is not a result of microbes pre-digestion, they found lack of an emergent microbial communities co-evolved to digest stored pollen and consider BSP as "preservative environment for beneficial bacteria".

11.3.3 Enterobacteriaceae and Related Families

Presence of Enterobacteriaceae family was recorded at max. 4.10 log CFU/g (with average value 2.80 ± 1.10 log CFU/g) in fresh and frozen BCP and max. 3.10 log CFU/g (with average value 1.80 ± 0.30 log CFU/g) in dried BCP [16]. De-Melo et al. [54] found coliform bacteria max. at level 3 log CFU/g in dried BCP from Brazil. Coliform bacteria is an indicator of the hygienic conditions and have a large capacity of forming colonies at different points of the processing plant, when sanitization is faulty [42]. The presence of Enterobacteriaceae family is limited in food products [28], because in this family, we can find various bacteria, at about twenty genera, including coliforms [23] and some of them are strict or facultative pathogens. Coliforms are normal residents of gastrointestinal tract of animals and humans as well as environments [23] and in general, they indicate faecal contamination in food.

Dinkov [23] tested Bulgarian fresh and dried BCP by cultivation method with identification through the system BioLog Gen III microplates and from Enterobacteriaceae family, he identified *Pantoea agglomerans*, *P. dispersa* (currently *Pantonea* spp. belong to Erwiniaceae family [64]), *Proteus mirabilis*, *P. vulgaris* (currently *Proteus* spp. belong to Morganellaceae family [64]), *Escherichia coli*, *Serratia fonticola* (currently *Serratia* spp. belong to Yersiniaceae family [64]), *Raoultella planticola/ornithinolytica* in fresh BCP and *Pantoea agglomerans*, *P. dispersa*, *Proteus mirabilis*, *Serratia fonticola*, *Raoultella planticola/ornithinolytica* and *Citrobacter freundii* in dried BCP stored for 1 year and total count of Enterobacteriaceae and related families was at level 3–4 log CFU/g regardless of pollen form [23]. Several genera from Enterobacteriaceae family were reclassified to other families [64]. However, if they are cultivated on the same medium, we have to evaluate then together. In Turkish BCP, *Pseudescherchia vulneris* (formerly *Escherichia vulneris*) and *Cronobacter malonaticus* were identified with high reads by metagenomic analysis and in BSP, *Mixta calida* (formerly *Pantoea calida*) were dominant [47].

Pantoea agglomerans create special structures "symplasmata", which are able to persist in the pollen by this form for a long time [23]. *P. agglomerans* is a plant pathogen, is usable in biocontrol of pests (against insect larvae of Lepidoptera and Coleoptera) and can cause infections, mainly wound infections with plant material and hospital-acquired infections in immunocompromised individuals [65, 66]. *P. agglomerans*, *P. dispersa* and *Mixta calida* are considered as human pathogens [67].

11.3.4 Others

Presence of staphylococci and enterococci was also observed in the pollen. From staphylococci, *Staphylococcus aureus* was not detected in dried BCP from Portugal [34]. Coagulase-positive staphylococci were negative also in forty five samples of dried BCP from Brazil [54]. *S. epidermidis* and *Mammaliicoccus sciuri* (formerly *S. sciuri*) were found in fresh BCP from Bulgaria [23]. In Turkish BCP, *Enterococcus casseliflavus* and *E. faecalis* were found with high reads by metagenomic analysis and in BSP only *E. faecalis* was dominant [47].

In dried BCP from Bulgaria, also *Pseudomonas oryzihabitans* (formerly *Flavimonas oryzihabitans*; Psedomonadaceae family), *Achromobacter denitrificans* (Alcaligenaceae family), *Arthrobacter globiformis* (Micrococaceae family), *Psychrobacter phenylpyruvicus* (Moraxellaceae family) were found [23]. These bacteria are interesting. *Pseudomonas oryzihabitans* is soil and saprophytic organism, which can cause human disease, particularly in immunocompromised patients [68]. Representatives of the families Alcaligenaceae, Micrococaceae and Moraxellaceae are typical for soil environment [69].

11.4 Microscopic Fungi (MF)

Microscopic fungi are eukaryotic microorganisms. They successfully colonize the terrestrial environment and efficiently utilize solid substrates by growing over their surfaces and penetrating into their matrices [70]. There are two main groups of microscopic fungi – yeasts and microscopic filamentous fungi (moulds). They are mainly divided by their cellular morphology. Microscopic fungi are often responsible for changes in organoleptic properties of the products [16]. Microscopic fungi were identified by method of cultivation in 60% of dried BCP samples from Portugal [34] and also in 60% of Italian BCP samples identified by metabarcoding and they were commonly associated to plants [43]. In German fresh and stored BCP, Friedle et al. [40] found fungal phyla Ascomycota (71%), Basidiomycota (21%), Motiellellomycota (0.15%) and unclassified were 3.1% of fungi.

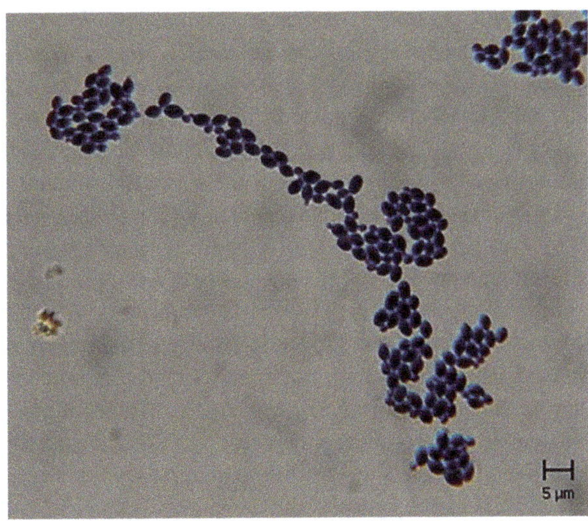

Fig. 11.3 Microscopic view of yeast cells, cultivated from fresh BCP. (Scale: 5 μm, photo: V. Kňazovická, 2011)

11.4.1 Yeasts

Counts of yeasts (Fig. 11.3) in fresh BCP, frozen BCP and BSP was found at level 4–5 log CFU/g [15]. Fresh BCP can contain higher yeast counts in some cases. In Slovenian fresh BCP, yeast count was found at level 3–7 log CFU/g [71]. Yeasts count in fresh pollen from wind-pollinated plants was lower, i.e. at level 2–4 log CFU/g [72]. Gilliam [18] tested plant pollen, BCP and BSP from the almond (*Prunus communis*) and found that *Starmerella magnoliae* (basionym *Torulopsis magnoliae*) was the most common isolate, present in all samples except plant pollen, what indicated its addition by bees. Later, Glushakova et al. [72] tested wind-pollinated plants – birch (*Betula pendula*), hazel (*Corylus avellana*), alder (*Alnus glutinosa*), cat grass (*Dactylis glomerata*) and timothy grass (*Phleum pratense*) in Russia and found twenty species in their pollen – *Rhodotorula mucilaginosa* (in 95% of samples), *Filobasidium magnum* (formerly *Cryptococcus magnus*, in 83.2% of samples), *Vishniacozyma victoriae* (formerly *Cryptococcus victoriae*, in 59.2% of samples), *Yarrowia lipolytica* (formerly *Candida oleophila*, in 43% of samples), *Candida zeylanoides* (in 42% of samples), *Debaryomyces hansenii* (in 33.5% of samples), *Candida friedrichii*, *Meyerozyma guilliermondii*, *Wickerhamomyces anomalus*, *Candida sake*, *Candida albicans* (these last five species: <30% > 20%), *Candida maltose*, *C. parapsilosis*, *C. glabrata*, *Metschnikowia pulcherrima*, *M. pimensis* (formerly *Candida pimensis*), *Ogataea cecidiorum* (these last six species: <20% > 10%) and less than 10%: *Candida tropicalis*, *Cryptococcus tephrensis* and *Filobasidium wieringae* (formerly *Cryptococcus wieringae*). In Brazilian dried BCP, yeasts were detected in 57.7% of samples and fourteen species were identified using biochemical system from yeast colonies on Petri plates: *Starmerella magnoliae*, *Candida parapsilopsis*, *Pichia norvegensis* (formerly *Candida norvegensis*), *Issatchenkia orientalis* (formerly *Candida krusei*),

Starmerella stellata (formerly *Candida stellata*), *Zygosaccharomyces bailii*, *Z. rouxii*, *Z. lentus*, *Rhodotorula mucilaginosa*, *Hanseniaspora uvarum*, *Vanrija humicola* (formerly *Cryptococcus humicola*), *Debaryomyces hansenii*, *Saccharomyces cerevisiae* and *Pichia membranifaciens* [54].

Some yeasts are natural part of plants including pollen and some of them are added to pollen by bees or from the hive environment. Besides bacteria, yeasts as *Saccharomyces* spp. are also responsible for fermentation during the transformation of BCP to BSP [20]. *Candida* spp. are widespread yeasts, often found and/or used in fermented products [73]. The ability of osmophilic yeasts from fermenting honey, brood comb, pollen and flower parts was proven in term of glycerol production from glucose [74]. Park et al. [75] tested biochemical characteristics of osmophilic yeasts isolated from pollen and honey and found that some of them are able to convert sucrose not only to polyols, but also to fructrooligosaccharides (FOS). FOS are considered as prebiotics, substances that should selectively stimulate the growth of desirable bacteria such as bifidobacteria and lactobacilli, however commercial available FOS may also stimulate other faecal bacteria such as clostridia, gram-negative bacteria, or *Escherichia coli* [13]. In Italian BCP, the most frequent fungi were *Metschnikowia* spp., widespread yeasts able to grow in floral nectar that is an environment of high sucrose concentrations (400 g/l), the next yeasts found were *Candida* spp. [43]. *Zygosaccharomyces* spp. dominated in BSP of German origin and was increased in BCP during the storage at 30 °C [40]. In general, number of yeasts isolates and species could decrease with time and optimal storage [18]. *Zygosaccharomyces* spp. can spoil BCP in warm and humid storage conditions, as they producing ethanol or carbon dioxide from sugar and their counts are not increasing at 4 °C [40]. Negative impact of yeast in pollen on humans is their potential ability of increasing the allergenic effect of pollen [72].

11.4.2 Microscopic Filamentous Fungi (MFF)

In Slovak BCP from rape (*Brassica napus*), common poppy (*Papaver somniferum*) and sunflower (*Helianthus annuus*) processed by freezing, drying or UV treatment, average counts of microscopic fungi at level 3–4 log CFU/g were found regardless of treatment [35]. In Slovak BCP from early spring (before *Brassica napus* blooming), from rape (*B. napus*) and from late spring (after *B. napus* blooming), MFF counts at level 4–5 log CFU/g were found and after fermentation, the counts were significantly lower as well as MFF diversity [33]. In fresh and dried BCP from Italy, a little bit lower mean values of microscopic fungi were found: 2.7 ± 1.0 and 2.4 ± 0.8 log CFU/g, respectively [16]. In sixty two dried BCP from Brazil, average values of MF were found at level 2 log CFU/g, ranged from <1.00 log CFU/g to 3 log CFU/g [42]. In dried BCP from Serbia, MFF counts were detected in 38.5% of samples at level 3–4 log CFU/g [76]. MFF counts are maintained in the freezer, because only small MFF decrease was recorded between the fresh BCP and BCP frozen for 1 year, counts in the samples stored at room temperature and cold store

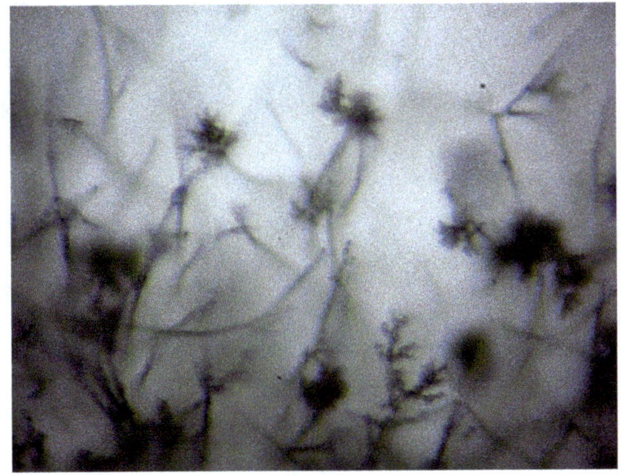

Fig. 11.4 View of the colony of *Cladosporium* sp., cultivated from pollen using magnifying lense. (Magnification 10 × 10, photo: V. Kňazovická, 2008)

decreased (from the level 3 log CFU/g to the level 1 log CFU/g); the same result can be seen in the diversity of fungi: 9–10 genera of fungi were observed in fresh and frozen BCP, 7 in pollen from cold store, 3 in pollen stored at room temperature and 2 in BSP [15].

In Slovak BCP, *Cladosporium* spp. (Fig. 11.4) occurred with the highest relative density, followed by *Penicillium* spp. and *Alternaria* spp., both with higher relative density; while *Arthrinium* spp., *Aspergillus* spp., *Aureobasidium* spp., *Botrytis cinerea*, *Epicoccum* spp. and *Mycelium sterilium* were also identified, but in low relative density [33]. In the similar study, the most frequent isolates in Slovak BCP were *Mucor mucedo*, *Alternaria alternata*, *Mucor hiemalis*, *Aspergillus fumigatus* and *Cladosporium cladosporioides* [35]. *Cladosporium* spp. were also detected as the most abundant fungus in German fresh BCP [40]. In fresh BCP samples from Lithuania, the most prevalent MFF were *Fusarium* spp., *Penicillium* spp., *Alternaria* spp. and *Mucor* spp. [77]. In next study [78], the most prevalent fungi in Lithuanian pollen were *Alternaria* spp., *Penicillium* spp., *Cladosporium* spp. and *Acremonium* spp. In Italian pollen, the most frequent fungi were the plant-pathogen *Mycosphaerella* spp., *Aureobasidium* spp. and *Alternaria* spp., less frequent were *Aspergillus* spp., all identified by DNA metabarcoding [43]. In Serbian pollen, *Mucor* spp., *Penicillium* spp. and *Alternaria* spp. were identified as the most frequent and *Rhizopus* spp., *Aspergillus flavus*, *Trichoderma* spp. and *Fusarium* spp. as less frequent [76].

Microscopic filamentous fungi could also represent a safety problem since they can produce allergenic substances and mycotoxins [16]. Mycotoxins can cause overt clinical mycotoxicosis, or supress immune functions, [79–81] or stimulate immune responses [80–82]. The reaction depends on concrete mycotoxin, its amount and host cell/organism. Key element for mycotoxin-free products is low water activity, which prevents moulds growth and mycotoxin production [83]. From mycotoxin-producing moulds, genera *Alternaria*, *Aspergillus*, *Fusarium* and *Penicillium* are the most common in the pollen [11, 33], where the *Aspergillus* and *Penicillium* genera are able to grow under conditions of lower water activity [33].

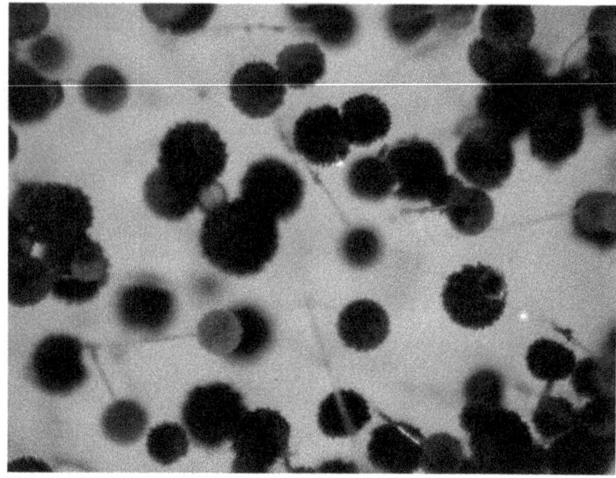

Fig. 11.5 View of the colony of *Aspergillus niger* complex, cultivated from pollen using magnifying lense. (Magnification 10 × 10, photo: V. Kňazovická, 2008)

From aspergilli, *Aspergillus flavus*, *A. fumigatus*, *A. nidulans* (formerly *Emericella nidulans*), *A. niger* complex (Fig. 11.5), *A. terreus*, *A. ochraceus*, *A. parasiticus* and *A. amstelodami* (formerly *Eurotium amstelodami*) were found in Slovak, Italian, Lithuanian and Serbian BCP [35, 76, 78, 84, 85]. From *Penicillium* spp., *P. brevicompactum*, *P. chrysogenum* and *P. verrucosum* were detected in Italian, Slovakian and Lithuanian BCP [35, 78, 85]. From *Fusarium* spp., *F. sporotrichioides*, *F. oxysporum*, *F. proliferatum*, *F. fujikuroi* (formerly *F. verticillioides*) and *F. graminearum* were found in Slovakian and Lithuanian BCP [35, 78]. González et al. [70] detected in Spanish and Argentinian BCP *Penicillium verrucosum*, *Aspergillus niger* complex, *A. carbonarius*, *A. ochraceus*, *A. flavus*, *A. parasiticus* and *Alternaria* spp. as potential producers of mycotoxins. Friedle et al. [40] found that location influenced the present mycobiota in German BCP and during the storage *Cladosporium* spp. and *Mycosphaerella* spp. decreased, while *Aspergillus* spp. increased during the storage at 30 °C, which could negatively influenced. In general, increased spread of MFF is influenced by intensive climate change [11, 86].

11.5 Viruses

Pollen acts as unique vehicle also for viruses transporting [87], including pollen-mediated viruses, which cause serious economic losses in fruit industry, because of viruses of plant diseases [88]. Plant-to-plant transmissions of viruses require vectors, e.g. insect [45, 87].

Fetters et al. [87] collected pollen directly from the flowers in several parts of USA and found pollen-associated viruses of viral families Bromoviridae, Secoviridae and Partitiviridae and several other viral families not previously known to pollen-associated.

Viruses, that cause bee diseases, are also present in bee colonies including healthy ones. Virome of healthy bee colonies, where *Varroa destructor* was under control and no other bee diseases was observed, DWV (*Deformed wing virus*) was highly prevalent [89]. DWV is considered as virus typical for pollen and later bee pathogen [87]. Caesar et al. [90] analysed healthy and unhealthy stingless bee colonies in Brazil and concluded that possible factor of viruses spread is the weakness of the colony.

11.6 Relations Between Pollen and Bee Pathogens

In scientific literature, there were published many studies about antimicrobial properties of pollen [5, 91–100]. On the other side, as we mentioned above, pollen contains relatively high amounts and large diversity of microorganisms. In some cases, presence of pathogens raises the concerns about serious bee diseases outbreaks. However, pollen as well as healthy bee colonies have effective mechanisms against pathogen development. Bacteria and fungi can affect as insect symbionts through the linking to their physiology and immature development or as a part of their defence system against parasites and diseases [101]. Fatty acids in the pollen have also antimicrobial properties, mainly against bee pathogens, as *Paenibacillus larvae*, causative agent of American foulbrood [102]. Current research is also aimed to interactions between microorganisms and role of their metabolites [103, 104]. Pollen processed by bees is also influences by microbiota from nectar, honey, bees, and hive environment. And these microorganisms produce a range of defence molecules [103]. *Apilactobacillus kunkeei* showed antibacterial activity against *Melissococcus plutonius*, one of causative agents of European foulbrood, probably due to special antibacterial peptides or proteins [61]. In general, lactobacilli play a central role in bee metabolism, nutrition and immune functions [27]. The lack of growth of pathogenic microorganisms that is observed in BSP may be the result of accumulation of various metabolites of fermentation process carried out also by LAB [5]. Hbs LAB, mainly *Apilactobacillus kunkeei,* are considered as perspective bee probiotics, however there were observed only low effectiveness in the field studies, e.g. against *Paenibacillus larvae* [105], probably due to various factors including level of pesticides and contaminants in bee surrounding [106]. In term of non-gut microbial communities associated with honeybees, many interactions and roles are still unclear [107].

Acknowledgements This publication was supported by the Operational Program Integrated Infrastructure within the project: Sustainable smart farming systems taking into account the future challenges 313011W112, co-financed by the European Regional Development Fund.

References

1. Linskens HF, Jorde W (1997) Pollen as food and medicine—a review. Econ Bot 51:78–86. https://doi.org/10.1007/BF02910407
2. Campos MGR, Frigerio C, Lopes J, Bogdanov S (2010) What is the future of bee-pollen? J ApiProduct ApiMedical Sci 2:131–144. https://doi.org/10.3896/IBRA.4.02.4.01
3. Komosinska-Vassev K, Olczyk P, Kaźmierczak J, Mencner L, Olczyk K (2015) Bee pollen: chemical composition and therapeutic application. Evid Based Complement Alternat Med 2015:e297425. https://doi.org/10.1155/2015/297425
4. Denisow B, Denisow-Pietrzyk M (2016) Biological and therapeutic properties of bee pollen: a review. J Sci Food Agric 96:4303–4309. https://doi.org/10.1002/jsfa.7729
5. Kieliszek M, Piwowarek K, Kot AM, Błażejak S, Chlebowska-Śmigiel A, Wolska I (2018) Pollen and bee bread as new health-oriented products: a review. Trends Food Sci Technol 71:170–180. https://doi.org/10.1016/j.tifs.2017.10.021
6. Kostić AŽ, Milinčić DD, Barać MB, Ali Shariati M, Tešić ŽL, Pešić MB (2020) The application of pollen as a functional food and feed ingredient—the present and perspectives. Biomol Ther 10:84. https://doi.org/10.3390/biom10010084
7. Bakour M, Laaroussi H, Ousaaid D, El Ghouizi A, Es-Safi I, Mechchate H, Lyoussi B (2022) Bee bread as a promising source of bioactive molecules and functional properties: an up-to-date review. Antibiotics 11:203. https://doi.org/10.3390/antibiotics11020203
8. Bakour M, Laaroussi H, Ferreira-Santos P, Genisheva Z, Ousaaid D, Teixeira JA, Lyoussi B (2022) Exploring the palynological, chemical, and bioactive properties of non-studied bee pollen and honey from Morocco. Molecules 27:5777. https://doi.org/10.3390/molecules27185777
9. Bogdanov S (2006) Contaminants of bee products. Apidologie 37:1–18. https://doi.org/10.1051/apido:2005043
10. Ačkar Đ, Flanjak I (2020) Safety aspects of bee pollen use in nutrition. Hrana U Zdr Boles Znan-Stručni Časopis Za Nutr Dijetetiku 9:63–68
11. Végh R, Csóka M, Sörös C, Sipos L (2021) Food safety hazards of bee pollen – a review. Trends Food Sci Technol 114:490–509. https://doi.org/10.1016/j.tifs.2021.06.016
12. Zafeiraki E, Kasiotis KM, Nisianakis P, Manea-Karga E, Machera K (2022) Occurrence and human health risk assessment of mineral elements and pesticides residues in bee pollen. Food Chem Toxicol 161:112826. https://doi.org/10.1016/j.fct.2022.112826
13. Bunešová V, Vlková E, Rada V, Kňazovická V, Ročková Š, Geigerová M, Božik M (2012) Growth of infant fecal bacteria on commercial prebiotics. Folia Microbiol (Praha) 57:273–275. https://doi.org/10.1007/s12223-012-0123-8
14. Ambika Manirajan B, Ratering S, Rusch V, Schwiertz A, Geissler-Plaum R, Cardinale M, Schnell S (2016) Bacterial microbiota associated with flower pollen is influenced by pollination type, and shows a high degree of diversity and species-specificity: the bacterial microbiota of flower pollen. Environ Microbiol 18:5161–5174. https://doi.org/10.1111/1462-2920.13524
15. Kňazovická V, Kačániová M, Miluchová M, Gábor M, Dovičičová M, Melich M, Trakovická A (2011) Honey and microorganisms: monitoring of microorganisms in Slovak honey by molecular-biological methods in relation to physico-chemical properties, 1st edn. Lambert Academic Publishing, Saarbrucken
16. Nuvoloni R, Meucci V, Turchi B, Sagona S, Fratini F, Felicioli A, Cerri D, Pedonese F (2021) Bee-pollen retailed in Tuscany (Italy): labelling, palynological, microbiological, and mycotoxicological profile. LWT 140:110712. https://doi.org/10.1016/j.lwt.2020.110712
17. Alippi AM, Fernández LA, López AC (2022) Diversity of aerobic spore-forming bacteria isolated from fresh bee pollen intended for human consumption in Argentina. J Apic Res 61:392–399. https://doi.org/10.1080/00218839.2021.1960747
18. Gilliam M (1979) Microbiology of pollen and bee bread: the yeasts. Apidologie 10:43–53. https://doi.org/10.1051/apido:19790106

19. Krell R (1996) Value-added products from beekeeping. Food and Agriculture Organization of the United Nations, Rome
20. Barta DG, Cornea-Cipcigan M, Margaoan R, Vodnar DC (2022) Biotechnological processes simulating the natural fermentation process of bee bread and therapeutic properties—an overview. Front Nutr 9. https://doi.org/10.3389/fnut.2022.871896
21. Zhang Z, Cao H, Chen C, Chen X, Wei Q, Zhao F (2017) Effects of fermentation by *Ganoderma lucidum* and *Saccharomyces cerevisiae* on rape pollen morphology and its wall. J Food Sci Technol 54:4026–4034. https://doi.org/10.1007/s13197-017-2868-1
22. Menezes C, Paludo CR, Pupo MT (2018) A review of the artificial diets used as pot-pollen substitutes. In: Vit P, Pedro SRM, Roubik DW (eds) Pot-pollen in stingless bee melittology. Springer International Publishing, Cham, pp 253–262
23. Dinkov D (2021) Bacterial contamination of vacuum stored flower bee pollen. J Microbiol Biotechnol Food Sci 2018:387–391
24. Oliveira D d J, Rodrigues dos Santos D, Andrade BR, Nascimento AS d, Oliveira da Silva M, da Cruz Mercês C, Lucas CIS, Cavalcante da Silva SMP, Dib de Carvalho P, Silva F d L, Estevinho LM, Carvalho CAL d (2021) Botanical origin, microbiological quality and physicochemical composition of the *Melipona scutellaris* pot-pollen ("samburá") from Bahia (Brazil) region. J Apic Res 60:457–469. https://doi.org/10.1080/00218839.2020.1797271
25. de Sousa BFS, Santos GG, Mesquita JA, Nascimento CADJ, Freire GC, Santos VTF, Carvalho-Zilse GA, Barros J d RS (2022) Physicochemical composition and microbiological quality of honey produced from Tiúba, *Melipona fasciculata* (Apidae, Meliponini) in Brazil. J Apic Res 1–9. https://doi.org/10.1080/00218839.2022.2143145
26. MaD B-A, Fraberger V, Schreiner M, Domig KJ, D'Amico S (2020) Safety aspects of stingless bee pot-pollen from the Philippines. Bodenkult J Land Manag Food Environ 71:87–100. https://doi.org/10.2478/boku-2020-0009
27. Dharampal PS, Diaz-Garcia L, Haase MAB, Zalapa J, Currie CR, Hittinger CT, Steffan SA (2020) Microbial diversity associated with the pollen stores of captive-bred bumble bee colonies. Insects 11:250. https://doi.org/10.3390/insects11040250
28. Campos MGR, Bogdanov S, de Almeida-Muradian LB, Szczesna T, Mancebo Y, Frigerio C, Ferreira F (2008) Pollen composition and standardisation of analytical methods. J Apic Res 47:154–161. https://doi.org/10.1080/00218839.2008.11101443
29. Bogdanov S (2017) Pollen: collection, harvest, compostion, quality. In: The bee pollen book
30. Kňazovická V, Mašková Z, Vlková E, Švejstil R, Salmonová H, Ivanišová E, Gažarová M, Gamráthová I, Repková M, Tokár M, Ducková V, Kročko M, Čanigová M, Kačániová M (2018) Pollen can – testing of bee pollen fermentation in model conditions. J Microbiol Biotechnol Food Sci 8:805–811. https://doi.org/10.15414/jmbfs.2018.8.2.805-811
31. Luo X, Dong Y, Gu C, Zhang X, Ma H (2021) Processing technologies for bee products: an overview of recent developments and perspectives. Front Nutr 8:727181. https://doi.org/10.3389/fnut.2021.727181
32. Anjos O, Paula VB, Delgado T, Estevinho LM (2019) Influence of the storage conditions on the quality of bee pollen. https://doi.org/10.13080/z-a.2019.106.012
33. Mašková Z, Kňazovická V, Tančinová D, Panáková S (2019) Production of pollen cans by fermentation of bee pollen in model conditions with regard to filamentous micromycetes occurrence. J Microbiol Biotechnol Food Sci 8:1223–1227. https://doi.org/10.15414/jmbfs.2019.8.5.1223-1227
34. Estevinho LM, Rodrigues S, Pereira AP, Feás X (2012) Portuguese bee pollen: palynological study, nutritional and microbiological evaluation. Int J Food Sci Technol 47:429–435. https://doi.org/10.1111/j.1365-2621.2011.02859.x
35. Kačániová M, Juráček M, Chlebo R, Kňazovická V, Kadasi-Horáková M, Kunová S, Lejková J, Haščík P, Mareček J, Šimko M (2011) Mycobiota and mycotoxins in bee pollen collected from different areas of Slovakia. J Environ Sci Health Part B 46:623–629. https://doi.org/10.1080/03601234.2011.589322

36. Di Cagno R, Filannino P, Cantatore V, Gobbetti M (2019) Novel solid-state fermentation of bee-collected pollen emulating the natural fermentation process of bee bread. Food Microbiol 82:218–230. https://doi.org/10.1016/j.fm.2019.02.007
37. Zhang H, Zhu X, Huang Q, Zhang L, Liu X, Liu R, Lu Q (2023) Antioxidant and anti-inflammatory activities of rape bee pollen after fermentation and their correlation with chemical components by ultra-performance liquid chromatography-quadrupole time of flight mass spectrometry-based untargeted metabolomics. Food Chem 409:135342. https://doi.org/10.1016/j.foodchem.2022.135342
38. Tolba O, Earle JAP, Millar BC, Rooney PJ, Moore JE (2007) Speciation of *Bacillus* spp. in honey produced in Northern Ireland by employment of 16S rDNA PCR and automated DNA sequencing techniques. World J Microbiol Biotechnol 23:1805–1808. https://doi.org/10.1007/s11274-007-9425-4
39. Kňazovická V, Gábor M, Miluchová M, Bobko M, Medo J (2019) Diversity of bacteria in Slovak and foreign honey, with assessment of its physico-chemical quality and counts of cultivable microorganisms. J Microbiol Biotechnol Food Sci 9:414–421. https://doi.org/10.15414/jmbfs.2019.9.special.414-421
40. Friedle C, D'Alvise P, Schweikert K, Wallner K, Hasselmann M (2021) Changes of micro-organism composition in fresh and stored bee pollen from southern Germany. Environ Sci Pollut Res 28:47251–47261. https://doi.org/10.1007/s11356-021-13932-4
41. Campos MG, Anjos O, Chica M, Campoy P, Nozkova J, Almaraz-Abarca N, Barreto LMRC, Nordi JC, Estevinho LM, Pascoal A, Paula VB, Chopina A, Dias LG, Tešić ŽL, Mosić MD, Kostić AŽ, Pešić MB, Milojković-Opsenica DM, Sickel W, Ankenbrand MJ, Grimmer G, Steffan-Dewenter I, Keller A, Förster F, Tananaki CH, Liolios V, Kanelis D, Rodopoulou M-A, Thrasyvoulou A, Paulo L, Kast C, Lucchetti MA, Glauser G, Lokutova O, de Almeida-Muradian LB, Szczęsna T, Carreck NL (2021) Standard methods for pollen research. J Apic Res 60:1–109. https://doi.org/10.1080/00218839.2021.1948240
42. de Arruda VAS, Vieria dos Santos A, Figueiredo Sampaio D, da Silva AE, de Castro Peixoto AL, Estevinho MLF, Bicudo de Almeida-Muradian L (2017) Microbiological quality and physicochemical characterization of Brazilian bee pollen. J Apic Res 56:231–238. https://doi.org/10.1080/00218839.2017.1307715
43. De Jesus IL, Merlanti R, Lucatello L, Bisutti V, Carraro L, Larini I, Vitulo N, Cardazzo B, Capolongo F (2021) Natural contaminants in bee pollen: DNA metabarcoding as a tool to identify floral sources of pyrrolizidine alkaloids and fungal diversity. Food Res Int 146:110438. https://doi.org/10.1016/j.foodres.2021.110438
44. Salmonová H, Killer J, Bunešová V, Geigerová M, Vlková E (2018) Cultivable bacteria from *Pectinatella magnifica* and the surrounding water in South Bohemia indicate potential new Gammaproteobacterial, Betaproteobacterial and Firmicutes taxa. FEMS Microbiol Lett 365:fny118. https://doi.org/10.1093/femsle/fny118
45. Dolja VV, Krupovic M, Koonin EV (2020) Deep roots and splendid boughs of the global plant virome. Annu Rev Phytopathol 58:23–53. https://doi.org/10.1146/annurev-phyto-030320-041346
46. Gilliam M (1979) Microbiology of pollen and bee bread: the genus *Bacillus*. Apidologie 10:269–274. https://doi.org/10.1051/apido:19790304
47. Arserim Ucar DK, Yurt MNZ, Tasbasi BB, Acar EE, Yegín Z, Ozalp VC, Sudagidan M (2022) Identification of bacterial diversity of bee collected pollen and bee bread micro-biota by metagenomic analysis. Acta Vet Eurasia 48:189–199. https://doi.org/10.5152/actavet.2022.22031
48. Leonhardt SD, Peters B, Keller A (2022) Do amino and fatty acid profiles of pollen provisions correlate with bacterial microbiomes in the mason bee *Osmia bicornis*? Philos Trans R Soc B Biol Sci 377:20210171. https://doi.org/10.1098/rstb.2021.0171
49. Elke DC, Paul DV (2004) Genotypic diversity among *Bacillus licheniformis* strains from various sources. FEMS Microbiol Lett 231:91–98. https://doi.org/10.1016/S0378-1097(03)00935-2

50. Eppinger M, Bunk B, Johns MA, Edirisinghe JN, Kutumbaka KK, Koenig SSK, Huot Creasy H, Rosovitz MJ, Riley DR, Daugherty S, Martin M, Elbourne LDH, Paulsen I, Biedendieck R, Braun C, Grayburn S, Dhingra S, Lukyanchuk V, Ball B, Ul-Qamar R, Seibel J, Bremer E, Jahn D, Ravel J, Vary PS (2011) Genome sequences of the biotechnologically important *Bacillus megaterium* strains QM B1551 and DSM319. J Bacteriol 193:4199–4213. https://doi.org/10.1128/JB.00449-11
51. Palmisano MM, Nakamura LK, Duncan KE, Istock CA, Cohan FM (2001) *Bacillus sonorensis* sp. nov., a close relative of *Bacillus licheniformis*, isolated from soil in the Sonoran Desert. Arizona Int J Syst Evol Microbiol 51:1671–1679. https://doi.org/10.1099/00207713-51-5-1671
52. Mohandas A, Raveendran S, Parameswaran B, Abraham A, Athira RSR, Mathew AK, Pandey A (2018) Production of pectinase from *Bacillus sonorensis* MPTD1. Food Technol Biotechnol 56:110–116. https://doi.org/10.17113/ftb.56.01.18.5477
53. Sinacori M, Francesca N, Alfonzo A, Cruciata M, Sannino C, Settanni L, Moschetti G (2014) Cultivable microorganisms associated with honeys of different geographical and botanical origin. Food Microbiol 38:284–294. https://doi.org/10.1016/j.fm.2013.07.013
54. De-Melo AA, Estevinho MLMF, Almeida-Muradian LBD (2015) A diagnosis of the microbiological quality of dehydrated bee-pollen produced in Brazil. Lett Appl Microbiol 61:477–483. https://doi.org/10.1111/lam.12480
55. Lindström M, Heikinheimo A, Lahti P, Korkeala H (2011) Novel insights into the epidemiology of *Clostridium perfringens* type a food poisoning. Food Microbiol 28:192–198. https://doi.org/10.1016/j.fm.2010.03.020
56. Khouri JM, Payne JR, Arnon SS (2018) More clinical mimics of infant botulism. J Pediatr 193:178–182. https://doi.org/10.1016/j.jpeds.2017.09.044
57. Stefka AT, Feehley T, Tripathi P, Qiu J, McCoy K, Mazmanian SK, Tjota MY, Seo G-Y, Cao S, Theriault BR, Antonopoulos DA, Zhou L, Chang EB, Fu Y-X, Nagler CR (2014) Commensal bacteria protect against food allergen sensitization. Proc Natl Acad Sci 111:13145–13150. https://doi.org/10.1073/pnas.1412008111
58. Grieger C, Vařejka F (1990) Mikrobiológia poživatín živočíšneho pôvodu. Príroda, Bratislava
59. König H, Fröhlich J (2017) Lactic acid bacteria. In: König H, Unden G, Fröhlich J (eds) Biology of microorganisms on grapes, in must and in wine. Springer International Publishing, Cham, pp 3–41
60. Olofsson TC, Vásquez A (2008) Detection and identification of a novel lactic acid bacterial flora within the honey stomach of the honeybee *Apis mellifera*. Curr Microbiol 57:356–363. https://doi.org/10.1007/s00284-008-9202-0
61. Endo A, Salminen S (2013) Honeybees and beehives are rich sources for fructophilic lactic acid bacteria. Syst Appl Microbiol 36:444–448. https://doi.org/10.1016/j.syapm.2013.06.002
62. Lamei S, Hu YOO, Olofsson TC, Andersson AF, Forsgren E, Vásquez A (2017) Improvement of identification methods for honeybee specific lactic acid bacteria; future approaches. PLoS One 12:e0174614. https://doi.org/10.1371/journal.pone.0174614
63. Anderson KE, Carroll MJ, Sheehan T, Mott BM, Maes P, Corby-Harris V (2014) Hive-stored pollen of honey bees: many lines of evidence are consistent with pollen preservation, not nutrient conversion. Mol Ecol 23:5904–5917. https://doi.org/10.1111/mec.12966
64. Adeolu M, Alnajar S, Naushad S, Gupta RS (2016) Genome-based phylogeny and taxonomy of the 'Enterobacteriales': proposal for Enterobacterales Ord. Nov. divided into the families Enterobacteriaceae, Erwiniaceae fam. Nov., Pectobacteriaceae fam. Nov., Yersiniaceae fam. Nov., Hafniaceae fam. Nov., Morganellaceae fam. Nov., and Budviciaceae fam. Nov. Int J Syst Evol Microbiol 66:5575–5599. https://doi.org/10.1099/ijsem.0.001485
65. Dutkiewicz J, Mackiewicz B, Kinga Lemieszek M, Golec M, Milanowski J (2016) *Pantoea agglomerans*: a mysterious bacterium of evil and good. Part III. Deleterious effects: infections of humans, animals and plants. Ann Agric Environ Med 23:197–205. https://doi.org/10.5604/12321966.1203878

66. Lorenzi AS, Bonatelli ML, Chia MA, Peressim L, Quecine MC (2022) Opposite sides of *Pantoea agglomerans* and its associated commercial outlook. Microorganisms 10:2072. https://doi.org/10.3390/microorganisms10102072
67. Bartlett A, Padfield D, Lear L, Bendall R, Vos M (2022) A comprehensive list of bacterial pathogens infecting humans. Microbiology 168:001269. https://doi.org/10.1099/mic.0.001269
68. Decker CF, Simon GL, Keiser JF (1991) *Flavimonas oryzihabitans* (*Pseudomonas oryzihabitans*; CDC group Ve-2) bacteremia in the immunocompromised host. Arch Intern Med 151:603–604
69. Joseph SJ, Hugenholtz P, Sangwan P, Osborne CA, Janssen PH (2003) Laboratory cultivation of widespread and previously uncultured soil bacteria. Appl Environ Microbiol 69:7210–7215. https://doi.org/10.1128/AEM.69.12.7210-7215.2003
70. González G, Hinojo MJ, Mateo R, Medina A, Jiménez M (2005) Occurrence of mycotoxin producing fungi in bee pollen. Int J Food Microbiol 105:1–9. https://doi.org/10.1016/j.ijfoodmicro.2005.05.001
71. Šimunović K, Abramovič H, Lilek N, Angelova M, Podržaj L, Smole Možina S (2019) Microbiological quality, antioxidative and antimicrobial properties of Slovenian bee pollen. AGROFOR 4:82–92. https://doi.org/10.7251/AGRENG1901005Q
72. Glushakova AM, Kachalkin AV, Zheltikova TM, Chernov IY (2015) Yeasts associated with wind-pollinated plants—leading pollen allergens in Central Russia. Microbiology 84:722–725. https://doi.org/10.1134/S0026261715050082
73. Hommel RK (2014) CANDIDA | introduction. In: Batt CA, Tortorello ML (eds) Encyclopedia of food microbiology, 2nd edn. Academic, Oxford, pp 367–373
74. Spencer JFT, Sallans HR (1956) Production of polyhydric alcohols by osmophilic yeasts. Can J Microbiol 2:72–79. https://doi.org/10.1139/m56-011
75. Park YK, Koo MH, Oliveira IM d A (1996) Biochemical characteristics of osmophilic yeasts isolated from pollens and honey. Biosci Biotechnol Biochem 60:1872–1873. https://doi.org/10.1271/bbb.60.1872
76. Kostić AŽ, Petrović TS, Krnjaja VS, Nedić NM, Tešić ŽL, Milojković-Opsenica DM, Barać MB, Stanojević SP, Pešić MB (2017) Mold/aflatoxin contamination of honey bee collected pollen from different Serbian regions. J Apic Res 56:13–20. https://doi.org/10.1080/00218839.2016.1259897
77. Sinkevičienė J, Marcinkevičienė A, Baliukonienė V, Jovaišienė J (2019) Fungi and mycotoxins in fresh bee pollen. Proc Int Sci Conf Rural Dev:69–72
78. Sinkevičienė J, Burbulis N, Baliukonienė V (2021) The influence of storage conditions on bee pollen contamination by microscopic fungi and their mycotoxins. Zemdirb-Agric 108:159–164. https://doi.org/10.13080/z-a.2021.108.021
79. Corrier DE (1991) Mycotoxicosis: mechanisms of immunosuppression. Vet Immunol Immunopathol 30:73–87. https://doi.org/10.1016/0165-2427(91)90010-A
80. Sharma RP (1993) Immunotoxicity of Mycotoxins. J Dairy Sci 76:892–897. https://doi.org/10.3168/jds.S0022-0302(93)77415-9
81. Surai PF, Mézes M (2005) Mycotoxins and immunity: theoretical consideration and practical applications. Prax Vet 53:71–88
82. Liu B-H, Chi J-Y, Hsiao Y-W, Tsai K-D, Lee Y-J, Lin C-C, Hsu S-C, Yang S-M, Lin T-H (2010) The fungal metabolite, citrinin, inhibits lipopolysaccharide/interferon-γ-induced nitric oxide production in glomerular mesangial cells. Int Immunopharmacol 10:1608–1615. https://doi.org/10.1016/j.intimp.2010.09.017
83. Marin S, Ramos AJ, Cano-Sancho G, Sanchis V (2013) Mycotoxins: occurrence, toxicology, and exposure assessment. Food Chem Toxicol 60:218–237. https://doi.org/10.1016/j.fct.2013.07.047
84. Kňazovická V, Kačániová M, Dovičičová M, Melich M, Barboráková Z, Kadási-Horáková M, Mareček J (2011) Microbial quality of honey mixture with pollen. Potravinárstvo Slovak Repub 5:27–32

85. Nardoni S, D'Ascenzi C, Rocchigiani G, Moretti V, Mancianti F (2016) Occurrence of moulds from bee pollen in Central Italy – a preliminary study. Ann Agric Environ Med 23. https://doi.org/10.5604/12321966.1196862
86. Kostić AŽ, Milinčić DD, Petrović TS, Krnjaja VS, Stanojević SP, Barać MB, Tešić ŽL, Pešić MB (2019) Mycotoxins and mycotoxin producing fungi in pollen: review. Toxins 11:64. https://doi.org/10.3390/toxins11020064
87. Fetters AM, Cantalupo PG, Wei N, Robles MTS, Stanley A, Stephens JD, Pipas JM, Ashman T-L (2022) The pollen virome of wild plants and its association with variation in floral traits and land use. Nat Commun 13:523. https://doi.org/10.1038/s41467-022-28143-9
88. Lee H-J, Jeong R-D (2022) Metatranscriptomic analysis of plant viruses in imported pear and kiwifruit pollen. Plant Pathol J 38:220–228. https://doi.org/10.5423/PPJ.OA.03.2022.0047
89. Kadlečková D, Tachezy R, Erban T, Deboutte W, Nunvář J, Saláková M, Matthijnssens J (2022) The virome of healthy honey bee colonies: ubiquitous occurrence of known and new viruses in bee populations. mSystems 7:e00072–e00022. https://doi.org/10.1128/msystems.00072-22
90. Caesar L, Cibulski SP, Canal CW, Blochtein B, Sattler A, Haag KL (2019) The virome of an endangered stingless bee suffering from annual mortality in southern Brazil. J Gen Virol 100:1153–1164. https://doi.org/10.1099/jgv.0.001273
91. Kačániová M, Vuković N, Chlebo R, Haščík P, Rovná K, Cubon J, Džugan M, Pasternakiewicz A (2012) The antimicrobial activity of honey, bee pollen loads and beeswax from Slovakia. Arch Biol Sci 64:927–934
92. Morais M, Moreira L, Feás X, Estevinho LM (2011) Honeybee-collected pollen from five Portuguese natural parks: Palynological origin, phenolic content, antioxidant properties and antimicrobial activity. Food Chem Toxicol 49:1096–1101. https://doi.org/10.1016/j.fct.2011.01.020
93. Fatrcová-Šramková K, Nôžková J, Kačániová M, Máriássyová M, Rovná K, Stričík M (2013) Antioxidant and antimicrobial properties of monofloral bee pollen. J Environ Sci Health Part B 48:133–138. https://doi.org/10.1080/03601234.2013.727664
94. Kačániová M, Vatľák A, Vukovic N, Petrová J, Brindza J, Nôžková J, Fatrcová-Šramková K (2014) Antimicrobial activity of bee collected pollen against clostridia. Sci Pap Anim Sci Biotechnol 47:362–365
95. Fatrcová-Šramková K, Nôžková J, Máriássyová M, Kačániová M (2016) Biologically active antimicrobial and antioxidant substances in the *Helianthus annuus* L. bee pollen. J Environ Sci Health Part B 51:176–181. https://doi.org/10.1080/03601234.2015.1108811
96. Bentrad N, Gaceb-Terrak R, Benmalek Y, Rahmania F (2017) Studies on chemical composition and antimicrobial activities of bioactive molecules from date palm (*Phoenix dactylifera* L.) pollens and seeds. Afr J Tradit Complement Altern Med 14:242–256. https://doi.org/10.4314/ajtcam.v14i3
97. AbdElsalam E, Foda HS, Abdel-Aziz MS, Abd FK (2018) Antioxidant and antimicrobial activities of Egyptian bee pollen. Middle East J Appl Sci 8:1248–1255
98. Bridi R, Atala E, Pizarro PN, Montenegro G (2019) Honeybee pollen load: phenolic composition and antimicrobial activity and antioxidant capacity. J Nat Prod 82:559–565. https://doi.org/10.1021/acs.jnatprod.8b00945
99. Didaras NA, Karatasou K, Dimitriou TG, Amoutzias GD, Mossialos D (2020) Antimicrobial activity of bee-collected pollen and beebread: state of the art and future perspectives. Antibiotics 9:811. https://doi.org/10.3390/antibiotics9110811
100. Soares de Arruda VA, Vieria dos Santos A, Figueiredo Sampaio D, da Silva AE, de Castro Peixoto AL, Estevinho LM, de Almeida-Muradian LB (2021) Brazilian bee pollen: phenolic content, antioxidant properties and antimicrobial activity. J Apic Res 60:775–783. https://doi.org/10.1080/00218839.2020.1840854
101. Westreich LR, Westreich ST, Tobin PC (2022) Bacterial and fungal symbionts in pollen provisions of a native solitary bee in urban and rural environments. Microb Ecol. https://doi.org/10.1007/s00248-022-02164-9

102. Manning R (2001) Fatty acids in pollen: a review of their importance for honey bees. Bee World 82:60–75. https://doi.org/10.1080/0005772X.2001.11099504
103. Brudzynski K (2021) Honey as an ecological reservoir of antibacterial compounds produced by antagonistic microbial interactions in plant nectars, honey and honey bee. Antibiotics 10:551. https://doi.org/10.3390/antibiotics10050551
104. Martín JF, Liras P (2019) Harnessing microbiota interactions to produce bioactive metabolites: communication signals and receptor proteins. Curr Opin Pharmacol 48:8–16. https://doi.org/10.1016/j.coph.2019.02.014
105. Lamei S, Stephan JG, Nilson B, Sieuwerts S, Riesbeck K, de Miranda JR, Forsgren E (2020) Feeding honeybee colonies with honeybee-specific lactic acid bacteria (Hbs-LAB) does not affect Colony-level Hbs-LAB composition or *Paenibacillus larvae* spore levels, although American foulbrood affected Colonies Harbor a more diverse Hbs-LAB community. Microb Ecol 79:743–755. https://doi.org/10.1007/s00248-019-01434-3
106. Kňazovická V, Lidiková J, Jakabová S, Kročko M, Bellova S, Staroň M, Benčaťová S, Hernández SV, Sanchez AC, Ramos Y (2021) Application of microbial agents to control diseases in agriculture with a focus to beekeeping: a review. Slovak J Anim Sci 54:99–105
107. Smutin D, Lebedev E, Selitskiy M, Panyushev N, Adonin L (2022) Micro "bee" ota: honey bee normal microbiota as a part of superorganism. Microorganisms 10:2359. https://doi.org/10.3390/microorganisms10122359

Chapter 12
Physical and Bioprocessing Techniques for Improving Nutritional, Microbiological, and Functional Quality of Bee Pollen

Carlos Alberto Fuenmayor, Carlos Mario Zuluaga-Domínguez, and Martha Cecilia Quicazán

12.1 Bee Pollen for Human Nutrition: Value and Limitations

The pollen grains are the male gametophyte in the sexual reproduction of flowering plants and correspond to microscopic bodies of highly variable morphology [1]. Bee pollen, or corbicular pollen, is obtained when worker bees gather pollen grains from flowers, adhering them to their corbiculae and forming pellets of different colors that are then transported to the hive for colony feeding purposes [2, 3]. This product can be collected by beekeepers using traps at the entrance of the beehive. In tropical countries, the productivity of bee pollen is usually higher due to sustained flowering seasons throughout the year. For instance, in the High-Andean forests of Colombia, harvests are commonly greater than 40 kg per hive per year [4]. However, its production is feasible and widespread everywhere beekeeping is practiced [1, 5].

The composition of bee pollen depends on the botanical and geographical origin; it is generally characterized by a high content of proteins, carbohydrates, dietary fiber, and lipids with a prevalence of polyunsaturated fatty acids, several vitamins, and minerals [6]. Bee pollen, regardless of its origin, contains also phenolic compounds, carotenoids, phytosterols, and other bioactive compounds [7–9].

The chemical composition of bee pollen has been the subject of various reviews [1, 6, 10–13]. From a macronutrient perspective, protein is perhaps the most interesting constituent of pollen. Roughly, the protein content varies between 10% and

C. A. Fuenmayor (✉) · M. C. Quicazán
Universidad Nacional de Colombia – Sede Bogotá – Instituto de Ciencia y Tecnología de Alimentos, Bogotá, D.C, Colombia
e-mail: cafuenmayorb@unal.edu.co; mcquicazand@unal.edu.co

C. M. Zuluaga-Domínguez
Universidad Nacional de Colombia – Sede Bogotá – Facultad de Ciencias Agrarias – Departamento de Desarrollo Rural y Agroalimentario, Bogotá, D.C, Colombia
e-mail: cmzuluagad@unal.edu.co

N. Ecem Bayram et al. (eds.), *Pollen Chemistry & Biotechnology*,
https://doi.org/10.1007/978-3-031-47563-4_12

nearly 50% [1, 6, 14]. The protein fraction of pollen includes many essential amino acids and some enzymes and enzymatic cofactors [6].

Carbohydrates are major constituents of pollen, ranging from approximately 15% to over 60% [1, 6], especially monosaccharides (glucose and fructose) and sucrose, which account for more than 90% of the sugar content [15] and are responsible for its sweet flavor. It also contains polysaccharides in variable amounts, including up to 20% of molecules classifiable as insoluble dietary fiber and starch-like compounds [15, 16]. Interestingly, there is evidence that suggests a potential prebiotic effect of bee pollen, mainly associated with the presence of dietary fiber and phenolic compounds [17, 18].

The lipid content of bee pollen is particularly variable according to its origin, representing between less than 1% and nearly 10% [19]. The main lipids are glycerol esters (oils) both saturated and unsaturated. Commonly, the most abundant fatty acids are palmitic, α-linolenic, linoleic, and oleic acid, but the fatty acid profile depends on the origin. For instance, a study by Gardana et al. [8], found that Colombian pollen was richer in ω–3 fatty acids, compared to European pollen, which contained higher proportions of ω–6 fatty acids. It also contains secondary and associated lipids, including phospholipids, carotenoids, tocopherols, and tocotrienols [6, 20–23], and phytosterols [11]. Indeed, due to the presence of carotenoids in some varieties of bee pollen, it has been proposed as a natural food colorant [3, 23].

The minerals present in bee pollen can also be nutritionally relevant. Total inorganic matter represents between 1% and 6% [14]. Potassium is usually the most concentrated mineral element, followed by phosphorous, sodium, calcium, magnesium, zinc, iron, and copper [1], in variable amounts that depend on the botanical origin.

Among bee pollen micronutrients, in addition to lipophilic vitamins (in particular vitamin E tocopherols and tocotrienols), vitamin C and complex B vitamins, including riboflavin, pantothenic acid, and nicotinic acid, have been widely reported [24]. Indeed, bee pollen typically presents contents of riboflavin (vitamin B2) higher than most plant-origin foods [25].

Besides macro- and micronutrients, bee pollen has a diversity of bioactive compounds, among which phenolic compounds and carotenoids stand out. Flavonoids, usually present in their glycoside forms, and phenolic acids are the main phenolics [6]. Carotenoids include carotenes, some of which have provitamin A activity, and xanthophylls commonly found esterified to fatty acids [8]. Other functional molecules that can be found in bee pollen are squalene and several phytosterols [26]. The complex mixture of vitamins C and E, carotenoids, phenolics, and sterols explain the antioxidant activity and other functionalities that have been associated with bee pollen consumption [9]. However, it is important to notice that these molecules can be labile towards oxidative reactions and thermal decomposition, and therefore can be partially lost as a consequence of heat treatments, drying processing, and storage.

Regardless of its rich composition, the digestibility and bioavailability of the nutrients contained in bee pollen are limited by the wall-like resistant microstructure that covers pollen grains, known as exine, especially for monogastric animals

[5, 27, 28]. It is important to notice that the bees consume pollen as bee bread, i.e., after solid state fermentation/ensiling processes that take place inside the hive [29]. Such processes would help release the pollen grain components, making substances more biologically available [30–32].

The following sections describe how physical treatments and bioprocessing techniques affect the nutritional features of bee pollen and how they may help to improve its value as a safe and nutritious food product for human consumption.

12.2 Bee Pollen Transformation by Physical Processing: Techniques and Effect on Nutritional Properties

Fresh bee pollen has high water activity and high microbial loads, which makes it susceptible to microbial spoilage and potentially compromises its adequacy for human consumption from a safety perspective. Different physical treatments can be applied to bee pollen to improve its quality as a food product, usually involving heat transfer. Some of these treatments may enhance nutrient availability by promoting the breakdown of the exine or wall of pollen grains, but they can also affect the micronutrient or bioactive profile.

There are still few reports dealing with the physical treatment of this food. Drying is the most common operation, however, other processes that could be useful to improve bee pollen quality have been proposed, such as sterilization or high hydrostatic pressure. In this section, a description of some physical treatments to which bee pollen can be subjected is presented.

12.2.1 Bee Pollen Drying

The bee pollen grain, once harvested by the beekeeper, has a moisture content that oscillates between 20% and 30% and a water activity between 0.66 and 0.82 [33], and consequently, it must be subjected immediately to moisture reduction processes until the free water content is reduced to values between 5% and 8% to prevent microbiological contamination and the triggering of enzymatic reactions. Zuluaga-Domínguez et al. [34] studied the effect of bee pollen drying at different temperatures on physicochemical quality, nutritional value, and bioactive features, using multifloral high-Andean bee pollen mainly from *Hypochaeris radicata* (43%) and *Brassica* sp. (34%). Comparative physicochemical characterization of bee pollen subjected to drying at 40 °C, 50 °C, and 60 °C is shown in Table 12.1.

These results showed that simple air-drying treatments reduce moisture to safe values, i.e., moisture content lower than 10%, however, the final stable moisture content is only achieved at 50 °C or 60 °C. As the drying temperature increases, the pH and acidity values approach those reported for fresh bee pollen. The increase in

Table 12.1 Physicochemical features of fresh and air-dried bee pollen [34]

Physicochemical parameter	Fresh bee pollen	Drying temperature 40 °C	50 °C	60 °C
Moisture (g/100 g)	20.4 ± 1.4	7.9 ± 1.0[b]	6.1 ± 0.6[a]	5.2 ± 0.6[a]
Water activity	0.762 ± 0.047	0.431 ± 0.028[b]	0.285 ± 0.036[a]	0.215 ± 0.042[a]
pH	4.16 ± 0.07[b]	4.02 ± 0.03[a]	4.04 ± 0.02[a]	4.11 ± 0.05[b]
Acidity* (meq/kg)	245.6 ± 31.8[a]	305.8 ± 9.3[b]	283.2 ± 12.5[b]	256.9 ± 30.1[a]
Digestibility* (g/100 g)	65.8 ± 10.1[a]	71.3 ± 10.4[a]	73.9 ± 11.7[a]	76.4 ± 9.8[a]
Total phenolics (mg GAE/g)	15.0 ± 2.1[a]	23.5 ± 2.5[b]	23.1 ± 2.8[b]	23.2 ± 2.3[b]
Total carotenoids (mg β-carotene equivalents/kg)	684.5 ± 82.8[b]	583.1 ± 54.0[a]	549.6 ± 68.0[a]	521.8 ± 84.4[a]
Total flavonoids (mg QE/g)	4.99 ± 1.79[a]	7.42 ± 0.53[b]	7.40 ± 0.33[b]	7.70 ± 0.37[b]
FRAP (μmol Trolox/g)	58.5 ± 15.6[a]	82.5 ± 14.1[b]	78.1 ± 10.0[b]	87.2 ± 15.6[b]
TEAC (μmol Trolox/g)	60.8 ± 9.9[a]	68.7 ± 8.3[a]	67.3 ± 8.9[a]	70.3 ± 9.7[a]

*Dry basis. Mean ± standard deviation. Different letters for the same row represent significant differences at a confidence level of 95%. The moisture and water activity values of the fresh bee pollen were not included in the statistical analysis due to their notorious difference from the treated grains. *FRAP* Ferric reducing antioxidant power assay, *TEAC* Trolox equivalent antioxidant capacity, *GAE* equivalents of gallic acid, *QE* equivalents of quercetin

acidity and the reduction in pH found at drying temperatures below 60 °C evidence the action of spoilage microorganisms that, as previously discussed, have a larger population as the temperature decreases. Moreover, a slight effect (although statistically non-significant) of drying temperature on protein digestibility was observed. It is important to mention that such an effect cannot be attributed to a change in the morphology, as air-drying or solar-drying conditions of bee pollen do not appear to affect the exine integrity [35, 36].

Many of the compounds with functional characteristics present in bee pollen, such as vitamins and antioxidants, are thermolabile. Therefore, the increase in temperature during dehydration is expected to induce their degradation [37]. Given the importance of these components, this is an important quality parameter to follow during drying [24]. Notwithstanding, the results presented in Table 12.1 demonstrated that, under appropriate conditions, increasing the drying temperature from 40 to 60 °C may not have a significant effect on several micronutrients and antioxidants. The total phenolics content of dried bee pollen in the study by Zuluaga-Domínguez et al. [34] was comparable to Algerian (21.90–26.68 mg GAE/g), Brazilian (19.28–48.90 mg GAE/g), Austrian (24.6 mg GAE/g) or Turkish (15.27 mg GAE/g) bee pollens [38–41]. The flavonoid content of dried bee pollen was comparable to that reported by Carpes et al. [38] for Brazilian bee pollen (8.92 ± 0.55 mg QE/g).

Carotenoid content of dried bee pollen experiences a significant reduction compared to fresh bee pollen, however, the losses of these compounds during air-drying may vary. In the work by Zuluaga-Domínguez et al. [34] drying at higher air temperatures (60 °C) did not entail a significantly higher loss of carotenoids with respect to lowering temperatures (40–50 °C). On the contrary, Barajas et al. [42]

studied the variation in the carotene content of fresh and dried bee pollen when temperatures above 45 °C were used, finding a significant reduction as the temperature rises. Other studies mention losses of vitamins after drying above 42 °C, being close to 31% for vitamins A, E, and C [33, 43].

Interestingly, antioxidant activity does not seem to be negatively affected by air-drying processes. Zuluaga-Domínguez et al. [34] (Table 12.1) found that FRAP antioxidant activity showed a significant 50% increase in dried bee pollen, while no significant differences were found using the TEAC method. Similar results have been found for certain fruits and vegetables [44]. This means that antioxidant activity can even be higher in dried samples compared to fresh pollen, due to partial oxidation of polyphenols or the formation of certain Maillard reaction products during drying, which ultimately contribute to the anti-radical activity of bee pollen [45].

After drying, bee pollen can be easily treated to separate extraneous matter, by operations such as sieving and/or cycloning, which removes insect parts, decomposed materials, sand, soil, or other foreign substances. Sieving operations can be also used for the separation of specific pollen grains according to composition or botanical origin in multi-floral bee pollen [23].

12.2.2 Bee Pollen Sterilization

In sterilization, microorganisms are completely removed from a food product by heat. It is a process that occurs more rapidly as the moisture content of the matrix is higher and requires temperatures of 121 °C or higher. Pressure must be higher than atmospheric pressure so exposure of the product to superheated steam can reach the thermal destruction of all the microbial loads [46]. This type of treatment has been proposed for bee pollen as a means to achieve a microbiologically safer product for human consumption [47, 48]. However, it may induce drastic changes in the composition and morphology of bee pollen to a larger extent than air-drying treatments. Moreover, unlike drying, the conditions of sterilization are enough to induce wall-breaking of pollen grains. Table 12.2 shows a comparative physicochemical characterization of bee pollen samples treated at 121 °C, for 5, 10, and 15 min [49].

Bee pollen showed a slight reduction in moisture content from 20% to 18% without a significant effect of time. The pH and acidity values did not differ between treatments, meaning that their variation is due to alterations caused by microorganisms. Additionally, there was no significant variation in digestibility.

No significant change was found in the total flavonoids of sterilized bee pollen compared to fresh bee pollen. Conversely, the content of phenolic compounds presented a significant increase in sterile bee pollen, where those treated for 15 min had a significantly higher value than the rest. Also, a decrease was observed in the total carotenoid content when subjected to thermal treatment, however, no significant variation was observed between bee pollen sterilized at different times. These results indicate that degradation of the exine was probably achieved in a greater proportion than in drying, potentially eliciting a greater release of phenolic

Table 12.2 Physicochemical features of bee pollen sterilized at 121 °C and different times compared to fresh bee pollen [49]

Physicochemical parameter	Fresh bee pollen	Sterilization time		
		5 min	10 min	15 min
Moisture (g/100 g)	20.4 ± 1.4[b]	18.3 ± 0.4[a]	18.1 ± 0.9[a]	18.2 ± 0.7[a]
pH	4.16 ± 0.07[a]	4.19 ± 0.12[a]	4.19 ± 0.13[a]	4.12 ± 0.15[a]
Acidity* (meq/kg)	245.6 ± 31.8[a]	251.0 ± 30.0[a]	251.7 ± 22.4[a]	266.1 ± 29.7[a]
Digestibility* (g / 100 g)	65.8 ± 10.1[a]	66.8 ± 1.7[a]	69.4 ± 8.2[a]	69.4 ± 10.9[a]
Total phenolics (mg GAE/g)	15.0 ± 2.1[a]	20.7 ± 1.0[b]	21.8 ± 0.6[b]	23.7 ± 0.7[b]
Total carotenoids (mg β-carotene equivalents/kg)	684.5 ± 82.8[b]	504.9 ± 81.5[a]	483.5 ± 60.1[a]	404.5 ± 64.4[a]
Total flavonoids (mg QE/g)	4.99 ± 1.79[a]	5.26 ± 0.19[a]	5.10 ± 0.44[a]	5.49 ± 0.59[a]
FRAP (μmol Trolox /g)	58.5 ± 15.6[a]	75.8 ± 11.3[a]	71.2 ± 9.6[a]	73.8 ± 14.4[a]
TEAC (μmol Trolox/g)	60.8 ± 19.9[a]	78.0 ± 17.4[a]	80.5 ± 17.3[a]	81.9 ± 38.3[a]

*Dry basis. Mean ± standard deviation. Different letters for the same row represent significant differences at a confidence level of 95%. *FRAP* Ferric reducing antioxidant power assay. *TEAC* Trolox equivalent antioxidant capacity. *GAE* equivalents of gallic acid. *QE* equivalents of quercetin

compounds. This can be evidenced in Fig. 12.1, which shows scanning electron microscopy (SEM) images of bee pollen grains subjected to drying and sterilization. It is clear how sterilization degrades the bee pollen surface, which would be indicative of greater availability of different nutritional and bioactive compounds.

While phenolics appeared to increase in concentration, the carotenoids, less stable to high temperatures, were gradually degraded. The statistical analysis showed antioxidant activity and digestibility of sterilized bee pollen did not have significant differences with respect to fresh bee pollen. A similar effect was found by Adaškevičiūtė et al. [50] for the content of phenolic compounds, which increased from 18.98 to 36.44 mg rutin equivalent/g, as well as the antioxidant activity measured by DPPH from 9.81 to 35.6 mg rutin equivalent/g, at 95 °C with a treatment of 40 minutes.

According to these results, sterilization of 5 min is an effective physical treatment to enhance the microbiological quality of bee pollen, with a moderate effect on carotenoids but no effect on antioxidant activity or digestibility.

12.2.3 Bee Pollen High-Pressure Treatments

High hydrostatic pressure (HHP) is one of the most promising non-thermal treatments to improve the microbiological safety of foods without altering their nutritional and sensory quality [51]. The principles governing HHP assume that food follows the isostatic principle, which mentions that the pressure is instantaneously and uniformly transmitted to the entire sample over time [52]. Unlike thermal processing, HHP is independent of sample size [53].

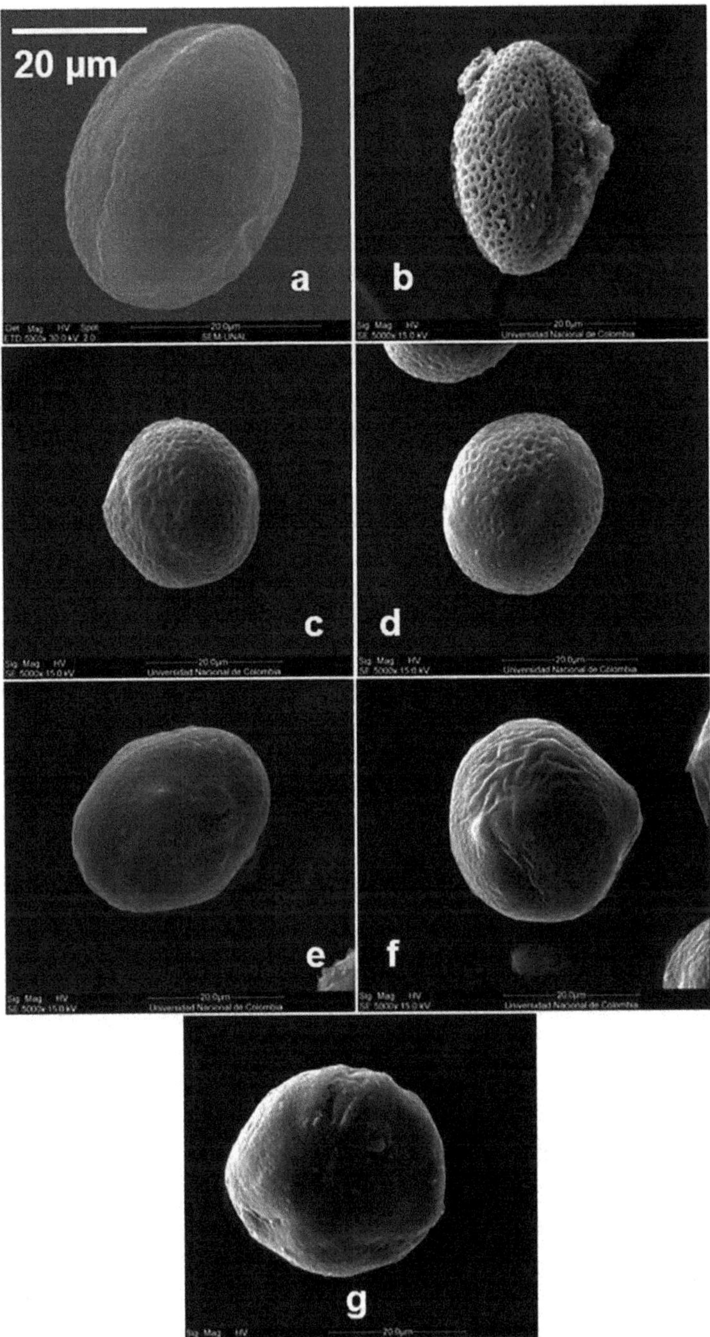

Fig. 12.1 SEM micrographs of fresh, dried, and sterilized bee pollen (5000×). (**a**) Fresh bee pollen, (**b**) Dried bee pollen at 40 °C, (**c**) 50 °C, (**d**) 60 °C, (**e**) sterilized bee pollen 121 °C, 5 min, (**f**) 10 min, (**g**) 15 min

Bee pollen [54] has been treated with high pressure to evaluate the effect of processing on the microstructure of plant tissue, the extractability of bioactive components, and the effectiveness of eliminating the microbiological load. A face-centered composite central design was used with three levels of each independent variable: pressure (200–400 MPa) and time (5–15 min), for the processing of a semi-solid product of bee pollen mixed with peptone water. This latter was used because preliminary tests showed that bee pollen subjected to high pressure by itself did not present modifications in its physicochemical and bioactive characteristics nor reduced the microbiological load. Regarding the effects of high-pressure treatment, both pressure and time had a positive effect in increasing the availability of phenolics and carotenoids. In the case of phenolic compounds, the extractability increased when the maximum pressure and time outlined in the experimental design (400 MPa, 15 min) were used, which means a total increase in the availability of these compounds of about 36%, compared to the control. Untreated bee pollen showed a mean content of total phenolics of 24.83 mg GAE/g, and after HHP this level increased up to 33.69 mg GAE/g. Regarding the content of carotenoids and antioxidant activity, no significant differences ($p > 0.05$) were observed at pressures of 300 or 400 MPa. As for the time factor, a trend was observed toward reaching higher values of carotenoid extraction and greater antioxidant activity with longer periods. Mean carotenoid content and FRAP antioxidant activity reported for untreated bee pollen were 552.3 mg β-carotene/kg and 100.6 μmol Trolox/g, respectively. Maximum pressure and time resulted also in the highest level of carotenoid extraction (778.0 mg β-carotene/kg) and antioxidant activity (126.5 μmol Trolox/g).

Microbial cell population reduction values (Log S) showed that the most effective treatments were only capable of inactivating at most three orders of magnitude, starting from inoculation of ca. 5×10^6 CFU/mL for *Salmonella* and ca. 5.1×10^6 CFU/mL for *Zigosaccharomyces rouxii*. In the case of *Salmonella*, the highest inactivation cycles were obtained after 15 min, regardless of the pressure submitted. For *Z. rouxii*, the highest values were obtained with a treatment of 400 MPa and 15 min.

HPP is therefore a promising emerging technology for preserving bee pollen and its nutritional and bioactive potential, however, more research is needed to provide guidelines for its commercial use.

12.3 Bee Pollen Transformation by Fermentative Bioprocessing

Bees induce a solid-state fermentation process to bee pollen, after which it is known as bee bread [29]. It is a complex process, involving several genera of bacteria and yeast [16, 25]. The nutrient digestibility of bee bread seems to be improved compared to fresh bee pollen, especially regarding protein, due to the metabolic action of yeasts and bacteria, mainly anaerobic lactic acid bacteria (LAB), the reduction of pH due to fermentation, and the multi-enzymatic activity that takes place [30–32,

55]. Also, bee bread is more stable to microbiological spoilage due to the increased content of lactic acid and other organic acids [29, 56], as well as a potential source of viable probiotic bacteria [57].

Inducing LAB-mediated fermentation is a cost-effective approach to improve the nutritional quality of foods for human nutrition [58]. The frequent consumption of LAB-fermented foods can be beneficial in terms of gut-microbiome modulation, if (a) fermentations are carried out by health-related LAB species or probiotics and, especially, if (b) the fermented food possesses enough viable cells of such species at the moment of ingestion; it is also beneficial in terms of the increased uptake of bioactive compounds if the metabolism of the fermenting LAB species can produce, or to make more bioavailable, nutrients and other chemical compounds with health functionalities, such as bacteriocins, organic acids, vitamins, bioactive peptides, among others [59]. Moreover, LAB fermentation of foods, in particular plant-based foods, can increase their sensory quality and safety [60]. Yeast- or Yeast/LAB-mediated fermentation with certain yeast strains has been extensively used in many cultures to prepare food with longer shelf life and better sensory features, but it can also exert a positive impact on food quality by improving nutritional and bio-functional quality [61].

Therefore, fermentation can be an outstanding strategy to improve the nutritional quality of bee pollen. Several attempts have been made to ferment pollen with lactic acid bacteria, yeasts, and their combination. Recently, Zuluaga-Domínguez & Fuenmayor [16] and Barta et al. [25] provided in-depth reviews of these attempts. Bee pollen fermentation is particularly challenging due to several reasons. Its initial environment microbial loads might lead to uncontrolled fermentation. On the other hand, its high concentration of sugars, which leads to a high osmotic pressure environment, and the presence of compounds with specific antimicrobial activities, can be hostile to LAB growth [6]. However, a proper selection of strains, an adequate treatment of the substrate, and the use of suitable bioprocessing conditions, can successfully promote safe LAB-fermentation of bee pollen, thus enhancing its features as a food product.

Table 12.3 shows some of the bioprocessing conditions that have been followed to achieve bee pollen fermentation, the LAB and yeasts species used, and the reported benefits in the nutritional or bioactive characteristics. Most of the developed pollen fermentation protocols involve the addition of sanitized, distilled, or sterilized water (or aqueous solutions) as part of the preparation of pollen as a substrate. A higher or similar amount of pollen with respect to added water is usually maintained, causing the mixture to rheological behave as a dense slurry, and therefore classifying the bioprocess as a solid-state fermentation. In these cases, the reported pollen: water ratios vary from 4:1 [62–64] to 1:1 [65–67], but the most commonly preferred ratio is 2:1 [30, 47, 48, 68–70]. Utoui et al. [71] developed a liquid fermentation using a Kombucha-fermented tea infusion as the medium, in which pollen was used as an additional ingredient rather than as the main substrate, at concentrations of 50 g/L. Besides water, supplementary carbohydrates, including honey [62–64], sugars [71], or prebiotic compounds such as inulin or lactulose [62–64], have been added to the substrate.

Table 12.3 Fermentation conditions for bee pollen and reported nutritional improvements

Fermenting microorganisms	Substrate formulation	Bioprocessing conditions	Nutritional and functional benefits at the best fermentation conditions	Reference
Lactobacillus sp. (possibly *L. acidophilus*)	20 g bee pollen, 3 g honey, 5 g distilled water	Temperature: 30–42 °C Time: 72 h	Bioaccumulation of lactic acid helps preserve the product naturally	[62]
Consortium of *L. plantarum* and *L. acidophilus*	20 g bee pollen, 3 g honey, 5 g distilled water	Temperature: 37 °C Time: 48–72 h	Reduction of cholesterol levels in an animal model (Wistar rats)	[63]
L. acidophilus La-14 (Danisco®)[a]	Bee pollen:water (2:1)	Substrate pre-treatment at 121 °C, 15 min. Temperature: 35 °C Time: 24 h	Bee pollen wall disruption; bioaccumulation of lactic acid; presence of viable probiotic LAB with cell counts of 8 log CFU/g in the fermented product	[47]
L. paracasei subsp. *paracasei* Lpc 37 (Danisco®)				
YOMIX 205 (consortium of *Streptococcus thermophilus, L. delbrueckii subsp. bulgaricus, L. acidophilus* and *Bifidobacterium lactis*) (Danisco®)				
CHOOZIT MY 800 (consortium of *S. thermophilus, L. delbrueckii* subsp. *lactis* and *L delbrueckii* subsp. *bulgaricus*) (Danisco®)				
Consortium of *Lactococcus fermentum, L. plantarum, L. paracasei* and *Bifidobacterium*	20 g ground bee pollen, 3 g honey, 5 g distilled water, 1% supplementation with prebiotic (inulin or lactulose)	Temperature: 37 °C Time: 7–14 days	The presence of inulin (prebiotic) favors bioaccumulation of lactic acid and the viability of LAB with possible probiotic features with cell counts of 8 log CFU/g in the fermented product	[64]
L. acidophilus	Bee pollen:water (1:1)	Substrate pre-treatment at 121 °C, 15 min Temperature: 35 °C Time: 72 h	Bioaccumulation of lactic acid; presence of viable probiotic LAB with cell counts of 8 log CFU/g in the fermented product; fermentation increases antioxidant activity and content of total phenolics and carotenoids	[65]
L. plantarum				
Saccharomyces cerevisiae				
Consortium of *L. plantarum* and *S. cerevisiae*[a]				

(continued)

Table 12.3 (continued)

Fermenting microorganisms	Substrate formulation	Bioprocessing conditions	Nutritional and functional benefits at the best fermentation conditions	Reference
Consortium of *L. plantarum* MRS3 and *Bacillus subtilis natto* ATCC15245	Bee pollen and MRS medium (35% moisture content)	Temperature: 33 °C Time: 9 days	Increased concentration of bioactive compound with thrombolytic activity (nattokinase).	[68]
SCOBY/ Kombucha consortium (LAB, acetobacteria, and yeasts)	Green tea infusion (5 g/L), sucrose (70 g/L), fermented Kombucha tea (10%), bee pollen (50 g/L)	Temperature: 28 °C Time: 30 days (simultaneous addition of Kombucha and pollen) or 10 days (pollen addition after 20 days of Kombucha inoculation)	Bee pollen increases LAB counts in the fermented Kombucha pollen; bee pollen wall disruption; increase in bioactive concentration and bioavailability (polyphenols and short-chain fatty acids); moderate antitumoral effect (Caco-2 cells).	[71]
Consortium of *L. kunkeei* (*Apilactobacillus kunkeei*) strains and *Hanseniaspora uvarum*	Bee pollen (60%) and water (40%)	Temperature: 30 °C Time: 216 h	Increased protein digestibility and bioavailability of phenolics compared to controls (raw pollen and spontaneously fermented pollen); improved microbiological stability towards molds growth; increased amount of bioaccessible bioactive compounds (phenolics, including rutin, luteolin, kaempferol, quercetin, p-coumaric acid, caffeic acid, 4-ethyl catechol, dihydrocaffeic acid, and dihydroferulic acid); increased concentration and complexity of volatile aroma compounds	[30, 70]

(continued)

Table 12.3 (continued)

Fermenting microorganisms	Substrate formulation	Bioprocessing conditions	Nutritional and functional benefits at the best fermentation conditions	Reference
L. plantarum ATCC 8014	Bee pollen:water (1:1)	Substrate pre-treatment at 121 °C, 15 min temperature: 37 °C Time: 72 h	Bee pollen wall disruption; bioaccumulation of lactic acid; presence of viable probiotic LAB with cell counts of >8 log CFU/g in the fermented product; increased (>30%) content of bioactive compounds (total phenolics) and antioxidant activity compared to fresh bee pollen.	[66]
CHOOZIT MY 800 (consortium of *S. thermophilus*, *L. lactis* and *L. bulgaricus*) (Danisco®)				
S. cerevisiae (ATCC 9763 and a commercial bakery starter)				
Consortia of *L. plantarum* ATCC 8014 and *S. cerevisiae* (ATCC 9763 and a commercial bakery starter)[a]				
Consortium of *L. delbrueckii* subsp. *bulgaricus* and *S. thermophilus*	Bee pollen (25 g) and sterile water (25 g)	Bee pollen can previously ground pulverization for wall breaking. Temperature: 42 °C (LAB); 37 °C (yeast or yeast-LAB) Time: 48 h Fermented bee pollen is freeze-dried.	Increased concentration of nutrients (oligopeptides, free amino acids, polyunsaturated fatty acids, and B-complex vitamins: riboflavin, nicotinic acid, nicotinamide). Bee pollen wall disruption.	[67]
Active dry yeast[a]				
Consortium of *L. delbrueckii* subsp. *bulgaricus*, *S. thermophilus* and active dry yeast				
L. lactis	Bee pollen (100 g) and pure culture of LAB (20 ml) adjusting moisture content to 35–40%	Temperature and time: 35 °C for 96 h followed by 20 °C for 72 h	Bioaccumulation of lactic acid; a slight increased concentration of protein and free amino acids; slight increase of polyphenol concentration and increase of antioxidant activity.	[69]

(continued)

Table 12.3 (continued)

Fermenting microorganisms	Substrate formulation	Bioprocessing conditions	Nutritional and functional benefits at the best fermentation conditions	Reference
Commercial consortium of *S. thermophilus*, *L. delbrueckii* ssp. *lactis*, *L. delbrueckii* ssp. *bulgaricus* (Danisco®)	Bee pollen:water (2:1); batches of 6 kg	Substrate pre-treatment at 115 °C, 80 kPa, 10 min, and pH adjustment to 5.8 with NaOH Temperature: 37 °C Time: 30 h	Bee pollen wall disruption; bioaccumulation of lactic acid; presence of viable probiotic LAB with cell counts of 8 log CFU/g in the fermented product	[48]
Commercial consortium of *L. delbrueckii* ssp. *lactis*, *L. delbrueckii* ssp. *cremoris*, L. *delbrueckii* ssp. biovar. d*iacetylactis* (Danisco®)				
Commercial *L. acidophilus* (Danisco®)[a]				

[a]Microorganisms/inoculum with best reported results

Due to the spontaneous high microbial loads of fresh raw bee pollen, several works recommend pretreating the substrate fermentation to avoid the growth of undesired microorganisms during this process. Fuenmayor et al. [47], Zuluaga-Domínguez et al. [66], Salazar-González [65], and Mora et al. [48] proposed sterilization protocols at 115–121 °C for 10–15 min (see Sect. 12.2.2). While such pretreatments dramatically reduce or eliminate the microbial loads, they do not have a drastic negative impact on most of the micronutrients and, instead, enhance wall breaking and thus the nutrient availability for the fermenting microorganisms. For the same purpose, Yan et al. [67] recommended ground pulverization of bee pollen as part of the pretreatment before fermentation observing better results as a consequence of wall-breaking.

LAB are, by far, the most widely used fermenting microorganisms for bee pollen, either in axenic starters or in a consortium of several species (Table 12.3). Also, consortia of LAB with yeasts [30, 65–67, 70, 71], or even yeast alone [65–67] have been investigated. Since bee pollen is a biochemically complex matrix, the proper selection of microorganism strains or consortia, capable to survive and thrive in this substrate, is critical. Most axenic LAB starters reported belong to the *Lactobacillus* genera (i.e., *L. acidophilus*, *L. plantarum*, and *L. lactis*), whereas other LAB species, such as *A. kunkeei*, *L. delbrueckii* subsp. *bulgaricus*, *B. lactis*, *L. paracasei*, *S. termophilus*, and *L. fermentum*, have only been used in consortia with other LAB or yeast species. There are fewer reports of the use of yeast for bee pollen fermentation, being *S. cerevisiae* and *H. uvarum* as the most commonly used species. *S. cerevisiae* in combination with the fungi *Ganoderma lucidum* was also investigated as a fermentative mechanism of wall rupture of bee pollen [72]. The effects and potential benefits of fermentation using different microorganisms can be seen in Table 12.3.

For an effective fermentation, particular attention must be paid to the fermentation conditions. These include inoculum preparation, incubation temperature and time, and anaerobicity, among others, all of which are highly dependent on microbial physiology and are strain-specific. Fermentation temperature must favor the growth of the desired microorganism. Temperature for LAB-mediated fermentation bee pollen is usually 35 °C–37 °C, however, temperatures as high as 42 °C have been used (Table 12.3). When the fermentation is mediated by LAB/yeast consortia or pure yeast, lower temperatures (28–30 °C) should be used [30, 70, 71]. The fermentation time ranges usually between 30 h and 72 h, but it can vary from 24 h [47] to 14 days [64] in solid-state fermentations and 30 days in liquid fermentations [71]. This time depends on the microbial growth kinetic, and it must be optimized to ensure the maximum possible microorganism survival and/or the desired level of fermentation (e.g., maximum acidity or specific metabolites).

Fermentation causes the acidification of bee pollen due to the bioaccumulation of lactic and other organic acids, which depends on the specific LAB or yeast metabolism. This helps preserve the product naturally. When the bioprocess is mediated by probiotic microorganisms, as far as the viable cell count is above a minimum threshold (usually 8 log CFU/g), the fermented bee pollen becomes a potentially probiotic food. The presence of cells and their bioactive metabolites confers several functional features to the product, including the maintenance of gut microbiota, the modulation of immune responses, and protection against pathogens [25].

Moreover, the induction of fermentation enhances the bee pollen wall (exine) disruption [47, 48, 67, 71] (Fig. 12.2). This, in turn, has been associated with an improvement in the bioavailability or concentration of nutrients [30, 65, 67, 69–71], bioactive compounds [30, 68, 69, 71], and antioxidant activity [30, 65, 66, 69, 70]. Other potential benefits of bee pollen fermented with certain probiotic microorganisms may include protection against gastrointestinal disorders, antiviral activity, and hypocholesterolemic effect [25], however, there is still very limited evidence of these effects.

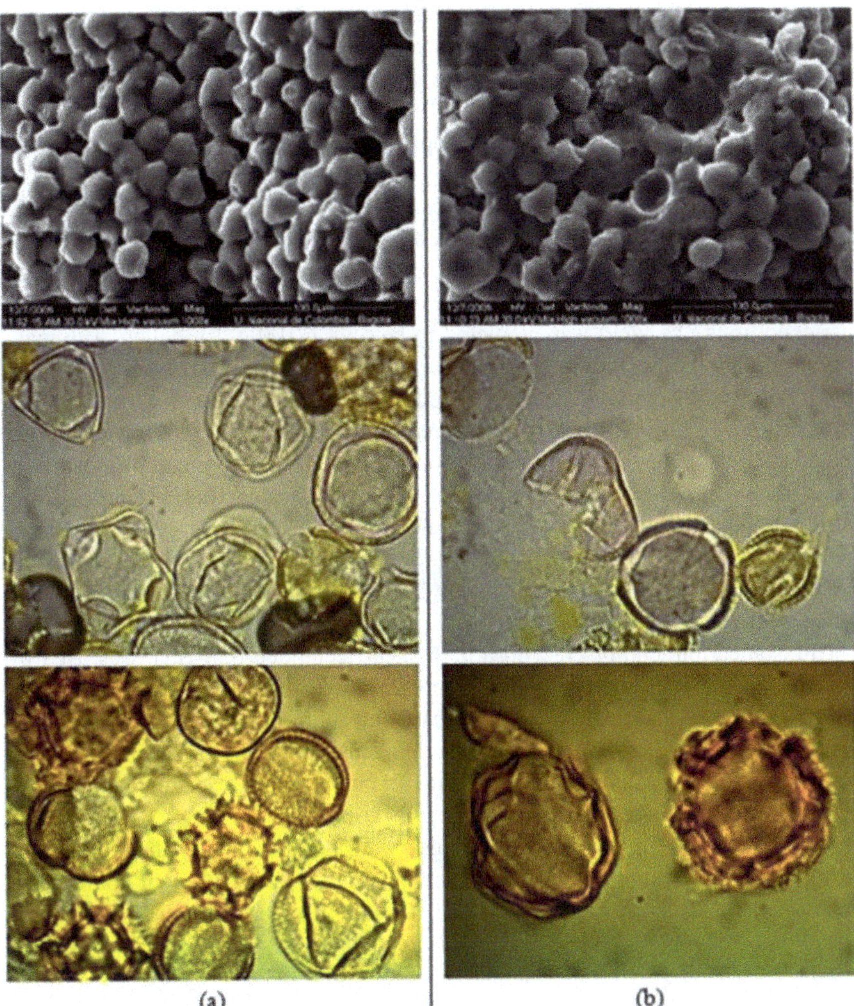

Fig. 12.2 Micrographs showing the typical appearance of pollen microstructure before (**a**) and after (**b**) LAB-fermentative processes. (Retrieved from Mora et al. [48] with permission from Elsevier)

12.4 Other Techniques to Improve the Nutritional Value of Bee Pollen That Focus on Wall-Breaking

As mentioned above, the exine and the intine i.e., the external and internal membranes that cover and protect pollen grains are so resistant that they can retain their morphology for extremely long periods. The resistance of the exine is mainly due to sporopollenin, a complex and highly inert biopolymer with several oxygenated and non-oxygenated functional groups. To take full advantage of the high nutritional value of pollen, it is therefore essential that the coating of each granule is broken, without damaging their nutrients, at some point before or after ingestion. Several modern techniques that rely on high-energy and/or pressure, the use of enzymes, or a combination of multiple approaches, have been applied to bee pollen to achieve wall-breaking.

Xu et al. [73] applied supercritical carbon dioxide (SC-CO_2) to successfully break the cell wall of rape (*Brassica campestris*) bee pollen and consecutively extract lyzed bee pollen oil. The process of consecutive extracting pollen oil was significantly affected by the SC-CO_2 pressure, temperature, and flow rate. In the best conditions, the yield was predicted to be 5.98 g/100 g dry pollen. The main polyunsaturated fatty acids in the extracted oil were linolenic acid (approximately 39%), followed by palmitic acid (20%) and linoleic acid (13%).

Dong et al. [74] investigated the degree of wall-disruption of rape bee pollen treated with protamex hydrolysis, ultrasonication, and a combination of the two. Protamex hydrolysis degraded the pollen coat and disintegrated the intine at the germinal apertures; ultrasonication treatment cracked the pollen exine into fragments but seemed to have little effect on the intine. It was the combination of the two techniques that could entirely disrupt both the exine and the intine of rape bee pollen, with the exine layer ruptured into three fragments along germinal apertures.

Ultrasonication, combined with a high-shear technique (US-HS), was used by Wu et al. [12] to break up bee pollen grains from five different species: lotus, rape, apricot, wuweizi, and camellia. After treatment, nutrients inside the pollen were largely released, including fatty acids, amino acids, proteins, crude fats, reducing sugars, carotenoids, calcium, iron, zinc, and selenium, possibly leading to the use of wall-disrupted pollen to improve nutrient utilization.

Enzymatic hydrolysis is another interesting strategy to modify the structure of bee-pollen and promote the release of compounds [75]. Six different commercial enzymes were used, and treatments were evaluated by calorimetry and microscopy, as well as in terms of protein, amino acids, phenolics, flavonoids content, and antioxidant activity. The calorimetric analysis showed an increase in the heat flow of hydrolyzed products compared to bee-pollen. Proteases improved the protein content by about 13–18%, phenolics 83–106%, flavonoids 85–96%, antioxidant activity up to 68%, and increased all essential amino acids.

Xie et al. [76] explored the effects of three techniques on the wall disruption, antioxidant activity, sensory qualities, and nutrients in tea pollen: freeze–thaw processing, a combination of enzymatic hydrolysis and ultrasonication, and superfine grinding. The broken-wall ratio was, respectively: 59.86%, 79.14%, and 100%. Compared with untreated samples, the release of nutrient compounds, including polyphenols, flavonoids, carbohydrates, total free amino acids, proteins, and theanine, increased after treatment with the three wall-disruption methods.

More recently, Xue & Li [77] applied ultrasonication, freeze-thawing, enzymatic hydrolysis, and their combinations were used to disrupt the cell wall and extract protein isolates of camellia bee pollen protein. Not only the wall disruption positively affected the release of proteins from pollen (Fig. 12.3), but also their

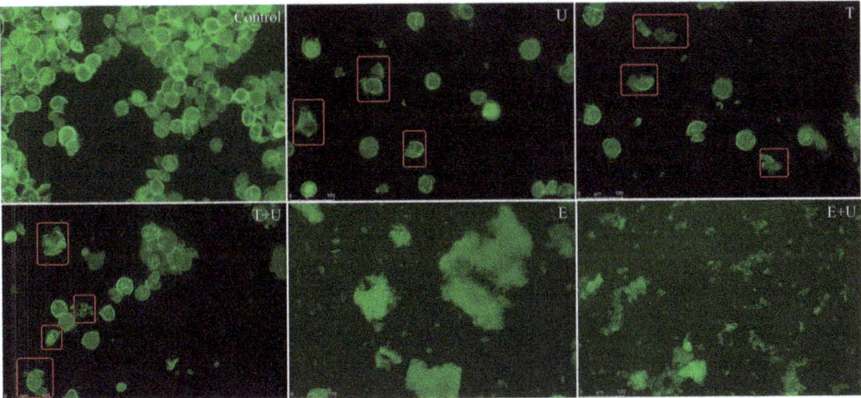

Fig. 12.3 Release of proteins from bee pollen samples by different wall-breaking techniques. U: pollen sample treated with ultrasonication at 400 W for 20 min; T: pollen sample treated with freeze (−20 °C for 24 h) and thawing (45 °C for 7 h); E: pollen sample treated with enzymatic hydrolysis at 45 °C for 6 h; T + U: pollen sample treated with a combination of freeze-thawing and ultrasonication; E + U: pollen sample treated with a combination of enzymatic hydrolysis and ultrasonication. (Retrieved from Xue and Li [77] with permission from Elsevier)

physiochemical properties and structure. As compared with physical wall disruption (ultrasonication, freeze-thawing, and their combination), enzymatic hydrolysis could significantly improve the yield of proteins. Interestingly, the extracted proteins exhibited better solubility, emulsifying properties, digestibility, and gelation property due to the partial hydrolysis of proteins induced by protease.

12.5 Use of Bee Pollen as an Ingredient in Food Products: Strategies and Nutritional Benefits

As bee pollen becomes more popular as a functional ingredient, its inclusion in the preparation of several types of food products is a promising strategy for the diversification of its use and to increase the nutritional value of traditional processed foods. The use of bee pollen as an ingredient has been tested in fruit pulps, bakery products, and milk drinks, among others.

12.5.1 Fruit Pulps

Table 12.4 shows the changes in nutritional and bioactive quality parameters of two types of fruit pulps (blackberry and mango) by the inclusion of 3% pollen. The moisture content in the pulps made from blackberry and mango decreased, and protein was increased by 80%, in addition to the fat content.

Table 12.4 Results of physicochemical characterization of pulps with the inclusion of bee pollen

	Fruit pulp			
Parameter	Blackberry	Blackberry – 3% bee pollen	Mango	Mango – 3% bee-polen
Moisture (g/100 g)	89.23 ± 1.05^{b}	86.83 ± 0.97^{a}	78.61 ± 1.12^{A}	76.72 ± 0.88^{A}
Protein (g/100 g)	1.0 ± 0.1^{a}	1.8 ± 0.2^{b}	0.7 ± 0.2^{A}	1.4 ± 0.1^{B}
Fat (g/100 g)	0.001 ± 0.000^{a}	0.120 ± 0.013^{b}	0.025 ± 0.009^{A}	0.150 ± 0.015^{B}
Ashes (g/100 g)	0.46 ± 0.07^{a}	0.53 ± 0.08^{a}	0.30 ± 0.05^{A}	0.34 ± 0.03^{A}
Carbohydrates (by difference)	9.31 ± 0.97^{a}	10.70 ± 1.01^{a}	19.60 ± 1.22^{A}	21.40 ± 1.33^{A}
pH	2.99 ± 0.01^{a}	3.07 ± 0.01^{b}	3.78 ± 0.01^{B}	3.37 ± 0.01^{A}
Soluble solids (°Bx)	9.5 ± 0.1^{a}	10.2 ± 0.1^{b}	19.7 ± 0.1^{A}	21.4 ± 0.3^{B}
Titrable acidity (% anhydrous citric acid)	2.590 ± 0.003^{a}	3.150 ± 0.040^{b}	0.520 ± 0.008^{A}	1.170 ± 0.070^{B}
TEAC (μmol Trolox/g)	11.0 ± 0.3^{a}	13.5 ± 0.1^{b}	3.80 ± 0.05^{A}	5.9 ± 0.02^{B}
FRAP (μmol Trolox/g)	10.3 ± 0.1^{a}	13.9 ± 1.2^{b}	1.700 ± 0.007^{A}	4.200 ± 0.070^{B}
Total phenolics (mg GAE/ 100 g)	167.9 ± 0.2^{a}	296.4 ± 0.2^{b}	48.10 ± 0.02^{A}	99.80 ± 0.06^{B}

Different letters in the same row indicate significant differences (uppercase letters describe differences for mango), with a confidence level of 95%. These experiments were carried out using High-Andean bee pollen from Colombia. *FRAP* Ferric reducing antioxidant power assay, *TEAC* Trolox equivalent antioxidant capacity, *GAE* equivalents of gallic acid, *QE* equivalents of quercetin

The inclusion of bee pollen in blackberry and mango pulps improved their nutritional composition since it is a source of nutrients [78]. The determination of the antioxidant capacity by TEAC increased by around 20% in the blackberry pulp with the inclusion of bee pollen, while by FRAP the increase was 30%. The antioxidant activity in the mango pulp including bee pollen increased by nearly 70%. Similarly, Silva et al. [79] found that the addition of up to 10% pollen in tropical fruit pulps had high acceptability values in color, flavor, aroma, and appearance.

12.5.2 Baked Goods

Some studies have been carried out on the inclusion of bee pollen in bakery products, especially bread. Table 12.5 presents possible formulations of soft bread with the inclusion of pollen (E3–4), modifications in the fat source, and the addition of spices.

Figure 12.4 shows the soft bread obtained through the baking process. The bread without the inclusion of bee pollen, made with margarine (E1) and oil (E2), presented a lighter crust color, as well as a larger size. Regarding the bread with the inclusion of bee pollen, aromatic herbs, and oil (E3 and E4), a darker crust and a yellower crumb were observed, even though the baking conditions were the same as E1 and E2. This result is associated with the presence of bee pollen in the formulation. The addition of carbohydrates from pollen caused a more accelerated crust

Table 12.5 Ingredients and formulation for the preparation of bread and with the inclusion of bee pollen and aromatic herbs

Ingredients	Quantity (g)			
	E1	E2	E3	E4
Wheat flour	250	500	500	500
Sugar	50	100	100	100
Margarine	38	–	–	–
Oil	–	75	72	72
Yeast	9	30	30	30
Salt	5	10	10	10
Eggs	0.5 units	1 unit	1 unit	1 unit
Water	75	150	145	145
Bee pollen	–	–	50	35
Rosemary	–	–	10	–
Oregano	–	–	–	5

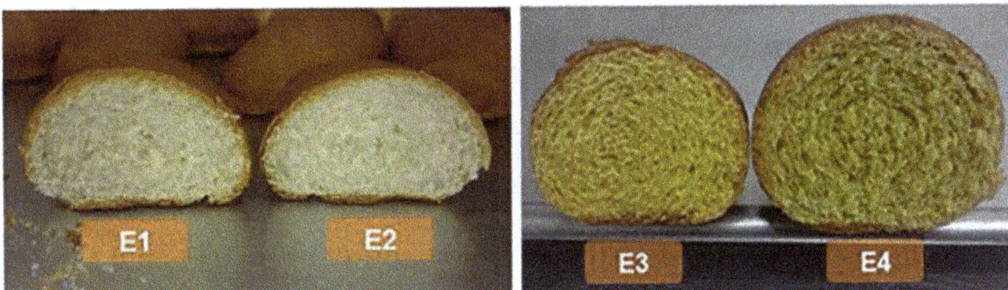

Fig. 12.4 Soft bread with (E3–4) and without (E1–2) the inclusion of bee pollen, spices, and modification of the fat source

browning, whereas the carotenoids in pollen naturally pigment the breadcrumb. The bread made with the inclusion of bee pollen and rosemary (E3) presented a lower growth than the bread with bee pollen and oregano (E4). This is directly related to the formulation because in the first case, the inclusion of bee pollen and rosemary was 10% and 2% respectively, while in E4 the inclusion of bee pollen and oregano was 8% and 1% respectively.

Concerning the crumb, low moisture, hardness, and a clear difference in size between both products were observed, as was previously discussed. It was also observed that E3, due to lower growth, presented a more compact crumb compared to E4, although both showed good alveolar distribution and adequate crust thickness.

Conte et al. [80] developed gluten-free bread with the inclusion of up to 5% bee pollen, without finding an effect on the rheology of the dough, however, some defects were detected in the crumb texture and staling kinetics. In another work, Conte et al. [81] found an increase in the content of phenolics in bread with the addition of bee pollen, going from 29.6 mg GAE/100 g in the control, to 84.1 mg GAE/100 g in the bread with 5% bee pollen. Likewise, significant increase of the lipid, ash, and protein content were observed.

12.5.3 Fermented Milk Drinks

One of the most attractive functional foods that can be obtained with the inclusion of bee pollen is fermented dairy beverages. The nutritional composition of yogurt and kumis obtained with the inclusion of bee pollen in a 10% ratio is presented in Table 12.6. In this study, a lyophilized mixed culture CHOOZIT MY800 (Danisco, Denmark) was used, made up of a mixture of selected strains (*S. thermophilus, L. delbrueckii* subsp. *lactis,* and *L. delbrueckii* subsp. *bulgaricus*) as well as *L. acidophilus* in a second treatment.

In general, in yogurt and kumis, the addition of bee pollen improved the nutritional characteristics, not only due to the increase in protein content but also to the contribution of other micronutrients such as minerals, particularly calcium. Moreover, Zlatev et al. [82] found an improvement in sensory characteristics and vitamin C content during the production of yogurt with the inclusion of up to 5% honey and 0.4% bee pollen.

Table 12.6 Nutrient composition of fermented milks with the inclusion of bee pollen

Parameter	Y1	Y2	K1	K2
Moisture (g/100 g)	83.71	83.19	83.71	82.87
Sodium (mg/100 g)	64.16	70.29	64.12	69.97
Potassium (mg/100 g)	0.02	0.02	0.02	0.02
Calcium (mg/100 g)	104.69	105.11	104.62	104.62
Iron (mg/100 g)	0.37	0.44	0.37	0.44
Zinc (mg/100 g)	0.15	0.19	0.15	0.19
Ashes (g/100 g)	0.70	0.73	0.70	0.73
Fat (g/100 g)	2.72	2.80	2.72	2.79
Protein (g/100 g)	3.24	3.49	3.24	3.48
Fructose (g/100 g)	0.60	0.79	0.60	0.79
Glucose (g/100 g)	0.40	0.51	0.40	0.50
Saccharose (g/100 g)	4.93	4.62	5.39	5.45
Lactose (g/100 g)	3.07	3.08	2.62	2.62
Dietary fiber (g/100 g)	0.44	0.59	0.44	0.59
Cholesterol (g/100 g)	0.01	0.01	0.01	0.01
Vitamin A (mg/100 g)	0.09	0.09	0.09	0.09

Y1 (yogurt with the inclusion of bee pollen and fermented with mixed cultures), Y2 (yogurt with the inclusion of bee pollen and fermented with *Lactobacillus acidophilus*), K1 (kumis with the inclusion of bee pollen and fermented with mixed cultures), and K2 (kumis with the inclusion of bee pollen and fermented with *Lactobacillus acidophilus*)

12.6 Conclusions

The adequacy of bee pollen as a source of nutrients and bioactive molecules is limited by the resistant microstructure of the pollen grains and its microbial loads. However, its nutritional quality, safety, and functional features can be noticeably improved by the application of several processing technologies.

The removal of moisture via air-drying is the most common unit operation to stabilize pollen against microbial spoilage. Usually, air temperatures around 60 °C are enough to remove water below safe and steady values. During air-drying, the content of carotenoids and some vitamins is partially reduced (around 20–30%), but there is little or no loss of other nutrients. Sterilization of bee pollen (115–121 °C, 5–10 min) allows to eliminate the microbial loads without the need for moisture removal. Like air-drying, sterilization causes a reduction of some micronutrients including carotenoids (ca. 40%). Interestingly, both techniques enhance antioxidant activity and increase the content of phenolic compounds. HPP is another promising technology for preserving pollen without compromising or even improving its nutritional content. However, more studies are needed to enable its commercial application.

Fermentation can be an outstanding strategy to improve the nutritional quality of bee pollen, especially by the use of lactic acid bacteria. Bee pollen has been successfully fermented using either axenic starters or consortiums of several LAB species and sometimes yeasts. The benefits of fermentation include its natural preservation and, depending on the bacterial species, the increase of certain bioactive metabolites and the addition of probiotic functionality.

Other technologies for bee pollen nutritional improvement focus on the application of energy, pressure, or enzymatic hydrolysis (or their combination) to achieve wall-braking and enhance the availability of micronutrients. These techniques include supercritical fluids (i.e., CO_2), the use of broad-spectrum proteases, ultrasonication, high-shear grinding, and freeze-thawing, and offer a wide range of potential effects in terms of the increased bioavailability of pollen nutrients.

Finally, the inclusion of bee pollen as a functionalizing ingredient is an exciting perspective for the development of novel foods and the diversification of the use of this valuable apicultural product.

References

1. Thakur M, Nanda V (2020) Composition and functionality of bee pollen: a review. Trends Food Sci Technol 98:82–106. https://doi.org/10.1016/j.tifs.2020.02.001
2. Mărgăoan R, Cornea-Cipcigan M, Topal E et al (2020) Impact of fermentation processes on the bioactive profile and health-promoting properties of bee bread, Mead and Honey Vinegar. Processes 8:1081. https://doi.org/10.3390/pr8091081
3. Salazar-González CY, Rodríguez-Pulido FJ, Terrab A et al (2018) Analysis of multifloral bee pollen pellets by advanced digital imaging applied to functional food ingredients. Plant Foods Hum Nutr 73:328–335. https://doi.org/10.1007/s11130-018-0695-9

4. Ortiz AMG, Tangarife MPO, Grosso GS (2010) Implementación de un modelo de calidad y trazabilidad en el proceso de agroindustrialización del polen. Rev Asoc Colomb Ciencias Biológicas 1:18–30
5. Fuenmayor C, Zuluaga C, Díaz C et al (2014) Evaluación de las propiedades fisicoquímicas y funcionales del polen apícola colombiano. Rev MVZ Córdoba 19:4003–4014
6. Salazar-González C, Díaz-Moreno C (2016) The nutritional and bioactive aptitude of bee pollen for a solid-state fermentation process. J Apic Res 55:161–175. https://doi.org/10.1080/00218839.2016.1205824
7. Leja M, Mareczek A, Wyżgolik G et al (2007) Antioxidative properties of bee pollen in selected plant species. Food Chem 100:237–240. https://doi.org/10.1016/j.foodchem.2005.09.047
8. Gardana C, Bo C, Quicazán MC et al (2018) Nutrients, phytochemicals and botanical origin of commercial bee pollen from different geographical areas. J Food Compos Anal 73:29–38. https://doi.org/10.1016/j.jfca.2018.07.009
9. Khalifa SAM, Elashal MH, Yosri N et al (2021) Bee pollen: current status and therapeutic potential. Nutrients 13:1876. https://doi.org/10.3390/nu13061876
10. Campos MGR, Bogdanov S, Almeida-Muradian LB et al (2008) Pollen composition and standardisation of analytical methods. J Apic Res Bee World 47:156–163. https://doi.org/10.1080/00218839.2008.11101443
11. Komosinska-Vassev K, Olczyk P, Kaźmierczak J et al (2015) Bee pollen: chemical composition and therapeutic application. Evidence Based Complement Altern Med. https://doi.org/10.1155/2015/297425
12. Li QQ, Wang K, Marcucci MC et al (2018) Nutrient-rich bee pollen: a treasure trove of active natural metabolites. J Funct Foods 49:472–484. https://doi.org/10.1016/j.jff.2018.09.008
13. Baky MH, Abouelela MB, Wang K et al (2023) Bee pollen and bread as a super-food: a comparative review of their metabolome composition and quality assessment in the context of best recovery conditions. Molecules 28:715. https://doi.org/10.3390/molecules28020715
14. Kostić AŽ, Barac MB, Stanojevic SP et al (2015) Physicochemical composition and techno-functional properties of bee pollen collected in Serbia. LWT Food Sci Technol 62:301–309. https://doi.org/10.1016/j.lwt.2015.01.031
15. Bertoncelj J, Polak T, Pucihar T et al (2018) Carbohydrate composition of Slovenian bee pollens. Int J Food Sci Technol 53:1880–1888. https://doi.org/10.1111/ijfs.13773
16. Zuluaga-Dominguez CM, Fuenmayor CA (2022) Bee bread and gut microbiota. In: Boyacioglu D (ed) Bee products and their applications in the food and pharmaceutical industries. Elsevier, London, pp 315–345
17. Kačániová M, Rovná K, Arpasová H et al (2013) The effects of bee pollen extracts on the broiler chicken's gastrointestinal microflora. Res Vet Sci 95:34–37. https://doi.org/10.1016/j.rvsc.2013.02.022
18. Aabed K, Bhat RS, Mouyabed N et al (2019) Ameliorative effect of probiotics (Lactobacillus paracaseii and Protexin®) and prebiotics (propolis and bee pollen) on clindamycin and propionic acid-induced oxidative stress and altered gut microbiota in a rodent model of autism. Cell Mol Biol 65:1–7. https://doi.org/10.14715/cmb/2019.65.1.1
19. Sattler JAG, Melo ILP, Granato D et al (2015) Impact of origin on bioactive compounds and nutritional composition of bee pollen from southern Brazil: a screening study. Food Res Int 77:82–91. https://doi.org/10.1016/j.foodres.2015.09.013
20. Sánchez EGT, Fuenmayor CA, Mejía SMV et al (2020) Effect of bee pollen extract as a source of natural carotenoids on the growth performance and pigmentation of rainbow trout (Oncorhynchus mykiss). Aquaculture 514:734490. https://doi.org/10.1016/j.aquaculture.2019.734490
21. Salazar-González C, Díaz-Moreno C, Fuenmayor CA (2019) Green extraction of carotenoids from bee pollen using sunflower oil: evaluation of time and matrix-solvent ratio. Chem Eng 75:541–546. https://doi.org/10.3303/CET1975091
22. Salazar-González CY, Rodríguez-Pulido FJ, Stinco CM et al (2020) Carotenoid profile determination of bee pollen by advanced digital image analysis. Comput Electron Agric 175:105601. https://doi.org/10.1016/j.compag.2020.105601

23. Salazar-González CY, Stinco CM, Rodríguez-Pulido FJ et al (2022) Characterization of carotenoid profile and α-tocopherol content in Andean bee pollen influenced by harvest time and particle size. LWT 170:114065. https://doi.org/10.1016/j.lwt.2022.114065
24. Arruda VAS, Pereira AAS, Estevinho MLMF et al (2013) Presence and stability of B complex vitamins in bee pollen using different storage conditions. Food Chem Toxicol 51:143–148. https://doi.org/10.1016/j.fct.2012.09.019
25. Barta DG, Cornea-Cipcigan M, Margaoan R et al (2022) Biotechnological processes simulating the natural fermentation process of bee bread and therapeutic properties—an overview. Front Nutr 9:871896. https://doi.org/10.3389/fnut.2022.871896
26. Xu X, Dong J, Mu X et al (2011) Supercritical CO_2 extraction of oil, carotenoids, squalene and sterols from lotus (Nelumbo nucifera Gaertn) bee pollen. Food Bioprod Process 89:47–52. https://doi.org/10.1016/j.fbp.2010.03.003
27. Zhou W, Yan Y, Mi J et al (2018) Simulated digestion and fermentation in vitro by human gut microbiota of polysaccharides from bee collected pollen of Chinese wolfberry. J Agric Food Chem 66:898–907. https://doi.org/10.1021/acs.jafc.7b05546
28. Wu W, Qiao J, Xiao X et al (2020) In vitro and in vivo digestion comparison of bee pollen with or without wall-disruption. J Sci Food Agric 101:2744–2755. https://doi.org/10.1002/jsfa.10902
29. Anderson KE, Sheehan TH, Mott BM et al (2013) Microbial ecology of the hive and pollination landscape: bacterial associates from floral nectar, the alimentary tract and stored food of honey bees (Apis mellifera). PLoS One 8:e83125. https://doi.org/10.1371/journal.pone.0083125
30. Di Cagno R, Filannino P, Cantatore V et al (2019) Novel solid-state fermentation of bee-collected pollen emulating the natural fermentation process of bee bread. Food Microbiol 82:218–230. https://doi.org/10.1016/j.fm.2019.02.007
31. Khalifa SAM, Elashal M, Kieliszek M et al (2020) Recent insights into chemical and pharmacological studies of bee bread. Trends Food Sci Technol 97:300–316. https://doi.org/10.1016/j.tifs.2019.08.021
32. Kieliszek M, Piwowarek K, Kot AM et al (2018) Pollen and bee bread as new health-oriented products: a review. Trends Food Sci Technol 71:170–180. https://doi.org/10.1016/j.tifs.2017.10.021
33. Campos M, Frigerio C, Lopes J et al (2010) What is the future of bee-pollen? J ApiProduct ApiMedical Sci 2:131–144. https://doi.org/10.3896/IBRA.4.02.4.01
34. Zuluaga-Domínguez C, Serrato-Bermudez J, Quicazán M (2018) Influence of drying-related operations on microbiological, structural and physicochemical aspects for processing of bee-pollen. Eng Agric Environ Food 11:57–64. https://doi.org/10.1016/j.eaef.2018.01.003
35. Zuluaga CM, Serrato JC, Quicazan MC (2015) Bee-pollen structure modification by physical and biotechnological processing: influence on the availability of nutrients and bioactive compounds. Chem Eng Trans 43:79–84. https://doi.org/10.3303/CET1543014
36. Castellanos-Paez BA, Durán-Jiménez A, Fuenmayor CA et al (2022) Effects of solar drying on the structural and thermodynamic characteristics of bee pollen. Vitae 29:350572. https://doi.org/10.17533/udea.vitae.v29n3a350572
37. Duran A, Quicazán MC, Zuluaga CM (2019) Effect of solar drying process on bioactive compounds and antioxidant activity in vitro of high Andean region bee pollen. Chem Eng Trans 75:91–96. https://doi.org/10.3303/CET1975016
38. Carpes S, Mourao G, Alencar S et al (2009) Chemical composition and free radical scavenging activity of Apis mellifera bee pollen from Southern Brazil. Brazilian J Food Technol 12:220–229
39. Kroyer G, Hegedus H (2001) Evaluation of bioactive properties of pollen extracts as functional dietary food supplement. Innov Food Sci Emerg Technol 2:171–174. https://doi.org/10.1016/S1466-8564(01)00039-X
40. Rebiai A, Lanez T (2012) Chemical composition and antioxidant activity of Apis mellifera bee pollen from northwest Algeria. J Fundam Appl Sci 4:26–35. https://doi.org/10.4314/jfas.v4i2.5
41. Keskin M, Özkök A (2020) Effects of drying techniques on chemical composition and volatile constituents of bee pollen. Czech J Food Sci 38:203–208. https://doi.org/10.17221/79/2020-CJFS

42. Barajas J, Cortes-Rodriguez M, Rodríguez-Sandoval E (2012) Effect of temperature on the drying process of bee pollen from two zones of Colombia. J Food Process Eng 35:134–148. https://doi.org/10.1111/j.1745-4530.2010.00577.x
43. Oliveira K (2006) Caracterização do pólen apícola e utilização de vitaminas antioxidantes como indicadoras do processo de desidratação. University of São Paulo, Brazil
44. Vidinamo F, Fawzia S, Karim MA (2021) Effect of drying methods and storage with agro-ecological conditions on phytochemicals and antioxidant activity of fruits: a review. Crit Rev Food Sci Nutr 62:353–361. https://doi.org/10.1080/10408398.2020.1816891
45. Aličić D, Flanjak I, Ačkar Đ et al (2020) Physicochemical properties and antioxidant capacity of bee pollen collected in Tuzla Canton (B&H). J Cent Eur Agric 21:42–50. https://doi.org/10.5513/JCEA01/21.1.2533
46. Wang J, Sun B, Cao Y et al (2009) Enzymatic preparation of wheat bran xylooligosaccharides and their stability during pasteurization and autoclave sterilization at low pH. Carbohydr Polym 77:816–821. https://doi.org/10.1016/j.carbpol.2009.03.005
47. Fuenmayor CA, Quicazán MC, Figueroa J (2011) Desarrollo de un suplemento nutricional mediante la fermentación en fase sólida de polen de abejas empleando bacterias ácido lácticas probióticas. Aliment Hoy 20:17–39
48. Mora-Adames WI, Fuenmayor CA, Benavides-Martín MA et al (2021) Bee pollen as a novel substrate in pilot-scale probiotic-mediated lactic fermentation processes. LWT 141:110868. https://doi.org/10.1016/j.lwt.2021.110868
49. Zuluaga CM, Serrato JC, Quicazán MC (2015) Chemical, nutritional and bioactive characterization of Colombian bee-bread. Chem Eng Trans 43:175–180. https://doi.org/10.3303/CET1543030
50. Adaškevičiūtė V, Kaškonienė V, Mickienė R et al (2018) The impact of sterilization process on bee pollen antioxidant activity. Vytautas Magnus University, Kaunas
51. Huang HW, Hsu CP, Wang CY (2020) Healthy expectations of high hydrostatic pressure treatment in food processing industry. J Food Drug Anal 28:1–13. https://doi.org/10.1016/j.jfda.2019.10.002
52. Sevenich R, Bark F, Kleinstueck E et al (2015) The impact of high pressure thermal sterilization on the microbiological stability and formation of food processing contaminants in selected fish systems and baby food puree at pilot scale. Food Control 50:539–547. https://doi.org/10.1016/j.foodcont.2014.09.050
53. Rastogi NK, Raghavarao KSMS, Balasubramaniam VM et al (2007) Opportunities and challenges in high pressure processing of foods. Crit Rev Food Sci Nutr 47:69–112. https://doi.org/10.1080/10408390600626420
54. Zuluaga C, Martínez A, Fernández J et al (2016) Effect of high pressure processing on carotenoid and phenolic compounds, antioxidant capacity, and microbial counts of bee-pollen paste and bee-pollen-based beverage. Innov Food Sci Emerg Technol 37:10–17. https://doi.org/10.1016/j.ifset.2016.07.023
55. Aylanc V, Falcão SI, Vilas-Boas M (2023) Bee pollen and bee bread nutritional potential: chemical composition and macronutrient digestibility under in vitro gastrointestinal system. Food Chem 413:135597. https://doi.org/10.1016/j.foodchem.2023.135597
56. Disayathanoowat T, Li HY, Supapimon N et al (2020) Different dynamics of bacterial and fungal communities in hive-stored bee bread and their possible roles: a case study from two commercial honey bees in China. Microorganisms 8:264. https://doi.org/10.3390/microorganisms8020264
57. Moreno Galarza L (2012) Aislamiento y Selección de Lactobacillus sp con potencial probiótico a partir de pan de abejas. Universidad Nacional de Colombia, Colombia
58. Mathur H, Beresford TP, Cotter PD (2020) Health benefits of lactic acid bacteria (LAB) fermentates. Nutrients 12:1679. https://doi.org/10.3390/nu12061679
59. Castellone V, Bancalari E, Rubert J et al (2021) Eating fermented: health benefits of LAB-fermented foods. Foods 10:2639. https://doi.org/10.3390/foods10112639

60. Garcia C, Guerin M, Souidi K et al (2020) Lactic fermented fruit or vegetable juices: past, present and future. Beverages 6:8. https://doi.org/10.3390/beverages6010008
61. Rai AK, Jeyaram K (2017) Role of yeasts in food fermentation. In: Satyanarayana T, Kunze G (eds) Yeast diversity in human welfare. Springer, Singapur, pp 83–113
62. Vamanu A, Vamanu E, Drugulescu M et al (2006) Identification of a lactic bacterium strain used for obtaining a pollen-based probiotic product. Turk J Biol 30:75–80
63. Vamanu A, Vamanu E, Popa O et al (2008) Obtaining of a symbiotic product based on lactic acid bacteria, pollen and honey. Pak J Biol Sci 11:613–617. https://doi.org/10.3923/pjbs.2008.613.617
64. Vamanu E, Vamanu A, Popa O et al (2010) The antioxidant effect of a functional product based on probiotic biomass, pollen and honey. Sci Pap Anim Sci Biotechnol 43:331–336
65. Salazar-González CY, Díaz-Moreno C (2014) Características bioactivas y sensoriales en procesos de fermentación en fase sólida de Polen apícola. Universidad Nacional de Colombia, Colombia
66. Zuluaga-Dominguez CM, Quicazan MC (2019) Effect of fermentation on structural characteristics and bioactive compounds of bee-pollen based food. J Apic Sci 63:209–222. https://doi.org/10.2478/jas-2019-0016
67. Yan S, Li Q, Xue X et al (2019) Analysis of improved nutritional composition of bee pollen (Brassica campestris L.) after different fermentation treatments. Int J Food Sci Technol 54:2169–2181. https://doi.org/10.1111/ijfs.14124
68. Duan C, Feng Y, Zhou H et al (2015) Optimization of fermentation condition of man-made bee-bread by response surface methodology. In: Zhang TC, Nakajima M (eds) Advances in applied biotechnology. Springer, Berlin, pp 353–363
69. Shirsat DV, Kad SK, Wakhle DM (2019) Solid state fermentation of bee-collected pollen. Int J Curr Microbiol App Sci 8:1557–1563
70. Filannino P, Di Cagno R, Gambacorta G et al (2021) Volatilome and bioaccessible phenolics profiles in lab-scale fermented bee pollen. Foods 10:286. https://doi.org/10.3390/foods10020286
71. Uţoiu E, Matei F, Toma A et al (2018) Bee collected pollen with enhanced health benefits, produced by fermentation with a Kombucha consortium. Nutrients 10:1365. https://doi.org/10.3390/nu10101365
72. Zhang Z, Cao H, Chen C et al (2017) Effects of fermentation by Ganoderma lucidum and Saccharomyces cerevisiae on rape pollen morphology and its wall. J Food Sci Technol 54:4026–4034. https://doi.org/10.1007/s13197-017-2868-1
73. Xu X, Sun L, Dong J et al (2009) Breaking the cells of rape bee pollen and consecutive extraction of functional oil with supercritical carbon dioxide. Innov Food Sci Emerg Technol 10:42–46. https://doi.org/10.1016/j.ifset.2008.08.004
74. Dong J, Gao K, Wang K et al (2015) Cell Wall disruption of Rape Bee pollen treated with combination of Protamex hydrolysis and ultrasonication. Food Res Int 75:123–130. https://doi.org/10.1016/j.foodres.2015.05.039
75. Zuluaga-Domínguez C, Castro-Mercado L, Quicazán MC (2019) Effect of enzymatic hydrolysis on structural characteristics and bioactive composition of bee-pollen. J Food Process Preserv 43:e13983. https://doi.org/10.1111/jfpp.13983
76. Xie J, Wei F, Luo L et al (2022) Effect of cell wall-disruption processes on wall disruption, antioxidant activity and nutrients in tea pollen. Int J Food Sci Technol 57:3361–3374. https://doi.org/10.1111/ijfs.15612
77. Xue F, Li C (2023) Effects of ultrasound assisted cell wall disruption on physicochemical properties of camellia bee pollen protein isolates. Ultrason Sonochem 92:106249. https://doi.org/10.1016/j.ultsonch.2022.106249
78. ICTA-UNC (2011) Informe Técnico. Proyecto Estrategias para Establecer la Denominación de Origen de Productos de las Abejas en Colombia. Universidad Nacional de Colombia, Colombia
79. Silva CEF, Abub AKS, da Silva ICC et al (2019) Acceptability of tropical fruit pulps enriched with vegetal/microbial protein sources: viscosity, importance of nutritional information and

changes on sensory analysis for different age groups. J Food Sci Technol 56:3810–3822. https://doi.org/10.1007/s13197-019-03852-0

80. Conte P, del Caro A, Balestra F et al (2018) Bee pollen as a functional ingredient in gluten-free bread: a physical-chemical, technological and sensory approach. LWT 90:1–7. https://doi.org/10.1016/j.lwt.2017.12.002
81. Conte P, Urgeghe PP, Petretto GL et al (2020) Nutritional and aroma improvement of gluten-free bread: is bee pollen effective? LWT 118:108711. https://doi.org/10.1016/j.lwt.2019.108711
82. Zlatev Z, Taneva I, Baycheva S et al (2018) A comparative analysis of physico-chemical indicators and sensory characteristics of yogurt with added honey and bee pollen. Bulg J Agric Sci 24:132–144

Chapter 13
Good Practice of Pollen Collection-What Pollen Traps Are Better Choice

Nebojša M. Nedić

13.1 Introduction

Pollen collecting by honey bees is a well organised collective behaviour. Bees collect pollen from the flowers, then moisture it by nectar and secretions from their mouth, producing in this way "bee pollen". Then they pack it in the form of pollen loads in pollen baskets (corbiculae) of their hind legs and transport it to the hives [1].

Nutrition of honey bee is based on two main sources, nectar and pollen [2, 3]. Nectar, which is in a bee hive being converted into honey, supplies bees with their nutritional needs for carbohydrates while pollen provides them with protein, lipids, vitamins and minerals [4].

Bee colony spends about 120 kg of nectar and 20 kg of pollen for its development at an annual level. Bee colonies do not keep a large quantity of pollen stored in a hive but actively regulate the quantity of pollen according to their current needs [5]. Pollen reserves in a hive are small and they amount to about 1 kg [6]. The requirements of bee colony for pollen are closely associated with quantity of larval brood and with pollen stored in a hive. Pollen is important both for the nutrition of bee brood and for a development and functioning of adult bees [7].

Bee colonies in certain periods during the year or in some localities may be exposed to a shortage of pollen from nature. In this case, the bees suffer from a lack of nutrients, often reduce the amount of brood grown, which leads to the weakening of the bee colony, weaker production results or a decrease in its efficiency if it is used in pollination of plant crops [8]. Therefore, an efficient way of collecting pollen when there is in excess during the season and its application in additional nutrition when it is needed helps bee colonies to overcome these unfavorable periods.

N. M. Nedić (✉)
University of Belgrade, Faculty of Agriculture, Chair for Breeding and Reproduction of Domestic and Bred Animals, Belgrade, Serbia
e-mail: nedicn@agrif.bg.ac.rs

N. Ecem Bayram et al. (eds.), *Pollen Chemistry & Biotechnology*,
https://doi.org/10.1007/978-3-031-47563-4_13

The diet of bees with the addition of different levels of pollen can have a favorable effect on the physiological characteristics of workers and on the queen fecundity [9].

Additionally to its importance for insects nutrition, bee pollen is used to a certain extent as a source of protein or an alternative antibiotic in animal feed [10].

Bee pollen is a nutritionally rich food and contains a number of bioactive ingredients that have shown various positive effects on the improvement of human health [11, 12]. Therefore, this natural product is considered as one of the promising functional supplements in human nutrition, and from this aspect, an efficient method of its obtaining is very important [13, 14].

13.2 A Brief History of the Development of Pollen Traps

The trait of a bee colony to collect pollen in order to satisfy its need for food has been used by a man to make pollen traps. Their construction and place of installation on hive may vary a lot (Fig. 13.1 and 13.2). Bee pollen can be artificially removed from the bee legs and collected by using special traps for bee pollen [15–17].

The first use of a pollen trap called "pollen guards" was done in order to prevent bees to take the pollen into the hive [18]. It was followed by a modification with a 5-mesh hardware cloth that replaced the screen made of perforated metal [19]. Next the pollen trap was described installed in the base of the hive so that the hive entrance could be used normally [20]. However, installed pollen screen in time became congested with dead bees and drones, so the extensions of hive had to be lifted in order to clean the trap. In order to enable the drones to leave the hive unhindered, in a modification of the bottom type trap the holes were made and there was no congestion of grid area with dead bees [21]. Over the time different authors worked on a modification of pollen traps [22–25]. Modern pollen traps have mostly plastic tubes that enable workers and drones to leave the hive (Fig. 13.3).

During the seventies of the twentieth century there appeared a pollen trap for obtaining small samples in a short time period that could be simply mounted on the front side of the hive without any additional manipulation [26]. Thereafter a pollen

Fig. 13.1 Cross section of an external front entrance pollen trap. (Figure by Beljanski 2023)

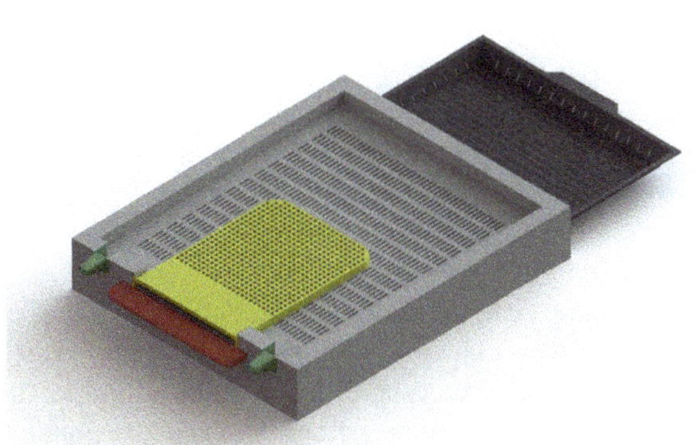

Fig. 13.2 Bottom pollen trap. (Figure by Beljanski N. 2023)

Fig. 13.3 Pollen trap with plastic tubes for bees and drones to fly out of hive. (Figure by Nedić N.)

trap with cleaning grid and trash collector was contrived [27]. This trap is made in such a way so as to be mounted at the front side of the hive between two brood chambers. An already existing entrance to the hive would close and bees could enter it through entrance on the trap. Bees had to pass through two double grids for removing pollen under which pollen drawer was placed. In the back of the trap there was a grid with trash collector beneath it. Pollen drawer and trash drawer could be drawn from the side of the trap in order not to disturb bees and even some apitechnical interventions could be performed without removing the trap. The next step was pollen trap that could be installed in the bottom, in the middle or at the top of the hive [28]. The improvements made in variously redesigned pollen traps and their efficacy has continued to be investigated to the present day [29, 30].

13.3 Types of Pollen Traps

Pollen trap is a mechanical device by which bees are forced to pass through 5 mm diameter-holes grid. These openings are large enough for foragers to pass through but at the same time they are small enough for a certain quantity of pollen loads from pollen baskets be removed when entering the hive [30, 31]. A well constructed pollen trap enables some bees to carry pollen pellets into the nest even past the pollen traps mounted at the hive entrance.

Pollen traps can be external and internal. External traps are mostly mounted at the level of bee's entrance (Fig. 13.4) or between boxes of hive on the front side.

Internal traps, according to the position of the mounting, can be bottom or top mounted, such as Sundance I and Sundance II pollen traps [32]. Most often, the internal pollen trap is mounted instead of the bottom board on the bottom of the hive (Fig. 13.5).

Pollen traps can be made of wood, plastic, metal or a combination of these materials (Figs. 13.6, 13.7, and 13.8).

The external pollen trap is being easily mounted by ears eyelets at the entrance of the hive. The trap mounted in this way forces the bees to pass through the pollen screen (Fig. 13.4), whereby the pollen loads are removed from the pollen baskets and fall through the fine mesh into the drawer. When the pollen trap is mounted on

Fig. 13.4 External front entrance pollen trap mounted on the hive. (Figure by Nedić N.)

Fig. 13.5 Wooden bottom pollen trap mounted instead bottom board. (Figure by Nedić N.)

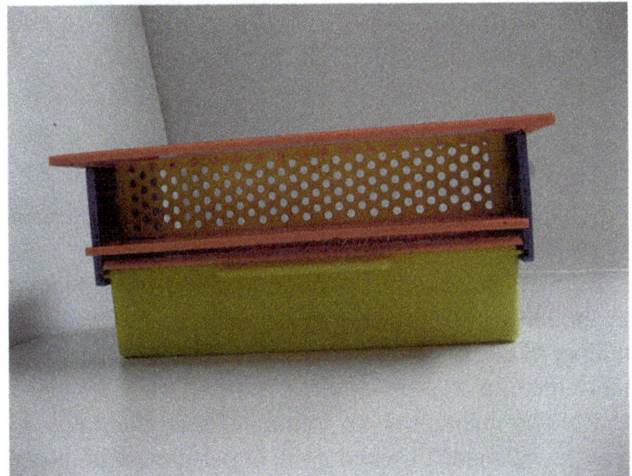

Fig. 13.6 External front entrance pollen trap made of plastic. (Figure by Nedić N.)

the hive it should take care the bees have only one entrance into the hive and that they have to pass the mesh of pollen trap in order to get to their nest.

External front entrance pollen trap (Fig. 13.9) consists of: 1. pollen screen, 2. pollen screen adjustment, 3. mesh screen with alighting platform, 4. mesh screen, 5. drawer, 6 pollen trap cover, 7. eyelets for hanging on the beehive, 8. side wall of the pollen trap, 9. metal shade screen on front side of trap (optional).

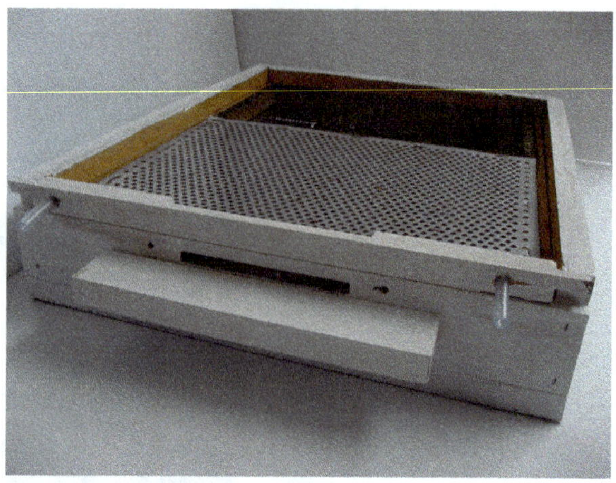

Fig. 13.7 Bottom wooden pollen trap with metal plate pollen screen. (Figure by Nedić N.)

Fig. 13.8 Bottom pollen trap made of plastic „Tehnoplast Gligorijević". (Figure by Nedić N.)

Between the alighting platform that is extended into a mesh (Fig. 13.9 marked by the number 3) and fine mesh that covers collecting drawer (Fig. 13.9 marked by the number 4) there is a 9 mm space which enables bee foragers and drones to fly freely out of the hive.

The two main parts are common for all the pollen traps. One is a pollen screen (Fig. 13.10) made of wired mesh or perforated tin of plastic or metal through which openings bees which carry pollen loads must pass. Openings are 5 mm in diameter. When designing trap with wired mesh for removing pollen the two grids mounted at the distance of around 7 mm from one another are often used. Openings on these meshes do not match one another so that forager bees are forced, when passing, to crawl through and thus pollen loads get stripped off their hind legs.

In practice the traps where pollen screen is a perforated strip of sheet plastic or metal are more often used. By a small wheel (1) on a side wall of a pollen trap it is possible to adjust position of pollen screen (Fig. 13.9). The first day after mounting

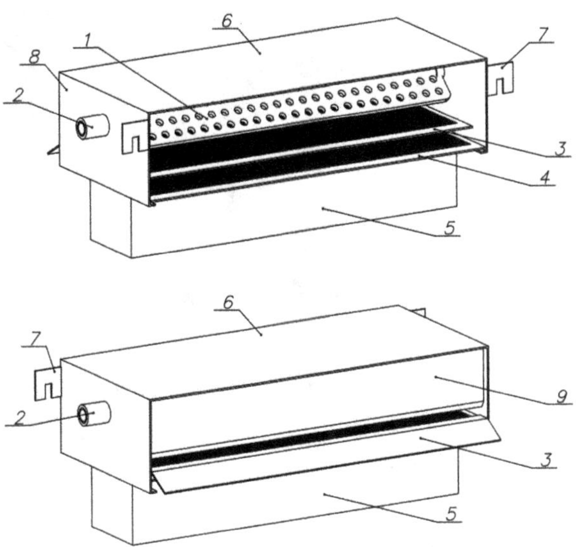

Fig. 13.9 Technical view of external front entrance pollen trap. (Figure by Beljanski N. 2023)

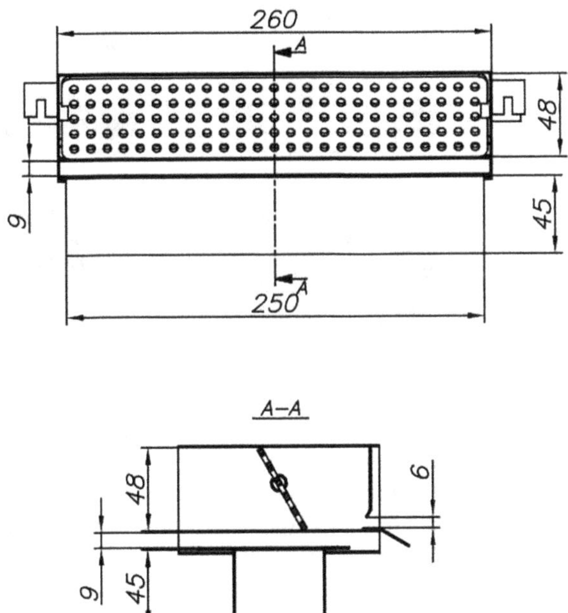

Fig. 13.10 Technical cross-section view of pollen screen and its position in the external pollen trap. (Figure by Beljanski N. 2023)

Fig. 13.11 Passing the forager bee through the plate pollen screen. (Figure by Nedić N.)

Fig. 13.12 Plastic drawer for collecting pollen covered by mesh. (Figure by Nedić N.)

the trap the pollen screen can be adjusted so that bees can freely enter the hive without stripping pollen loads. In this way bees get used to mounted traps. On the following day pollen screen is mounted into a position which force bees, on their way back to the hive, to pass through openings on pollen screen (Fig. 13.11).

The other standardized part of every trap is a drawer or pollen tray covered by very fine mesh whose openings are of diameter of 3 mm (Fig. 13.12). Those mesh prevents bees to get into a drawer with collected pollen and can be made of wire or plastic. The drawer in which pollen is collected has a bottom of fine wire what enables ventilation of collected pollen (Fig. 13.12). Collected pollen contains high moisture what favours development of microorganisms so the pollen from the collecting drawer should be removed daily [33, 34].

Bottom pollen trap (Fig. 13.13) made of plastic consists of: 1. body of the pollen trap, 2. pollen drawer, 3. pollen screen (pollen plate with openings), 4. pipe for

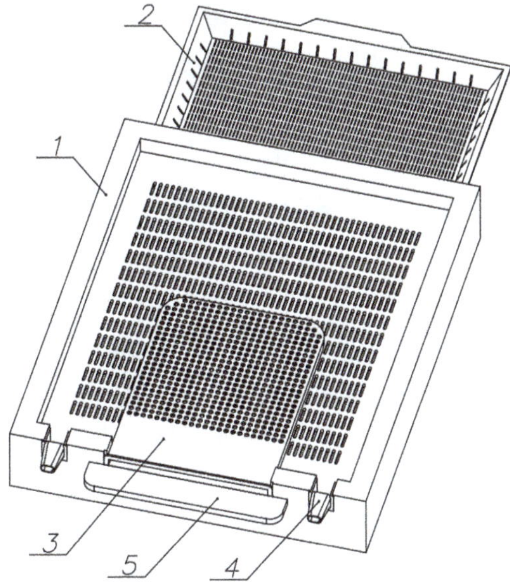

Fig. 13.13 Technical view of the bottom pollen trap. (Figure by Beljanski N. 2023)

Fig. 13.14 Passing the forager bee trough plastic pollen screen of bottom pollen trap. (Figure by Nedić N.)

letting out forager bees and drones from the hive, 5. alighting platform with hind hive entrance. The pollen plate with openings is the main part of this type of bottom trap and it can be easily removed, which regulates the process of pollen collection.

The bottom pollen trap is installed by lifting the extensions of the hive and it is placed instead of the bottom board. With this type of pollen trap, bees can freely enter the hive entrance, but they can only reach the nest by passing through a horizontally placed pollen screen (Fig. 13.8 and 13.14).

13.4 The Efficiency of Pollen Collectors and the Yield of Collected Pollen

In the description of different pollen traps, their efficacy is often emphasized. It is obtained by counting the number of bees that enter the trap carrying pollen pellets and by determining pollen percentage collected in a pollen drawer [31, 35, 36]. The efficacy of various traps to take off the pollen ranges from 1% to more than 60% [2, 37]. By use of a good trap at least 50% pollen loads from the legs of foragers can be taken off before they enter the hive [38].

Pollen collection begins with the mounting of pollen traps on the hive in the spring and during the season when there are more plants that produce considerable quantities of pollen. Forager bees collect larger quantity of pollen in the morning [1, 39]. It is recommended to remove pollen traps from the hive for a few days after 10 days of their use in order to enable bee colony to normally collect pollen from nature.

The mass of pollen pellets, which bees take into the hive, can vary depending on the time of day and plant species which pollen originates from. An average mass of a pollen pellet is 7.9 mg and bee foragers can bring two such loads on their hind legs [40]. A yield of pollen per hive depends on plant source which it is collected from, the strength of bee colony, quantity of open brood, type of pollen trap, weather conditions [41]. One bee colony can be expected to yield from 1 to 7 kg bee pollen annually [30].

An activity of foragers for obtaining pollen is intensified in order to satisfy bee colony's needs for pollen [42]. Foragers made 16.8% distance flights to collect pollen and 83.2% flights to reach the source of nectar showing that bees cover greater distances on account of reaching the source of nectar in relation to pollen. Middle distance covered by foragers to obtain pollen was 1.074 m while total middle distance covered by foragers to obtain nectar was 1.408 m [43].

Quantity of daily collected pollen can vary significantly from 50 g to 200 g [44]. It is closely associated with internal requirements of bee colony, climate conditions that affect the behaviour of pollen foragers, presence of pollen attractants, quantity of flowering plants in the zone where bee colonies are situated and the type of pollen trap [40, 45, 46]. The average daily pollen yield per colony obtained from the front entrance pollen trap and the bottom pollen trap was 12.55 ± 1.66 g and 22.5 ± 1.25 g, respectively [47].

Pollen drawer with collected pollen should be emptied every day because otherwise pollen quickly absorbs moisture from the air and is susceptible to decay. Upon collecting pollen from pollen traps it is taken to drying and further processing.

Pollen traps can be stressful to bee colonies that raise a great quantity of brood. By mounting the pollen traps on bee hives the foragers can be encouraged to more intense working activity to collect pollen from nature. Presence of open brood and especially of older larvae can act stimulatingly on foragers to collect pollen [48].

The use of highly efficient pollen traps over a longer time period can provoke decreasing of quantity of raised brood. Use of pollen traps can cause the decrease of yield of honey per bee colony. Reason for this phenomenon can be found in foragers` more difficult passing through the screen of pollen trap. It is thought that long-term use of pollen traps can be associated with the occurrence of some diseases like viruses of chronic paralysis or chalkbrood [49, 50]. A short-time mounting of pollen traps, during a period of high availability of pollen in nature does not negatively affect production of brood.

13.4.1 Characteristics of Different Types of Pollen Traps and Concluding Remarks

When selecting pollen traps, certain preconditions should be considered. Pollen trap should have some characteristics such as simple mounting, providing simple manipulation to colonies, simple removing of the collected pollen from the trap without disturbing the bees in their work, the size of pollen drawer should be large enough to collect pollen that bees bring within the period of 1–2 days, obtaining clean pollen, simple maintenance, simple dismantling and disinfecting.

Front entrance pollen trap and bottom pollen trap have their advantages and disadvantages. The advantage of an external front entrance pollen trap is in its easy mounting and removing. This trap is universal but its drawer capacity of daily collected pollen is smaller in relation to bottom traps. When removing pollen from the drawer of external front entrance trap, one passes in front of the hive, which disturbs the bees.

The advantage of a bottom pollen trap is in its larger surface of pollen screen for removing the pollen thus higher yield of pollen per day can be obtained. In external pollen traps (Fig. 13.4) number of openings on pollen screen is 149, while in bottom pollen trap (Fig. 13.8) number of openings on pollen screen amounts to 462. The bottom trap ensures better protection of pollen from moisture. Pollen removal in the bottom trap is done by pulling out the drawer from the back of the hive without disturbing the bees.

There are variants of bottom traps where by moving the position of the alighting platform, the same effect is achieved and the bees pass with pollen loads freely into their nest (Fig. 13.7). With some subtypes of bottom trap (Fig. 13.8), there is a possibility that by removing the pollen plate, the bees can freely enter the nest, without taking away the pollen loads, and thus the bottom trap turns into a bottom board.

Plastic bottom pollen trap shown in Fig. 13.8 can be completely dissembled and efficiently cleaned and disinfected.

The disadvantage of bottom trap is that during mounting, the hive boxes with bees must be separated from the bottom board that is being removed and pollen trap installed at that place. During dismantling the procedure is being repeated. Since the bottom pollen trap is placed on the bottom of the hive, particles from the nest fall and end up in the trap drawer.

The bee pollen removal technology is based on the use of different types of pollen traps. Many papers describe the chemical composition of bee pollen, its increasing importance as a functional food, but also its sensitivity to microbiological contamination. Taking into account good beekeeping practice and guidelines for bee pollen to maintain food safety, it is necessary to apply the type of pollen traps that has satisfactory efficiency, that can be easily installed and dismantled on the hive without significantly disturbing the bees, with an accessible way of removing the collected pollen and that their construction enables efficient and quick disinfection.

Acknowledgements The author gratefully acknowledge to BSc Nikola Beljanski for preparing and creating technical drawings and pictures.

References

1. Louveaux J (1990) Les relations abeilles-pollens. Bull Soc Bot Fr Actual Bot 137:121–131. https://doi.org/10.1080/01811789.1990.10827009
2. Rodney S, Purdy J (2020) Dietary requirements of individual nectar foragers, and colony-level pollen and nectar consumption: a review to support pesticide exposure assessment for honey bees. Apidologie 51:163–179. https://doi.org/10.1007/s13592-019-00694-9
3. Nedić N, Nešović M, Radišić P, Gašić U et al (2022) Polyphenolic and chemical profiles of honey from the Tara Mountain in Serbia. Front Nutr 24:941463. https://doi.org/10.3389/fnut.2022.941463
4. DeGrandi-Hoffman G, Eckholm JB, Huang HM (2013) A comparison of bee bread made by Africanized and European honey bees (*Apis mellifera*) and its effects on hemolymph protein titers. Apidologie 44:52–63. https://doi.org/10.1007/s13592-012-0154-9
5. Hoover SE, Ovinge LP (2018) Pollen collection, honey production, and pollination services: managing honey bees in an agricultural setting. J Econ Entomol 111:1509–1516. https://doi.org/10.1093/jee/toy125
6. Seeley TD (1995) The wisdom of the hive: the social physiology of honey bee colonies. Harvard University Press, Cambridge, MA/London
7. Brodschneider R, Crailsheim K (2010) Nutrition and health in honey bees. Apidologie 41:278–294. https://doi.org/10.1051/apido/2010012
8. Avni D, Dag A, Shafir S (2009) The effect of surface area of pollen patties fed to honey bee (*Apis mellifera*) colonies on their consumption, brood production and honey yields. J Apic Res 48(1):23–28. https://doi.org/10.3896/IBRA.1.48.1.06
9. Fine JD, Shpigler HY, Ray AM et al (2018) Quantifying the effects of pollen nutrition on honey bee queen egg laying with a new laboratory system. PLoS One 13:e0203444. https://doi.org/10.1371/journal.pone.0203444
10. Abdelnour SA, Abd El-Hack ME, Alagawany M et al (2019) Beneficial impacts of bee pollen in animal production, reproduction and health. J Anim Physiol Anim Nutr 103:477–484. https://doi.org/10.1111/jpn.13049
11. Hsu P-S, Wu T-H, Huang M-Y et al (2021) Nutritive value of 11 bee pollen samples from major floral sources in Taiwan. Foods 10:2229. https://doi.org/10.3390/foods10092229
12. Alshallash KS, Abolaban G, Elhamamsy SM et al (2023) Bee pollen as a functional product – chemical constituents and nutritional properties. J Ecol Eng 24:173–183. https://doi.org/10.12911/22998993/156611

13. Kostić AŽ, Milinčić DD, Barać MB et al (2020) The application of pollen as a functional food and feed ingredient-the present and perspectives. Biomol Ther 10:84. https://doi.org/10.3390/biom10010084
14. Kostić AŽ, Milinčić DD, Nedić N et al (2021) Phytochemical profile and antioxidant properties of bee-collected artichoke (*Cynara scolymus*) pollen. Antioxidants 10:1091. https://doi.org/10.3390/antiox10071091
15. Campos RGM, Bogdanov S, de Almeida-Muradian BL et al (2008) Pollen composition and standardisation of analytical methods. J Apic Res 47:154–161. https://doi.org/10.1080/00218839.2008.11101443
16. Siuda M, Jerzy W, Tomasz B (2012) The effect of various storage methods on organoleptic quality of bee pollen loads. J Apic Sci 56:71–79. https://doi.org/10.2478/v10289-012-0008-8.1
17. Mauriello G, De Prisco A, Di Prisco G et al (2017) Microbial characterization of bee pollen from the Vesuvius area collected by using three different traps. PLoS One 12:e0183208. https://doi.org/10.1371/journal.pone.0183208
18. Farrar CL (1934) Bees must have pollen. Glean Bee Cult 62:276–278
19. Todd EF, Bishop KR (1940) Trapping honeybee-gathered pollen and factors affecting yields. J Econ Entomol 33:866–870. https://doi.org/10.1093/jee/33.6.866
20. Schaefer CW, Farrar CL (1946) The use of pollen traps and pollen supplements in developing honeybee colonies. U.S. Dept Agr Bur Ent Pl Quar E 531:1–13
21. Smith MV, Adie A (1963) A new design in pollen traps. Can Bee J 74:4–5
22. Nye WP (1959) A modified pollen trap for honeybee hives. J Econ Entomol 52:1024–1025. https://doi.org/10.1093/jee/52.5.1024
23. Durante G (1960) Trappe à pollen à grille horizontale. Abeilles et Fleurs 81:5–8
24. Smith MV (1965) The O.A.C. pollen trap. Apic Dept Ontario Agr Col: 2 pp
25. Detroy BF, Harp ER (1976) Trapping pollen from honey bee colonies, USDA production research report 163. US Department of Agriculture, Agricultural Research Service
26. Stewart JD, Shimanuki H (1971) Rapid sample pollen trap for honey bees. J Econ Entomol 63:1350
27. Kauffeld MN (1976) Pollen trap with cleaning grid. US Patent 3,995,338, 7 Dec 1976
28. Brown R (1982) Pollen trap for beehives. US Patent 4,337,541, 6 July 1982
29. Mahmood R, Asad S, Sarwar G, Iftikhar F, Abdul Qadir Z (2013) Comparative study of pollen traps on improvement collecting technology. Pak J Sci 65:202–205
30. Eman M, Ali AM, Ghazala N (2022) Evaluation of the efficiency different types of bee pollen collection traps in honey bee colonies during summer season. Arab Univ J Sci 30:141–146. https://doi.org/10.21608/AJS.2022.111909.1448
31. Delaplane SK, Dag A, Danka GR et al (2013) Standard methods for pollination research with *Apis mellifera*. J Apic Res 52:1–28. https://doi.org/10.3896/IBRA.1.52.4.12
32. MacFawn D (2020) Collecting pollen. Bee Cult 10:70–73
33. Petrović T, Nedić N, Paunović D et al (2014) Natural mycrobiota and aflatoxin B_1 presence in bee pollen collected in Serbia. Biotechnol Anim Husb 30:731–741. https://doi.org/10.2298/BAH1404731P
34. Bogdanov S (2017) Chapter 1: The pollen book. In: Bogdanov S (ed) Pollen: collection, harvest, composition, quality. Available online: https://www.bee-hexagon.net/english/bee-products/downloads-pollen-book/. Accessed 15 Jan 2023
35. Levin MD, Loper GM (1984) Factors affecting pollen trap efficiency. Am Bee J 124:721–723
36. Goodwin MR, Perry HJ (1992) Use of pollen traps to investigate the foraging behaviour of honey bee colonies in kiwifruit orchards. N Z J Crop Hortic Sci 20:23–26
37. Ismail AHM, Owayss AA, Mohanny KM, Salem RA (2013) Evaluation of pollen collected by honey bee, Apis mellifera L. colonies at Fayoum governorate, Egypt. Part 1: botanical origin. J Saudi Soc Agric Sci 12:129–135. https://doi.org/10.1016/j.jssas.2012.09.003
38. Knäbe S et al (2014) Available methods for the sampling of nectar, pollen, and flowers of different plant species. In: Oomen AP, Pistorius J (eds) Hazards of pesticides to bees. 12th International Symposium of the ICP-PR Bee Protection Group, Ghent, September 2014.

Section II: Developments in laboratory, semi-field and field testing for honeybees. Julius Kühn-Institut, Berlin, p 131
39. Reyes-Carrillo JL, Eischen FA, Cano-Rios P et al (2007) Pollen collection and honey bee forage distribution in cantaloupe. Acta Zool Mex 23:29–36
40. García-García MC, Ortiz PL, Dapena MJD (2004) Variations in the weights of pollen loads collected by *Apis mellifera* L. Grana 43:183–192. https://doi.org/10.1080/00173130410020350
41. Crane E (1975) Pollen and its harvesting. Bee World 56(4):155–158
42. Free JB (1967) Factors determining the collection of pollen by honeybee foragers. Anim Behav 15:134–144
43. Weidenmuller A, Tautz J (2002) In-hive behavior of pollen foragers (*Apis mellifera*) in honey bee colonies under conditions of high and low pollen need. Ethology 108:205–221. https://doi.org/10.1046/j.1439-0310.2002.00759.x
44. Dimou M, Thrasyvoulou A, Tsirakoglou V (2006) Efficient use of pollen traps to determine the pollen flora used by honey bees. J Apic Res 45:42–46. https://doi.org/10.1080/00218839.2006.11101312
45. Dimou M, Thrasyvoulou A (2007) Seasonal variation in vegetation and pollen collected by honeybees in Thessaloniki, Greece. Grana 46:292–299. https://doi.org/10.1080/00173130701760718
46. Nikolova I, Georgieva N, Kirilov A, Mladenova R (2016) Dynamics of dominant bees – pollinators and influence of temperature, relative humidity and time of the day on their abundance in forage crops in Pleven region, Bulgaria. J Global Agr Ecol 5:200–209
47. Raja S, Waghchoure SE, Mahmood R et al (2010) Comparative study on improvement in pollen collection technology. Halteres 1:1–6
48. Hellmich R, Rothenbuhler WC (1986) Relationship between different amounts of brood and the collection and use of pollen by the honey bee (*Apis mellifera*). Apidologie 17:13–20. https://doi.org/10.1051/apido:19860102
49. Dubois E, Reis C, Schurr F et al (2017) Effect of pollen traps on the relapse of chronic bee paralysis virus in honeybee (*Apis mellifera*) colonies. Apidologie 49:235–242. https://doi.org/10.1007/s13592-017-0547-x
50. Smart M, Otto C, Cornman R et al (2018) Using colony monitoring devices to evaluate the impacts of land use and nutritional value of forage on honey bee health. Agriculture. https://doi.org/10.3390/agriculture8010002

Chapter 14
Techno-Functional Properties of Pollen

Danijel D. Milinčić, Aleksandar Ž. Kostić, Slađana P. Stanojević, and Mirjana B. Pešić

14.1 Pollen-"The Only Perfectly Complete Food"

From ancient times until the present day, (bee) pollen, also known as "the life-giving dust" has been widely used in human diets as a source of healthy compounds, while in medicine it is famous for its potential curative and therapeutic properties [1–3]. However, it wasn't applied until the 1970s when pollen was introduced as a dietary supplement, which significantly increased the scientists' attention to its detailed characterization, as well as evaluation of its nutritional and biological properties. Thanks to effective and advanced instrumental techniques, globally popular pollen from different botanical and geographical origins has been well examined and reviewed in recent years. Nowadays, pollen is known as "the only perfectly complete food" because it contains all the nutritionally and functionally important compounds necessary for human health, such as carbohydrates, proteins, peptides, amino acids, lipids (essential fatty acids, phospholipids, phytosterols), vitamins (primarily group-B vitamins), carotenoids, minerals, and phenolic compounds (primarily quercetin and its derivatives) [4–9]. For these reasons, different pollen formulations in the form of granules, capsules, pellets, powders, oral liquids or candy bars are widely commercially available becoming one of the most commonly consumed food supplements in the world [2, 10]. Moreover, some countries with high pollen producing rates (Argentina, Brazil, Switzerland, China and Poland) have officially recognized pollen as an additive and developed regulations (standard norms) for the evaluation of its sensory, physico-chemical and microbiological criteria [9]. Recommended doses of pollen consumption for adults (20–40 g) have also

D. D. Milinčić (✉) · A. Ž. Kostić · S. P. Stanojević · M. B. Pešić
Department of Chemistry and Biochemistry, Faculty of Agriculture, University of Belgrade, Belgrade, Serbia
e-mail: danijel.milincic@agrif.bg.ac.rs

N. Ecem Bayram et al. (eds.), *Pollen Chemistry & Biotechnology*,
https://doi.org/10.1007/978-3-031-47563-4_14

been established. However, the digestibility and absorption of important nutrients from pollen products are often low due to the structure of the pollen grain [9, 10]. Recently, due to people's need to consume functional and healthy foods, pollen is gaining more importance as a functional ingredient in different food products. So far, pollen has been successfully incorporated into different beverages, bakery, confectionery, meat and dairy products, with the aim of improving their nutritional and functional properties [10]. However, most studies focus on the functional and health properties of pollen as a dietary or food supplement [1, 2, 10, 11], while the data for the techno-functional properties of pollen are often limited. Only a few studies evaluate the techno-functional properties of pollen from different botanical origins [12–14], as well as the effect that added pollen has on textural, surface, rheological and sensory properties of various fortified functional products [15–18].

Given the increasing popularity of pollen as a food additive, this chapter aims to provide an overview and summary of all research studies related to the techno-functional properties of pollen. Special attention will be paid to the lipophilic/hydrophilic nature of pollen which is responsible for its unique characteristics among food additives. The chapter will also highlight how the addition of pollen affects the techno-functional, textural, and sensory properties of newly formulated food products. It is worth noting that all research studies related to techno-functional properties were conducted on bee pollen. Therefore, the rest of the chapter will refer to and describe pollen accordingly.

14.2 Lipophilic/Hydrophilic Nature of Pollen

Bee pollen (BP) is primarily recognized as a rich source of different antioxidants, including phenolic compounds, carotenoids, and certain vitamins. These microconstituents typically have a pronounced lipophilic character such as quercetin and its derivatives and carotenoids, [8, 19], which are easily positioned and diffunded in the high fat food matrix, and provide excellent antioxidant properties [20]. However, the main constituents of pollen are carbohydrates (18.50–84.25%), proteins (4.50–33.50%) and lipids (0.66–19.04%) (Table 14.1), which significantly contribute to its techno-functional properties [3, 12]. These macroconstituents most often determine the lipophilic/hydrophilic character of BP and its behavior in complex food matrices. Therefore, the content of these compounds should be taken into account when considering, evaluating or potentially applying BP as food ingredient. However, their content in BP is often variable and closely dependent on its botanical and geographical origin (Table 14.1), as well as seasonal environmental variations and processing methodologies.

Since carbohydrates are the predominant constituents of pollen (primarily due to honey or nectar incorporated in pollen pellets) it primarily contains hydrophilic monosaccharides such as glucose and sucrose, with other sugars found in significantly smaller quantities [2, 21]. Pollen also contains water soluble (starch and

Table 14.1 Physico-chemical composition (lipids, proteins and carbohydrates) of different bee pollen samples

Geographical origin of pollen#	Number analyzed pollen samples	Chemical composition (g/100 g)			Reference
		Carbohydrates*	Proteins*	Lipids*	
Spain and Portugal	8 commercial pollen samples	69.68–84.25	12.50–25.15	2.35–3.33	[35]
Portugal	22 organic bee pollen samples (dried pollen)	61.2–70.6	19.1–27.1	4.3–6.3	[36]
China	12 pollen commercial samples	59.43–77.82	14.26–28.95	0.66–6.56	[37]
Serbia	26 pollen samples (3 monofloral and 23 polyfloral pollen samples)	64.42–81.84	14.81–27.25	1.31–6.78	[12]
Italy	3 fresh polen samples	54.84–57.98	25.87–28.42	1.92–2.83	[38]
	2 dried pollen samples (from bees *M. seminigra* and *M. interrupta*)	25.66–44.27	37.63–24.00	6.47–10.81	[39]
Italy	5 organic fresh bee pollen samples	44.8–61.3	21.3–28.7	0.91–2.22	[40]
India	35 fresh bee pollen samples (25 monofloral and 10 polyfloral pollen samples)	42.32–46.15	19.62–25.39	7.13–12.38	[4]
Greece	1 Commercial bee-collected pollen originating from a mixture of flowers	61.00	17.60	7.00	[41]
Italy, Spain, Colombia	3 commercial bee pollen samples	37.7–44.1	12.3–21.6	2.5–6-0	[42]
Brazil	25 fresh bee pollen samples (from stingless bee)	18.5–45.00	4.5–9.9	2.00–6.00	[43]
Brazil	56 dried bee pollen samples	54.90–82.80	7.90–32.20	3.2–13.5	[44]
Turkey	1 fresh bee pollen sample	61.96	30.36	5.50	[45]
Serbia	1 monofloral bee pollen sample	82.01	14.36	1.62	[25]
Croatia	16 polyfloral and 48 monofloral bee pollen samples	26.20–52.58	13.90–27.26	4.51–19.04	[46]
Serbia	1 commercially bee collected pollen	71.00	19.54	8.26	[20]
Morocco	1 monofloral and 6 polyfloral bee pollen samples	18.52–46.44	19.18–33.50	2.11–5.20	[14]
range	**min-max**	**18.50–84.25**	**4.50–33.50**	**0.66–19.04**	

#Worldwide Literature Reports; * Crude lipids and crude proteins determined according to the standard methods proposed by Association of Official Analytical Chemists (AOAC). Total carbohydrates calculated according to equations (carbohydrates = 100 − (g proteins + g lipids + g ash)

pectin) and insoluble (callose, cellulose, lignin and sporopollenin) polysaccharides [5, 12], which can potentially act as thickeners, emulsion stabilizers or gelling agents [3, 12]. Due to the diversity of these polysaccharides, pollen can easily bind and adsorb water or oil, contributing to the sensory properties of some food products [12]. Partially charged or uncharged polysaccharides can be adsorbed at oil–water interfaces and exhibit stabilizing or destabilizing effects on emulsions [22]. Moreover, some non-surface active polysaccharides can form complexes with proteins or peptides, creating a steric layer that can affect interfacial stabilization [22, 23]. When heat dried, pollen can form Maillard type conjugates that may act as potential stabilizing agents due to their strong steric stabilizing ability [24].

Proteins from pollen have not been extensively analyzed thus far [12, 25]. Most investigations have focused on evaluating pollen's amino acid profile. Pollen contains amino acids in both free (glutamic acid, proline, aspartic acid, lysine, tryptophan, threonine) and bound forms [2, 5]. However, electrophoretic analysis of different pollen samples has confirmed the presence of soluble proteins, ranging from 10 to 80 kDa, that exhibit good emulsifying, oil adsorption and gelling properties [12, 25]. Low molecular weight proteins quickly diffuse and easily adsorb, with the ability to distribute themselves on oil/water interfaces, thus stabilizing the emulsion by steric and electrostatic mechanisms, and preventing aggregation or coalescence of oil droplets [22, 26]. During adsorption, proteins often change their native conformation to ensure a better distribution of hydrophilic and hydrophobic side chains of amino acids on oil/water interfaces, covering the maximum area [26]. Proteins can also easily form three-dimensional networks, where molecules of water, oil droplets, some flavors and bioactive compounds are entrapped [23, 27, 28], providing an added functional dimension of pollen proteins.

Lipids are the third largest group of macroconstituents in pollen grains characterized by great diversity. Fatty acids (FAs) of pollen of different botanical origin have been most frequently analyzed, showing the dominant presence of linoleic and saturated FAs [5, 19, 25, 29]. Furthermore, Conte et al. [30] showed the presence of different sterols (24-methylcholesterol, 24-methylencholestanol, β-sitosterol, 25-dehydrositostanol, isofucosterol and campesterol), diglycerides and phospholipids in chestnut and willow pollen. Several classes of polar phospholipides were detected in different pollen grains, such as phosphatidylcholine, phosphatidylethanolamine, phosphatidylglycerol, phosphatidylserine and lysophosphatidylcholine [31, 32]. Due to the mixture of lipids (glycerides, esters of lipid acids, and phospholipids), pollen shows good emulsifying and flavor retaining properties, as well as poor foaming properties [12, 25]. Polar lipids show higher surface activity in comparison to proteins, they easily squeeze out proteins from the air/water interface and weaken the cohesiveness of the film, causing the air bubbles to coalesce and rupture [33]. Furthermore, FAs are an integral part of pollen's sporopollenins, significantly contributing to its good oil adsorption properties [12, 34].

14.3 Techno-Functional Properties of Pollen

Application of pollen as a functional ingredient requires knowledge about its physical, chemical and techno-functional properties, which directly affect processing, texture, quality, rheological and sensory properties of different food formulations [23]. Data related to the techno-functional properties of pollen are scarce. The first research in this area was conducted by Kostić et al. [12], examining the techno-functional properties of different monofloral and polyfloral bee pollen samples collected in Serbia. From then until today, only a few papers have been published [13, 14, 25]. In that sense, further research about this topic is necessary in order to harmonize and explain the techno-functional properties of pollen as much as possible. The available results for solubility, wettability, dispersibility, emulsifying properties, foaming properties and oil/water absorption capacity of bee pollen from different botanical origin are presented in Table 14.2.

Wettability refers to the ability of water to spread over the surface of a pollen grain, which is important for its reconstitution properties. Dispersibility, on the other hand, determines how uniformly pollen can be distributed in solutions [13, 47]. To date, only Thakur and Nanda [13] have analyzed these properties for different types of monofloral and polyfloral pollen samples. The wettability time varied between 285.67 (rapeseed pollen) and 1909.46 s (coconut pollen), while the percentage of dispersibility of pollen ranged from 34.10 (coconut pollen) to 51.06% (rapeseed pollen). The different values for dispersibility and wettability time can be attributed to various factors such as the specific surface area, chemical composition, moisture content, particle structure, porosity, texture and capillary properties of pollen grains [13].

Protein and carbohydrate solubility are important parameters for analyzing the techno-functional properties of multicomponent food samples, such as pollen. Protein solubility is determined by the amino acid composition (content of amino acids with hydrophylic or hydrophobic residues), structure and conformation (e.g., surface hydrophobicity) of the proteins, complexation of proteins with other compounds and pH conditions of the environment [48]. It is known that soluble proteins of pollen are primarily composed of low molecular weight proteins (25–50 kDa), which significantly affect emulsification, foaming or gelling properties [12]. According to the literature, protein solubility varied from 2.79 (polyfloral pollen which predominantly contain *Fraxinus* (60%)) to 25.90% (bifloral pollen containing predominantly *Sophora* (46%) and *Helianthus annuus* (21%) pollen) [12, 14, 25]. On the other hand, soluble carbohydrates are commonly composed of sugars (glucose, fructose and sucrose) which are known as carriers of a sweet taste, and some polysaccharides (starch and pectin) that, in complexes with proteins, can exert the ability to stabilize the emulsion. Values for carbohydrate solubility are significantly higher compared to proteins, and vary from 31.2% to 77.09%, depending on the botanical (type of predominant and accompanying pollen grains) and geographical origin, as well as the year of pollen harvesting [12, 14, 25].

Emulsions play a key role in the formulation and stability of different food products, such as mayonnaise, dressings, infant formulas, coffee creamers, cream liqueurs and some meat products), as well as in their sensory and mouthfeel perception [26]. For these reasons, the emulsifying properties of different food ingredients, such as milk proteins [49], flour of some beans (black gram, kidney, soy) [50–52] or some protein isolates (adzuki, pea and soy) [53], are often evaluated. Pollen is frequently incorporated into various colloidal food products, so it is desirable to evaluate its emulsifying properties before application [15]. Based on literature data, the emulsifying properties of pollen are most often expressed through two parameters: emulsifying stability (ESI) and emulsifying activity (EAI) index, for 0.1% aqueous dispersion. The emulsifying activity of different pollen samples varied from 9.83 (polyfloral pollen from Morocco) to 80.54 g/m^2 (*Helianthus annuus* monofloral pollen from Serbia), while the emulsifying stability ranged from 16.52 to 49.3 min (polyfloral pollen samples) [12, 14, 25]. Good emulsifying properties of pollen are a consequence of its multicomponent nature. Proteins (primarily lower MW proteins) and polar lipids (monoglycerides and esters of lipid acids, phospholipids) from pollen dispersion, easily migrate and adsorb on oil/water interfacial surfaces, forming a compact layer around small oil droplets [12, 22, 26]. Charged/uncharged polysaccharides (pectin and starch) and their conjugates with proteins can also contribute to stabilizing the emulsion [24]. Furthermore, EAI and ESI values strongly depend on the concentration of analyzed pollen in prepared dispersions [15]. For this reason, it is desirable to estimate the optimal concentration of pollen before its food application, in order to avoid the destabilization of emulsions caused by an excess of some biomolecules that remain unabsorbed on oil/water interfaces.

Many food products have a foam form (angel food cakes, sponge cakes, divinity-type confections, candy, meringue, soufflés, various whipped toppings, icings, fudges, ice-creams), and their acceptability depends on the incorporation and stabilization of air bubbles in structures during processing and storage [27]. The mechanism of formation of foams and emulsions is similar, however, foams are much more difficult to maintain and stabilize, due to a more sensitive microstructure and higher interfacial tension on the air-water layer [54]. It is known that low molecular weight proteins are responsible for foam formation and stability, because they are densely packed in the continuous phase and at the interface [28]. Although they contain proteins, different bee pollen suspensions (0.1–2.5%) have shown pronounced antifoaming properties [12, 25]. These samples can be used as antifoaming agents in food products where foam formation is not desirable. Lipids from pollen obviously prevent foam formation and stability, because they have higher surface activity than proteins [33]. After the formation of the foam, the film separating the bubble drains into the Plateau border, because the Laplace pressure inside the Plateau border is lower in comparison to the adjacent films [28]. However, lipid compounds quickly diffuse from foam films to Plateau borders and form asymmetric oil-water-air films, which contributes to increasing the rate of water drainage from the foam and narrowing of the borders, causing ruptures of foam films [55, 56]. Unlike previous claims, Thakur and Nanda [13] showed low foaming capacity and stability for different Indian bee pollen samples. However, these differences in

Table 14.2 Summary of techno-functional properties of bee pollen from different botanical origins from over the world

Geographical origin of pollen	Bee-collected/Floral pollen sample(s)	Techno-functional properties									Reference
		Solubility (%)	Wettability (s)	Dispersibility (%)	Emulsifying properties		Foaming properties		OAC (g/g)	WAC (g/g)	
					ESI (min)	EAI (g/m³)	FS (%)	FC (%)			
Serbia	**3 monofloral*** pollen samples (>80% Brassicaceae; *Salix*; Fabaceae pollen grains);	**Monofloral:** 5.77–8.38 (PS) 45.9–71.1 (CS)	/	/	**Monofloral:** 27.9–43.0	**Monofloral:** 10.76–17.58	Absence of foaming	Absence of foaming	**Monofloral:** 2.01–3.17	**Monofloral:** 1.28–2.25	[12]
	23 polyfloral* pollen samples (*Fraxinus*; Ranunculaceae; *Zea mays*; *Rumex*; Pinaceae; *Juglans*; Asteraceae; Fabaceae; Brassicaceae; *Vitis*; *Helianthus*; *Cannabaceae*; Rosaceae; *Quercus*, *Plantago*; *Ambrosia*; Apiaceae; *Artemisia*; Fabaceae; *Tilia*; *Salix*; Moraceae; *Sambucus*; Chenopodiaceae; Poaceae; *Sophora*; *Cornus*)	**Polyfloral:** 2.79–25.90 (PS) 31.2–75.0 (CS)			**Polyfloral:** 19.6–49.3	**Polyfloral:** 10.40–24.52			**Polyfloral:** 1.00–3.53	**Polyfloral:** 0.92–1.99	
India	**25 monofloral** pollen samples (coconut; coriander; rapeseed pollen grains);	/	**Monofloral:** 285.67–1909.46	**Monofloral:** 34.10–51.06	/	/	**Monofloral:** 17.50–20.00	**Monofloral:** 6.21–8.69	**Monofloral:** 1.31–2.13	**Monofloral:** 0.47–0.72	[13]
	10 polyfloral pollen samples (maize, pearls millet, onionweed, pigeon pea and cotton)		**Polyfloral:** 861.45 (Average)	**Polyfloral:** 42.6 (Average)			**Polyfloral:** 18.56 (Average)	**Polyfloral:** 7.43 (Average)	**Polyfloral:** 1.35 (Average)	**Polyfloral:** 0.57 (Average)	

Table 14.2 (continued)

Geographical origin of pollen	Bee-collected/Floral pollen sample(s)	Techno-functional properties									Reference
		Solubility (%)	Wettability (s)	Dispersibility (%)	Emulsifying properties		Foaming properties		OAC (g/g)	WAC (g/g)	
					ESI (min)	EAI (g/m^3)	FS (%)	FC (%)			
Serbia	**1 monofloral** pollen sample (*Helianthus annuus* L.)	3.64 (PS) 77.09 (CS)	/	/	19.98	80.54	Absence of foaming	Absence of foaming	2.43	0.87	[25]
Serbia	**1 commercially pollen** (according to producers' declaration pollen predominantly originated from *Brassica napus* and *Fraxinus spp.*)	/	/	/	For different pollen concentration: 11.53–14.04	For different pollen concentration: 58.59–156.6	Absence of foaming	Absence of foaming	1.5	0.9	[15]
Morroco	1 **monofloral** pollen sample (*Reseda luteola*)	**Monofloral:** 15.79 (PS) 57.39 (CS)	/	/	**Monofloral:** 26.53	**Monofloral:** 18.28	/	/	**Monofloral:** 1.91	**Monofloral:** 1.36	[14]
	6 polyfloral pollen samples (Fagaceae; Rosaceae; Lamiaceae; Fabaceae; Asteraceae; Oleaceae; Papaveraceae; Brassicaceae; Rhamnaceae; Ericaceae; Salicaceae; Myrthaceae; Moraceae; Rutaceae; Poaceae)	**Polyfloral:** 7.28–23.31 (PS) 34.47–75.53 (CS)			**Polyfloral:** 16.52–45.38	**Polyfloral:** 9.83–25.05			**Polyfloral:** 1.15–3.50	**Polyfloral:** 1.06–2.19	

*__Monofloral pollen__ sample contain >80% of one type of pollen grains; * **Polyfloral pollen** sample contain mixture of different type of pollen grains. **Abbreviations**: *ESI* emulsifying stability index, *EAI* emulsifying activity index, *FS* foaming stability, *FC* foam capacity, *OAC* oil absorption capacity, *WAC* Water absorption capacity, *PS* protein solubility (determined according to Bradford procedure, g/100 g), *CS* carbohydrate solubility (determined according to the conventional anthrone method, g/100 g) comparison to the results of other studies can be explained by applied methodology (air introduction into pollen suspension versus its whipping in a mixer blender), time of foam production (maximum 1 min versus 5 min) different concentrations of pollen suspension (2% versus 3%) and differences in the chemical composition of the pollen samples.

The ability of some food ingredients to absorb water or oil is a very important characteristic, because it directly affects consistency, texture (hardness, brittleness, crispiness, dryness, cohesiveness) and sensory properties (flavor retention and mouthfeel perception) of food products, especially during storage [27]. Pollen is a specific food ingredient that has both, good water (WAC) and oil (OAC) adsorption capacity, due to its unique multicomponent composition and capillary action. According to literature, the WAC of different pollen samples varied from 0.47 (monofloral coriander pollen) to 2.25 g/g (81% Fabaceae pollen), while OAC ranged from 1.00 (polyfloral pollen containing predominantly Brassicaceae pollen (76%) to 3.53 g/g (polyfloral pollen containing predominantly *Sophora* pollen (42%)) [12–15, 25]. The major compounds contributing to WAC from pollen are insoluble proteins and polysaccharides which contain hydrophilic parts (side chain or groups) and some polar lipids. On the other hand, proteins and insoluble molecules with hydrophobic parts, as well as sporopollenin (complex polymer from exine of pollen) have a high ability to absorb oils. Mentioned macromolecules have three-dimensional structures which can bind (entrap) water or oil molecules by capillary actions [3, 12]. Finally, all pollen samples have unique OAC and WAC characteristics that depend on their protein, carbohydrate and lipid compositions. Based on the WAC and OAC values, it is possible to determine the water-oil absorption index (WOAI), as an important parameter for evaluating the lipophilic/hydrophilic character of different food ingredients. Most of the analyzed pollen samples of different botanical and geographical (Serbia, Morocco and India) origin showed better lipophilic characteristics (WOAI<1) [12–14].

14.4 Effect of Processing and Storage on Nutrient Composition of Pollen Grains

Pollen is rarely used immediately after collection, due to its high water activity, which influence microbial spoilage and chemical/enzymatic reactions, and decrease its shelf life [10, 57]. The applications of pollen as functional food ingredients depend on the processing (usually involving different drying and freezing treatments) and storage conditions (humidity, temperature, oxygen presence), which directly affect and can modify its nutrient composition, as well as its techno-functional properties. For examples, Isik et al. [45] showed that increasing the temperature (from 40 to 60 °C) of hot air used for pollen drying has a significant influence on the content of crude proteins, fat and total carbohydrates, as well as on changing the color, morphological structures and organoleptic characteristics of pollen grains. Additionally, increasing pollen drying temperatures affected the increasing browning index, content of 5-hydroxymethyl furfural (5-HMF) [58] and other products of Maillard reaction [59] and lipid oxidation [60]. On the other hand, Domínguez-Valhondo et al. [61] showed that the content of monosaccharides (glucose, fructose and sucrose), dietary fiber and proteins was not significantly different in freeze dried and fresh pollen samples (monofloral and polyfloral), except for lipid content which increased in freeze dried samples. In addition to the above, lyophilized pollen samples exhibited a significantly higher content of proteins and lipids than the same samples dehydrated by the electric oven [62]. Some advanced drying techniques based on microwave and/or pulsed vacuum treatments can also be used to process pollen. However, the use of these techniques significantly increased the content of hydroxymethylfurfural and products of lipid oxidation and decreased free/total amino acids in the pollen samples, probably due to their involvement in Maillard reactions [63–65]. The content of pollen nutrients also closely depends on the conditions and time of storage. Žilić et al. [59] highlighted the increased content of Maillard products in different maize pollen samples during storage (7 days at 4 °C/~45% moisture). The increasing amount of Maillard products during storage and heat treatment is associated with the decreasing content of proteins, peptides, amino acids, glucose and fructose, which will directly reflected in techno-functional properties of pollen.

14.5 Products of Fermented Pollen as Techno-Functional Ingredients

Pollen presents ideal raw material for solid state fermentation, and its transformation results in a completely new product known as bee bread with a unique chemical composition [66]. Fermentation of pollen (*Brassica campestris* L.) using yeast gives a highly valuable nutritional product with decreased content of glucose/fructose and increased content of oligopeptides, free amino acids, fatty acids and phenolic compounds [67]. However, bee bread is most often consumed as a highly available nutritional product, but there are no studies that analyze its techno-functional properties as a potential food ingredient. In the future studies, it would be very interesting to analyze the techno-functional properties and evaluate the differences between bee bread and its initial, unfermented pollen.

14.6 Pollen as a Functional and Techno-Functional Ingredient in Different Food Products

Recently, pollen has been often used as an ingredient in the production of specific beverages, bakery, confectionery, meat, and dairy products, with the aim to improve primarily their primarily functional, nutritional, and health-promoting properties.

However, most of the processed food products are multicomponent colloidal systems, containing a mixture of different ingredients such as biopolymers (proteins and carbohydrates) and small molecules (sugars, lipids, salts, phenolics, carotenoids etc.), as well as various types of dispersed entities such as oil droplets, gas bubbles, lipid crystals, starch granules, etc. [12, 23]. Consequently, the texture, rheology and physical stability of different food products depend on the nature and strength of interactions among food constituents and dispersed particles. Taking this into account, the effect of added pollen on textural and sensory properties of different food formulations is very important due to the multicompound nature of pollen and must be taken into consideration, because it directly affects processing, quality, and overall acceptance of products (Table 14.3). Therefore, in most of the studies, these technological properties are crucial for evaluating the optimal amount of pollen to be incorporated into food products.

The addition of pollen in the production of mead or white wines primarily stimulated fermentation, increased the content of volatile compounds and affected the sensory characteristics of the product as a carrier of a specific aroma [68, 69]. However, increasing pollen content in wines contributes to increased turbidity and color intensity, which can affect the final perception and acceptance of the product.

Pollen possesses excellent emulsifying properties and oil/water absorption capacity, which qualifies it for application in high fat and colloidal meat products such as meatballs [70], sausages [71] or frankfurters [15, 20], where these features are very desirable both when formulating the products and during their storage. Meatballs and frankfurters with pollen addition showed increased yellowness in color parameters, due to the presence of carotenoids and some flavonoids which migrated from pollen grains to high fat meat matrices [16, 20, 70]. Furthermore, pollen gave a consistent texture to these meat products and additionally stabilized this colloidal matrix, but on the other hand it weakened the protein network and decreased its resistance to compression. It can be concluded that increasing pollen content affected the decrease of hardness, gumminess, and chewiness of these meat products, but did not cause other negative changes in the texture [16, 20]. Meatballs with increased content of pollen also had higher cooking loss which can be due to the difference in the meatball's matrix and its ability to keep water or fat [16]. As sensory aspects, higher content of pollen in meat products gave floral odor and increased juiciness but decreased overall acceptability.

All bakery and confectionary products with bee pollen had a higher browning index, changed color, and had a more intensive color in comparison to control samples, due to the color of added pollen (usually yellow) and Maillard products (pyrazines, furfural, furans and melanoidins) formed during baking [17, 72–74]. In addition, increased furan content in bakery products with pollen gave pleasing sensory properties and a specific aroma [75]. Pollen content also affected the texture properties of bakery products, i.e. in most cases it reduced hardness, thickness, cohesiveness, springiness, fructurability and resilience, but increased crispiness and chewiness of cookies, biscuits and bread [73, 74, 76, 77]. Only the bread samples containing more than 20% pollen possessed increased hardness which could be connected with poorer water retention of bread crumbs [77]. During bread production,

Table 14.3 Techno-functional and sensory properties of bee-collected pollen-based bakery, confectionery, juice, and meat products

Products	Botanical source of pollen	Geographical origin of pollen	Main observations		Reference
			Techno-functional properties	Sensory properties	
Beverages					
Mead (honey-wine) with pollen addition (10–50 g/l)	Commercially produced pollen	Spain	After pollen addition: 1. Turbidity ↑ 2. Improves fermentation rates	Pollen addition improvement mead sensory properties 1. 30 g/hl of pollen in mead is the most acceptable, because it contributes a moderate aroma of beverages with pronounced notes of almonds, dried fruits, apple, caramel and sweets.	[68]
White wines with pollen addition (0.1–20 g/l)	Commercial bee-collected pollen	Spain	After pollen addition: 1. Density (not significantly changed) 2. Increases exponential velocity during fermentation 3. Color intensity ↑	/	[69]

(continued)

Table 14.3 (continued)

Products	Botanical source of pollen	Geographical origin of pollen	Main observations		Reference
			Techno-functional properties	Sensory properties	
Meat Products					
Meatballs formulated with pollen (0–6%)	Bee pollen (Fanus Gida ve Organik Urunler San. Tic. Ltd. Sti.,Turkey)	Turkey	Increasing pollen share in meatballs: 1. Texture (hardness, gumminess, chewiness) ↓ 2. Springiness (3%) ↑ 3. Cohesiveness (not significantly changed) 4. Cooking loss (%) ↓ 5. Redness and lightness ↓ 6. Yellowness (Hunter b) ↑	Increasing pollen share in meatballs: 1. Appearance (raw samples), flavor, hardness, juiciness, overall acceptability (cooked samples) ↓	[16]
Meatballs formulated with pollen (0–6%)	Bee pollen (Fanus Gida ve Organik Urunler San. Tic. Ltd. Sti.,Turkey)	Turkey	Increasing pollen share in meatballs: 1. Color intensity (redness ↓) 2. Yellowness, chroma and hue angle ↑	/	[70]
Pork sausages with the addition of lyophilized pollen extract (0.2 g/kg)	Polyfloral pollen (*Arecaceae*, Brassicaceae families, *Asteraceae baccharis*, *Asteraceae eupatorium*)	Brazil	/	1. TBARS values during storage as indicator of rancid odors ↑ 2. TBARS values similar for pollen extracts and sodium erythorbate at the end of storage	[71]

(continued)

Table 14.3 (continued)

Products	Botanical source of pollen	Geographical origin of pollen	Main observations: Techno-functional properties	Main observations: Sensory properties	Reference
Black pudding with fresh bee pollen and lyophilized extract addition	Bee.collected pollen (mainly composed by *Cistus ladanifer* pellets)	Portugal	/	1. TBARS values during storage as indicator of rancid odors ↓ 2. Product quality and consumer acceptance ↑	[85]
Frankfurters formulated with bee pollen powder (0–1.5%)	Commercially pollen (according to producers' declaration from *Brassica napus* and *Fraxinus* spp.)	Serbia	Increasing pollen share in frankfurters: 1. Texture (hardness and chewiness) ↓ 2. Springiness and cohesiveness (not significantly changed) 3. Color intensity (lightness) ↓ 4. Redness and yellowness ↑	1. Pollen addition did not cause any adverse sensory sensations 2. Floral sensation and juiciness at the end of mastication (1 and 1.5% pollen in frankfurters)	[20]
Frankfurters formulated with bee pollen powder (0–1.5%)	Commercially pollen (according to producers' declaration from *Brassica napus* and *Fraxinus* spp.)	Serbia	Increasing pollen share in frankfurters: 1. Warner-Bratzler sheer force of frankfurters (toughness/tenderness) (not significantly changed) 2. Stable emulsion and consistent meat product during storage	1. Pronounced floral odor and decreased pork flavor (1 and 1.5% pollen in frankfurters)	[15]

(continued)

Table 14.3 (continued)

Products	Botanical source of pollen	Geographical origin of pollen	Main observations		Reference
			Techno-functional properties	Sensory properties	
Bakery and confectionary products					
Cookies enriched with bee pollen (1 and 2 g bee pollen per cookie)	Monofloral rape bee pollen	Slovakia	Increasing pollen share in cookies: 1. Diameter and weight of cookies ↑ 2. Thickness of cookies ↓ 3. Color intensity ↑ 4. Crispiness/ chewiness ↑	1. Pleasant and easy chewiness, delicate taste 2. Overall impression ↓	[74]
Biscuits enriched with bee pollen (0–10%)	Dried bee collected pollen	Poland	Increasing pollen share in biscuits: 1. Color intensity (lightness) ↓ 2. Redness and yellowness ↑ 3. Texture (hardness) ↓ 4. Penetration work ↑	Increasing pollen share in biscuits: 1. Aroma (not significantly changed) 2. Taste (5%) ↑ 3. Total score (5%) ↑	[72]

(continued)

Table 14.3 (continued)

Products	Botanical source of pollen	Geographical origin of pollen	Main observations: Techno-functional properties	Main observations: Sensory properties	Reference
Gluten-free bread with pollen (1–5%)	Polyfloral dry bee pollen	Italy	Increasing pollen share in bread: 1. not influence on rheological characteristics (solid, elastic-like behavior of dough) 2. Leavening properties of dough (4%) ↑ 3. Textural properties (cohesiveness and resilience) (from 1% to 5%) ↓ 4. Textural properties (hardness)(4%) ↓ 5. Springiness (not significantly changed) 6. Crust color ↓ 7. Browning index ↑ 8. Whiteness index ↓	Increasing pollen share in bread: 1. Color (4%) ↑ 2. Odor, sweetness, flavor, overall acceptance ↑ 3. Hardness in mouth (not significantly changed)	[17]
Gluten-free bread with pollen (1–5%)	Polyfloral dry bee pollen	Italy	Increasing pollen share in bread: 1. Furans ↑ 2. Furfural ↑ 3. 5-Methyl-2-furaldehyde ↑ 4. Pyrazin derivatives (↓ or not significantly changed)	Increased content of some furans in bread with pollen significantly influence on some pleasant sensory properties	[75]

(continued)

Table 14.3 (continued)

Products	Botanical source of pollen	Geographical origin of pollen	Main observations		Reference
			Techno-functional properties	Sensory properties	
Bread with pollen (0–25%)	Dry rape bee pollen (Sichuan Kuake Technology Development Co., Ltd.)	China	Increasing pollen share in bread: 1. Specific volume ↓ 2. Texture properties (hardness, chewiness) ↑ 3. Springiness ↓	Increasing pollen share in bread: 1. Appearance, color, taste, flavor, inner texture ↓	[77]
Reduced-fat cookies (0–15%)	Bee pollen	Turkey	Increasing pollen share in cookies: 1. Diameter ↑ 2. Texture and thickness ↓ 3. Spread ration ↑ 4. Color (L*) ↓ 5. Color (a* and b*)↑	Increasing pollen share in cookies: 1. Appearance and texture ↓ 2. Taste ↑ 3. Eagerness to Buy (10%) ↑ 4. Overall acceptability (10%) ↑	[73]
Biscuits with pollen (0–10%)	Bee collected pollen	Egypt	Increasing pollen share in biscuits: 1. color (L* ↓; b*↓; a* ↑)	Pollen biscuits in comparison to control biscuits: 1.Overall acceptance, odor and taste (5 and 7.5%) ↑ 2. Texture ↓	[86]

(continued)

Table 14.3 (continued)

Products	Botanical source of pollen	Geographical origin of pollen	Main observations		Reference
			Techno-functional properties	Sensory properties	
Biscuits (2%, 5% and 10%)	Monofloral bee pollen: 1. rapeseed (*Brassica napus*) 2. phacelia (*Phacelia tanacetifolia*) 3. sunflower (*Helianthus annuus*)	Hungary	Increasing pollen share in biscuits: 1. Color (L*) and b* (phacelia) ↓ 2. Color (a* and b*)↑ 3. Diameter, volume, area and density (not significantly changed) 4. Baking loss (not significantly difference) 5. Hardness, adhesive force, fracturability, ↓ (except phacelia 10% ↑) 6. Cohesiveness (not significantly changed) 7. Springiness ↓ (except phacelia ↑) 8. Gumminess and chewiness (rapeseed (5%)↑; phacelia (5%) ↓; sunflower ↓)	1. Consumer acceptance of biscuits was strongly influenced by the botanical origin; 2. Biscuits enriched with sunflower pollen had more acceptable sensory properties compared to other samples; 3. Biscuits with rapeseed or phacelia pollens ("cabbage" and "cut honey" odors) 4. Biscuits with sunflower pollen did not have any off-flavors 5. Purchase intentions (biscuits with 2% sunflower) ↑	[76]
Milk products					
Fermented milk beverages (0–20 mg/mL)	Commercial bee pollen	Turkey	Increasing pollen share in fermented milk: 1. Viscosity ↑	Increasing pollen share in fermented milk: 1. Aroma and overall acceptability ↓ 2. Texture ↓ (10 mg/mL)	[79]

(continued)

Table 14.3 (continued)

Products	Botanical source of pollen	Geographical origin of pollen	Main observations		Reference
			Techno-functional properties	Sensory properties	
Bio-yogurt with probiotic bacteria, royal jelly and bee-collected pollen grains (0.8%)	Bee pollen grains	Egypt	Effect of pollen addition: 1. Coagulation time ↑ 2. Curd tension ↓ 3. Syneresis ↓	Effect of pollen addition: 1. Flavor, body and texture, appearance and overall acceptability (not significantly difference)	[81]
Probiotic yogurt with the bee-collected pollen grains (0.8%)	Bee pollen grains	Egypt	Effect of pollen addition: 1. Texture (hardness) ↑ 2. Springiness, cohesiveness, chewiness, gumminess (not significantly changed) 3. Syneresis ↓ 4. Microstructure (more comprehensive network was formed, improved consistency and water holding capacity)	/	[80]
Yoghurt with bee pollen (0.4%; 0.6%; 0.8%)	Commercial bee collected pollen	Bulgaria	Increasing pollen share in yoghurt (sensory panel): 1. Surface, color, type of coagulant, sectional structure and consistency after breaking ↓ (in comparison to control)	Increasing pollen share in yoghurt (sensory panel): 1. Taste and flavor ↓ (in comparison to control)	[87]

(continued)

Table 14.3 (continued)

Products	Botanical source of pollen	Geographical origin of pollen	Main observations: Techno-functional properties	Main observations: Sensory properties	Reference
Bio-functional, "bee-collected pollen yogurt "prepared from cow, goat and sheep milk (0.5%; 1.0%; 2.5%; 3.0% *w/v*)	Commercial bee-collected pollen originating from a mixture of flowers (pollen brand name "Greek Bee Pollen", (Attiki Bee Culturing Co.-Alex Pittas S.A., Athens, Greece))	Greece	Increasing pollen share in yoghurt (sensory panel): 1. Cohesion (cow (1%) ↑; goat ↓; sheep (3.0%) ↑)	Increasing pollen share in yoghurt (sensory panel): 1. Taste (cow (1%) ↑; goat ↓; sheep (0.5%) ↑) 2. Odor (cow (1%) ↓; goat (1%) ↑; sheep (1%) ↑) 3. Appearance (cow (1%) ↑; goat ↓; sheep (1%) ↑) 4. panelist impression ("wonderful"; sweet and pleasant taste"; "nice smell"; covers the odor of milk"; "attractive appearance"; "improved cohesion")	[41]
Bee pollen-skim milk powders (5–15%) (vacuum drying)	Fresh monofloral bee pollen (*Brassica napus*)	India	Optimized bee pollen-milk powders: 1. Solubility (74.86–83.73%) 2. Bulk density (0.37–0.63 g cm^{-3}) 3. Hygroscopicity (13.10–18.29%)	/	[83]

(continued)

Table 14.3 (continued)

Products	Botanical source of pollen	Geographical origin of pollen	Main observations		Reference
			Techno-functional properties	Sensory properties	
Bee pollen-skim milk powders (5–15%) (spray drying)	Fresh monofloral bee pollen (*Brassica napus*)	India	Optimized bee pollen-milk powders: 1. Solubility (93.38–97.82%) 2. Bulk density (0.31–0.52 g cm^{-3}) 3. Hygroscopicity (18.96–24.22%) 4. Color, whiteness (78.09–86.86)	/	[84]
Bee pollen-skim milk powder (8.04%)	Fresh monofloral bee pollen (*Brassica napus*)	India	Pollen-milk powder in comparison to commercial milk powder: 1. Cohesion properties (cohesion coefficient ↓; cohesion index ↑) 2. Caking properties (height ratio (showed the compressibility (%), 5 cycle) ↓; cake height ↑; cake strength ↑; mean cake strength ↑) 3. Flow Speed Dependence test (compaction coefficient ↑; flow stability (not significantly different); cohesion coefficient ↓)	/	[18]

(continued)

Table 14.3 (continued)

Products	Botanical source of pollen	Geographical origin of pollen	Main observations		Reference
			Techno-functional properties	Sensory properties	
Fermented milk with bee pollen and (pollen, royal jelly and bee bread) mix (1% *w/v*)	Bee collected pollen	Egypt	Fermented milk with pollen in comparison to control (sensory panel): 1. Texture ↓ 2. Color ↑	Fermented milk with pollen in comparison to control (sensory panel): 1. Appearance and overall acceptance ↓ 2. Odor ↑ 3. Taste (not significantly changed)	[82]

the addition of pollen to the formulation did not show any effects on the rheological characteristics, but it did improve the leavening properties of dough (up to 4% pollen content). In a complex matrix like gluten free bread, proteins obviously contributed to the formation of a stable dough network, which stabilized the formed bubbles and suppressed the antifoaming properties of pollen. However, the antifoam activity of surface active pollen lipids came to the fore only when the share of pollen in bread was >5%, as indicated by increased pores and reduced cell number in crumb grains, which was later reflected in decreased bread hardness [17]. Pollen can also act as a good crumb softener, because of its good emulsifying properties. During storage, biscuits with/without pollen showed similar texture due to pronounced water sorption and redistribution [72]. Finally, overall acceptance of bakery products with pollen is often variable and depends on the botanical origin, content of added pollen, technological parameters, and complex interactions in prepared bakery formulations [73, 76].

Dairy products must have specific techno-functional criteria, in order to satisfy quality and be accepted by consumers. Quality of fermented milk and yoghurts primarily depends on gel strength and syneresis, because it directly affects consumer perception [78]. The addition of pollen to fermented dairy products affects the rearrangements of the casein network as a consequence of numerous interactions that occur in the complex milk-pollen matrix, which directly affects changes by viscosity, texture (hardness) and microstructure of final products [79–81]. Fermented dairy products with pollen have more comprehensive network, improved consistency and water holding capacity, and therefore increased hardness of formed gel and decreased syneresis [80]. Gelling properties of pollen depend on the critical concentration of proteins capable of participating in the formation of the gel network and the interactions of proteins with other components in the food matrices [28]. Yoghurt with 0.8% pollen except hardness, did not show significant changes for other textural parameters (springiness, cohesiveness, chewiness, gumminess) in comparison to the control sample [80]. In general, acceptability of dairy products with pollen is often lower and closely dependent on pollen content in products. For example, Karabagias et al. [41] showed high sensory scores for sheep, cow and goat yogurt enriched with 0.5 and 1% (w/w) pollen, with improved cohesion of these products. Interaction of pollen compounds with lipid/protein from yoghurt, contributed to the formation of additional glazing surface coating, which prevented water loss, improved cohesion and acceptance of new fermented milk products. Color change of milk products caused by pollen addition also significantly affected the perception of final products [82]. In addition to the above, the production of milk-pollen powders, as new functional products, or additives, is being increasingly examined [11, 83]. Optimized and then spray/vacuum dried pollen-milk powders showed good properties such as solubility, bulk density, hygroscopicity, cohesion and caking properties [18, 83, 84]. Moreover, it was shown that functional skimmed thermally treated goat milk-pollen powder prepared by Kostić et al. [11] had excellent emulsifying properties and oil absorption capacity, but lower or absent foaming properties (data not published). The mentioned characteristics give the possibility for further production and application of milk-pollen powders as techno-functional agents in a wide range of different food products.

14.7 Conclusions and Future Directions

With respect to the techno-functional properties of (bee) pollen there is significant space for developing methods and collecting data. Since these properties are extremely important for formulation of novel food products it should be validated how (bee) pollen can influence by:

- Performing additional and more comprehensive studies on pollen samples with different botanical/geographical origin in order to observe any regularity
- Trying to unify results with application of monofloral samples since it only can have more or less consistent chemical contexture
- Proposing some recommended or limited values for the most important techno-functional properties
- Introducing pollen in national legislations as functional food additive with proposed values for several nutritional parameters, including techno-functional

References

1. Mărgăoan R, Stranț M, Varadi A, Topal E, Yücel B, Cornea-Cipcigan M et al (2019) Bee collected pollen and bee bread: bioactive constituents and health benefits. Antioxidants (Basel) 8(12):568
2. Li Q-Q, Wang K, Marcucci MC, Sawaya ACHF, Hu L, Xue X-F et al (2018) Nutrient-rich bee pollen: a treasure trove of active natural metabolites. J Funct Foods 49:472–484

3. Kostić AŽ, Milinčić DD, Tešić ŽL, Pešić MB (2022) Chapter 11 – Bee pollen in cosmetics: the chemical point of view. In: Boyacioglu D (ed) Bee products and their applications in the food and pharmaceutical industries. Academic, pp 261–282
4. Thakur M, Nanda V (2018) Assessment of physico-chemical properties, fatty acid, amino acid and mineral profile of bee pollen from India with a multivariate perspective. J Food Nutr Res 57:328
5. Thakur N, Raigond P, Singh Y, Mishra T, Singh B, Lal MK et al (2020) Recent updates on bioaccessibility of phytonutrients. Trends Food Sci Technol 97:366–380
6. Ares AM, Valverde S, Bernal JL, Nozal MJ, Bernal J (2018) Extraction and determination of bioactive compounds from bee pollen. J Pharm Biomed Anal 147:110–124
7. Kieliszek M, Piwowarek K, Kot AM, Błażejak S, Chlebowska-Śmigiel A, Wolska I (2018) Pollen and bee bread as new health-oriented products: a review. Trends Food Sci Technol 71:170–180
8. Kostić AŽ, Milinčić DD, Gašić UM, Nedić N, Stanojević SP, Tešić ŽL et al (2019) Polyphenolic profile and antioxidant properties of bee-collected pollen from sunflower (Helianthus annuus L.) plant. LWT Food Sci Technol 112:108244
9. Aylanc V, Falcão SI, Ertosun S, Vilas-Boas M (2021) From the hive to the table: nutrition value, digestibility and bioavailability of the dietary phytochemicals present in the bee pollen and bee bread. Trends Food Sci Technol 109:464–481
10. Kostić A, Milinčić DD, Barać MB, Ali Shariati M, Tešić ŽL, Pešić MB (2020) The application of pollen as a functional food and feed ingredient-the present and perspectives. Biomolecules 10(1):84
11. Kostić AŽ, Milinčić DD, Stanisavljević NS, Gašić UM, Lević S, Kojić MO et al (2021) Polyphenol bioaccessibility and antioxidant properties of in vitro digested spray-dried thermally-treated skimmed goat milk enriched with pollen. Food Chem 351:129310
12. Kostić AŽ, Barać MB, Stanojević SP, Milojković-Opsenica DM, Tešić ŽL, Šikoparija B et al (2015) Physicochemical composition and techno-functional properties of bee pollen collected in Serbia. LWT Food Sci Technol 62(1, Part 1):301–309
13. Thakur M, Nanda V (2020) Exploring the physical, functional, thermal, and textural properties of bee pollen from different botanical origins of India. J Food Process Eng 43(1):e12935
14. Laaroussi H, Ferreira-Santos P, Genisheva Z, Bakour M, Ousaaid D, El Ghouizi A et al (2023) Unveiling the techno-functional and bioactive properties of bee pollen as an added-value food ingredient. Food Chem 405:134958
15. Novaković S, Djekic I, Pešić M, Kostić A, Milinčić D, Tomasevic I (2021) Techno-functional, textural and sensorial properties of frankfurters as affected by the addition of bee pollen powder. Theory Pract Meat Process 6:135–140
16. Turhan S, Yazici F, Saricaoglu FT, Mortas M, Genccelep H (2014) Evaluation of the nutritional and storage quality of meatballs formulated with bee pollen. Korean J Food Sci Anim Resour 34(4):423–433
17. Conte P, Del Caro A, Balestra F, Piga A, Fadda C (2018) Bee pollen as a functional ingredient in gluten-free bread: a physical-chemical, technological and sensory approach. LWT Food Sci Technol 90:1–7
18. Thakur M, Nanda V (2022) Investigating the flow properties of bee pollen enriched milk powder during storage. J Stored Prod Res 96:101940
19. Mărgăoan R, Mărghitaş LA, Dezmirean DS, Dulf FV, Bunea A, Socaci SA et al (2014) Predominant and secondary pollen botanical origins influence the carotenoid and fatty acid profile in fresh honeybee-collected pollen. J Agric Food Chem 62(27):6306–6316
20. Novaković S, Djekic I, Pešić M, Kostić A, Milinčić D, Stanisavljević N et al (2021) Bee pollen powder as a functional ingredient in frankfurters. Meat Sci 182:108621
21. Thakur M, Nanda V (2020) Composition and functionality of bee pollen: a review. Trends Food Sci Technol 98:82–106
22. Dalgleish DG (2006) Food emulsions—their structures and structure-forming properties. Food Hydrocoll 20(4):415–422

23. Dickinson E (2013) Stabilising emulsion-based colloidal structures with mixed food ingredients. J Sci Food Agric 93(4):710–721
24. Dickinson E (2011) Double emulsions stabilized by food biopolymers. Food Biophys 6(1):1–11
25. Kostić AŽ, Milinčić DD, Trifunović BDŠ, Stanojević SP, Lević S, Nedić N et al (2020) Nutritional and techno-functional properties of monofloral bee-collected sunflower (Helianthus annuus L.) pollen. Emir J Food Agric:768–777
26. Dalgleish DG (1997) Adsorption of protein and the stability of emulsions. Trends Food Sci Technol 8(1):1–6
27. Kinsella J, Melachouris N (2009) Functional properties of proteins in foods: a survey. CRC Crit Rev Food Sci Nutr 7:219–280
28. Foegeding EA, Davis JP (2011) Food protein functionality: a comprehensive approach. Food Hydrocoll 25(8):1853–1864
29. Kostić AŽ, Mačukanović-Jocić MP, Špirović Trifunović BD, Vukašinović IŽ, Pavlović VB, Pešić MB (2017) Fatty acids of maize pollen – quantification, nutritional and morphological evaluation. J Cereal Sci 77:180–185
30. Conte G, Benelli G, Serra A, Signorini F, Bientinesi M, Nicolella C et al (2017) Lipid characterization of chestnut and willow honeybee-collected pollen: Impact of freeze-drying and microwave-assisted drying. J Food Compos Anal 55:12–19
31. Liang M, Zhang P, Shu X, Liu C, Shu J (2013) Characterization of pollen by MALDI-TOF lipid profiling. Int J Mass Spectrom 334:13–18
32. Li Q, Liang X, Zhao L, Zhang Z, Xue X, Wang K et al (2017) UPLC-Q-exactive orbitrap/MS-based lipidomics approach to characterize lipid extracts from bee pollen and their in vitro anti-inflammatory properties. J Agric Food Chem 65(32):6848–6860
33. Phillips LG, Davis MJ, Kinsella JE (1989) The effects of various milk proteins on the foaming properties of egg white. Food Hydrocoll 3(3):163–174
34. Li F-S, Phyo P, Jacobowitz J, Hong M, Weng J-K (2019) The molecular structure of plant sporopollenin. Nat Plant 5(1):41–46
35. Nogueira C, Iglesias A, Feás X, Estevinho LM (2012) Commercial bee pollen with different geographical origins: a comprehensive approach. Int J Mol Sci [Internet] 13(9):11173–11187
36. Feás X, Vázquez-Tato MP, Estevinho L, Seijas JA, Iglesias A (2012) Organic Bee pollen: botanical origin, nutritional value, bioactive compounds, antioxidant activity and microbiological quality. Molecules [Internet] 17(7):8359–8377
37. Yang K, Wu D, Ye X, Liu D, Chen J, Sun P (2013) Characterization of Chemical Composition of Bee Pollen in China. J Agric Food Chem 61(3):708–718
38. Domenici V, Gabriele M, Parri E, Felicioli A, Sagona S, Pozzo L et al (2015) Phytochemical composition and antioxidant activity of Tuscan bee pollen of different botanic origins. Italian J Food Sci 27:248–259
39. Rebelo KS, Ferreira AG, Carvalho-Zilse GA (2016) Physicochemical characteristics of pollen collected by Amazonian stingless bees. Ciência Rural 46:927–932
40. Sagona S, Pozzo L, Peiretti PG, Biondi C, Giusti M, Gabriele M et al (2017) Palynological origin, chemical composition, lipid peroxidation and fatty acid profile of organic Tuscanian bee-pollen. J Apic Res 56(2):136–143
41. Karabagias IK, Karabagias VK, Gatzias I, Riganakos KA (2018) Bio-functional properties of bee pollen: the case of "Bee Pollen Yoghurt". Coatings [Internet] 8(12):423
42. Gardana C, Del Bo' C, Quicazán MC, Corrrea AR, Simonetti P (2018) Nutrients, phytochemicals and botanical origin of commercial bee pollen from different geographical areas. J Food Compos Anal 73:29–38
43. Duarte A, Santos Vasconcelos M, Oda-Souza M, Oliveira F, Lopez AM (2018) Honey and bee pollen produced by Meliponini (Apidae) in Alagoas, Brazil: multivariate analysis of physicochemical and antioxidant profiles. Cienc Tecnol Aliment 38
44. De-Melo AAM, Estevinho LM, Moreira MM, Delerue-Matos C, Freitas AS, Barth OM et al (2018) A multivariate approach based on physicochemical parameters and biological potential for the botanical and geographical discrimination of Brazilian bee pollen. Food Biosci 25:91–110

45. Isik A, Ozdemir M, Doymaz I (2018) Effect of hot air drying on quality characteristics and physicochemical properties of bee pollen. Food Sci Technol 39:224
46. Prđun S, Svečnjak L, Valentić M, Marijanović Z, Jerković I (2021) Characterization of bee pollen: physico-chemical properties, headspace composition and ftir spectral profiles. Foods [Internet] 10(9):2103
47. Jinapong N, Suphantharika M, Jamnong P (2008) Production of instant soymilk powders by ultrafiltration, spray drying and fluidized bed agglomeration. J Food Eng 84(2):194–205
48. Boye J, Zare F, Pletch A (2010) Pulse proteins: processing, characterization, functional properties and applications in food and feed. Food Res Int 43(2):414–431
49. Shokri S, Javanmardi F, Mohammadi M, Mousavi KA (2022) Effects of ultrasound on the techno-functional properties of milk proteins: a systematic review. Ultrason Sonochem 83:105938
50. Wani IA, Sogi DS, Gill BS (2013) Physicochemical and functional properties of flours from three Black gram (Phaseolus mungo L.) cultivars. Int J Food Sci Technol 48(4):771–777
51. Wani IA, Sogi DS, Wani AA, Gill BS (2013) Physico-chemical and functional properties of flours from Indian kidney bean (Phaseolus vulgaris L.) cultivars. LWT Food Sci Technol 53(1):278–284
52. Heywood A, Myers D, Bailey T, Johnson L (2002) Functional properties of low-fat soy flour produced by an extrusion-expelling system. J Am Oil Chem Soc 79(12):1249
53. Barac MB, Pesic MB, Stanojevic SP, Kostic AZ, Bivolarevic V (2015) Comparative study of the functional properties of three legume seed isolates: adzuki, pea and soy bean. J Food Sci Technol 52(5):2779–2787
54. Damodaran S (2005) Protein stabilization of emulsions and foams. J Food Sci 70:R54–R66
55. Denkov ND (2004) Mechanisms of foam destruction by oil-based antifoams. Langmuir 20(22):9463–9505
56. Karakashev SI, Grozdanova MV (2012) Foams and antifoams. Adv Colloid Interface Sci 176–177:1–17
57. Sagona S, Bozzicolonna R, Nuvoloni R, Cilia G, Torracca B, Felicioli A (2017) Water activity of fresh bee pollen and mixtures of bee pollen-honey of different botanical origin. LWT Food Sci Technol 84:595–600
58. Bi Y-X, Zielinska S, Ni J-B, Li X-X, Xue X-F, Tian W-L et al (2022) Effects of hot-air drying temperature on drying characteristics and color deterioration of rape bee pollen. Food Chem X 16:100464
59. Žilić S, Vančetović J, Janković M, Maksimović V (2014) Chemical composition, bioactive compounds, antioxidant capacity and stability of floral maize (Zea mays L.) pollen. J Funct Foods 10:65–74
60. Song X-D, Mujumdar AS, Law C-L, Fang X-M, Peng W-J, Deng L-Z et al (2020) Effect of drying air temperature on drying kinetics, color, carotenoid content, antioxidant capacity and oxidation of fat for lotus pollen. Dry Technol 38(9):1151–1164
61. Domínguez-Valhondo D, Gil D, Hernández M, Gonzalez-Gomez D (2011) Influence of the commercial processing and floral origin on bioactive and nutritional properties of honeybee-collected pollen. Int J Food Sci Technol 46:2204–2211
62. De-Melo A, Estevinho L, Gasparotto Sattler J, Rodrigues de Souza B, Freitas A, Barth O et al (2015) Effect of processing conditions on characteristics of dehydrated bee-pollen and correlation between quality parameters. LWT Food Sci Technol 65:808
63. Canale A, Benelli G, Castagna A, Sgherri C, Poli P, Serra A et al (2016) Microwave-assisted drying for the conservation of honeybee pollen. Materials (Basel) 9(5):363
64. Kanar Y, Mazı BG (2019) HMF formation, diastase activity and proline content changes in bee pollen dried by different drying methods. LWT Food Sci Technol 113:108273
65. Wang S, Bi Y, Zhou Z, Peng W, Tian W, Wang H et al (2022) Effects of pulsed vacuum drying temperature on drying kinetics, physicochemical properties and microstructure of bee pollen. LWT Food Sci Technol 169:113966

66. Salazar-González C, Díaz-Moreno C (2016) The nutritional and bioactive aptitude of bee pollen for a solid-state fermentation process. J Apic Res 55(2):161–175
67. Yan S, Li Q, Xue X, Wang K, Zhao L, Wu L (2019) Analysis of improved nutritional composition of bee pollen (Brassica campestris L.) after different fermentation treatments. Int J Food Sci Technol 54(6):2169–2181
68. Roldán A, van Muiswinkel GCJ, Lasanta C, Palacios V, Caro I (2011) Influence of pollen addition on mead elaboration: physicochemical and sensory characteristics. Food Chem 126(2):574–582
69. Amores-Arrocha A, Roldán A, Jiménez-Cantizano A, Caro I, Palacios V (2018) Effect on white grape must of multiflora bee pollen addition during the alcoholic fermentation process. Molecules 23(6):1321
70. Turhan S, Sarıcaoğlu F, Mortaş M, Yazici F, Gençcelep H (2016) Evaluation of color, lipid oxidation and microbial quality in meatballs formulated with bee pollen during frozen storage: evaluation of quality of meatballs with pollen. J Food Process Preserv 41
71. Almeida JF, Reis AS, Heldt LFS, Pereira D, Bianchin M, Moura C et al (2017) Lyophilized bee pollen extract: a natural antioxidant source to prevent lipid oxidation in refrigerated sausages. LWT Food Sci Technol 76:299–305
72. Krystyjan M, Gumul D, Ziobro R, Korus A (2015) The fortification of biscuits with bee pollen and its effect on physicochemical and antioxidant properties in biscuits. LWT Food Sci Technol 63(1):640–646
73. Sokmen O, Ozdemir S, Dundar AN, Cinar A (2022) Quality properties and bioactive compounds of reduced-fat cookies with bee pollen. Int J Gastronomy Food Sci 29:100557
74. Solgajová M, Nôžková J, Kadáková M (2014) Quality of durable cookies enriched with rape bee pollen. J Cent Eur Agric 15:24–38
75. Conte P, Del Caro A, Urgeghe PP, Petretto GL, Montanari L, Piga A et al (2020) Nutritional and aroma improvement of gluten-free bread: is bee pollen effective? LWT Food Sci Technol 118:108711
76. Végh R, Csóka M, Stefanovits-Bányai É, Juhász R, Sipos L (2023) Biscuits enriched with monofloral bee pollens: nutritional properties, techno-functional parameters, sensory profile, and consumer preference. Foods [Internet] 12(1)
77. Yan S, Wan Y, Wang F, Xue X, Wu L (2021) Fortification of bread with bee pollen, and its effects on quality attributes and antioxidant activity. Int Food Res J 28(3):517–526
78. Kandylis P, Dimitrellou D, Moschakis T (2021) Recent applications of grapes and their derivatives in dairy products. Trends Food Sci Technol 114:696–711
79. Yerlikaya O (2014) Effect of bee pollen supplement on antimicrobial, chemical, rheological, sensorial properties and probiotic viability of fermented milk beverages. Mljekarstvo/Dairy 64(4):268
80. Morsy M (2017) Effect of Incorporating Royal Jelly and Bee Pollen Grains on Texture and Microstructure Profile of Probiotic Yoghurt. J Food Process Technol 8:1–4
81. Mabrouk A (2016) The production of bio-yoghurt with probiotic bacteria, royal jelly and bee pollen grains. J Nutr Food Sci 6
82. Darwish AMG, Abd El-Wahed AA, Shehata MG, El-Seedi HR, Masry SHD, Khalifa SAM et al (2022) Chemical profiling and nutritional evaluation of bee pollen, bee bread, and royal jelly and their role in functional fermented dairy products. Molecules 28(1):227
83. Thakur M, Nanda V (2019) Process optimization of polyphenol-rich milk powder using bee pollen based on physicochemical and functional properties. J Food Process Eng 42(6):e13148
84. Thakur M, Pant K, Naik RR, Nanda V (2021) Optimization of spray drying operating conditions for production of functional milk powder encapsulating bee pollen. Dry Technol 39(6):777–790
85. Anjos O, Fernandes R, Cardoso SM, Delgado T, Farinha N, Paula V et al (2019) Bee pollen as a natural antioxidant source to prevent lipid oxidation in black pudding. LWT Food Sci Technol 111:869–875

86. Abd-Elmegaly FM, Latif SS, Saleh SAM, Abdel-Hameed SM (2022) Nutritional quality and sensory attributes of biscuits fortified with bee pollen. Aswan Univ J Sci Technol 2(2):1–13
87. Zlatev Z, Taneva I, Baycheva S, Petev M (2018) A comparative analysis of physico-chemical indicators and sensory characteristics of yogurt with added honey and bee pollen. Bulg J Agric Sci 24:132–144

Chapter 15
Bee Pollen as a Source of Pharmaceuticals: Where Are We Now?

Rachid Kacemi and Maria G. Campos

15.1 Introduction

Bee products have been used for medical, nutritional, and cosmetic purposes since ancient times. Their use in human medicine dates back to thousands of years ago and was known in different ancient civilizations across the world including Egyptian, Greek, Indian, Roman and Islamic ones [1–3]. Archaeological paintings even give evidence for their use since the Stone Age. Apiculture and bee product collection were illustrated in rock paintings from around 10,000–5000 B.C. in figuring female collecting honey Cuevas de la Araña de Bicorp, Valencia, eastern Spain [4, 5]. Other hieroglyphic wall paintings in Egyptian temples from 3500 B.C. represent humans managing bees (possibly the Asian honeybee, *Apis cercana*) for honey and crop pollination [6]. These products have been used in ancient medicine to treat and prevent many ailments such as ulcers, burns, wounds, allergic rhinitis, hyperlipidemia, bowel problems, and rheumatoid arthritis [4, 7]. Going back in time, Ancient Egyptians described bee pollen (BP) as a "life-giving dust" [3], that resulted certainly from its known nutritional and medicinal benefits. Also Greeks believed that this hive product and honey gave youth to kings [8]. Data about ethnopharmacological use of BP is unfortunately very scarce. However, some of these uses are studied nowadays for robust data which support them. For instance, pollen use in prostate

R. Kacemi
Observatory of Drug-Herb Interactions/Faculty of Pharmacy, University of Coimbra, Coimbra, Portugal

M. G. Campos (✉)
Observatory of Drug-Herb Interactions/Faculty of Pharmacy, University of Coimbra, Coimbra, Portugal

Coimbra Chemistry Centre (CQC, FCT Unit 313) (FCTUC), University Coimbra, Coimbra, Portugal
e-mail: mgcampos@ff.uc.pt

N. Ecem Bayram et al. (eds.), *Pollen Chemistry & Biotechnology*,
https://doi.org/10.1007/978-3-031-47563-4_15

disease, for example, is largely known especially in non-infectious situations [9]. Historical evidence shows also that pollen was used for cosmetic purposes in the ancient China [10]. BP was used for medicinal purposes in many major known civilizations, as Ancient Egyptians, Greeks, native Americans, Chinese, and Indians which used it for provision and energy in long journeys as well as for other health benefits [11].

A simple search (done on February 15th, 2023) with the keyword "bee pollen" in the PubMed database returns more than 2200 publications in the last 50 years, with more than 1000 works published in the last 5 years and more than 500 in the last 2 years. This gives a clear preview on how research is going on studying BP implication in human medicine. Experimental and preclinical research is cumulating a large body of evidence to support pollen bioactivities as well as nutritional and preventive attributes. Numerous studies have shown that BP have a rich and well balanced composition to serve as a human food and supplement, while its richness of bioactive compounds, especially polyphenols, confer to it a large array of biological and pharmacological activities [12].

However, BP composition is extremely variable due to many biological, genetic, and environmental factors. Despite all these diversity factors, BP comprises many quasi-universal components which may provide a background for its pharmacological studies and bioprospection. More than 250 compounds have been reported in this matrix, including proteins (comprising all essential amino acids needed by humans), carbohydrates, lipids (long-chain essential fatty acids), phenolic compounds (with extensive variability), vitamins (tocopherols, calciferol, and B complex), minerals, organic acids, and many other bioactive elements (e.g., more than 100 enzymes and coenzymes, carotenoids, phytosterols, some hormones and growth regulators, terpenes) are among the most important and universally encountered molecules in pollens of different botanical origins [13–15]. In addition, mineral [16] and protein [17] contents were found to vary depending on the harvesting season mostly due to the diversity of flowering plants according to the season. These results are very likely to be true for other constituents despite the scarcity of studies.

Furthermore, the universal presence of phenolic compounds in pollen is also a key determinant of its most known bioactivities. Pollen importance in many worldwide spread diseases such as inflammatory, malignant, neurodegenerative, cardiovascular, and infectious diseases is mainly, or at least partly, due to phenolic compounds which are widely known for their anti-inflammatory, antioxidative and anti-infectious activities. However, the synergistic and complementary interplay between pollen compounds remain very poorly and insufficiently studied. Such inadequacy still hinders our understanding of its preventive and therapeutic potential and needs to be urgently addressed, at least partly by comparative studies between pollen as a whole product and its different types of extracts and molecules. Data analysis remains very scarcely exploited in understanding pollen bioactivities, whereas extensive studies exploring some of its key known molecules are available from other research works that studied either these molecules alone or derived from other natural sources such as medicinal plants. This great amount of data is unfortunately still underexploited and may, if well analyzed, give us unvaluable insights to guide preclinical and translational research and thus to fructify these works at the clinical level. In this chapter, it will be provided an overview of selected possibilities for drug discovery based on the research already available with this crude material.

15.2 Bioactivities & Pharmacognosy

In addition to the vast array of nutrients that are present in BP it also contains a myriad of other phytochemicals manifesting numerous biological activities correctly mostly to the phenolic compounds, such as phenolic acids and flavonoids, carotenoids, and diverse volatile compounds [18]. Phenolic acids and flavonoids are most likely responsible for the major part of antioxidant properties of BP [19].

Several bioactivities of BP were reported by numerous studies from different regions and environmental conditions worldwide and it has been reported to exert anesthetic, anti-allergic, anti-androgen, anti-atherosclerotic, anti-cancer (anticarcinogenic and anti-mutagenic), anti-inflammatory, antimicrobial (antibacterial, antifungal, and antiviral), antioxidant, antiulcer, and immunostimulant activities [18–26]. In addition, due to its multiple bioactivities, BP has been notably shown to possess anti-obesity [25], antidiabetic [25], hypocholesterolemic [26], and hepatoprotective effects [21, 25]. On the digestive system, BP has been notably shown to ameliorate [20], maintain [18] and regulate [26] gut functions. On the cardiovascular system, it was also found to reduce capillary fragility [18], and to overall improve the cardiovascular system [20, 26]. Among the preventive properties and due to possible positively impact is explored in neurodegeneration [20, 27], overall aging process [24, 26], and cellular apoptosis [28], as well as to promote recovery from chronic diseases and to have chemo-preventive properties [21]. This product has also been shown to promote skin health and reparation and to have many cosmetic qualities as well (such as protecting cells from abnormal melanogenesis, and eliminating age and freckly spots, and melasma) [18, 29].

BP has possible advantageous properties that could be exploited in nutritional, preventive, therapeutic and recovery indications. This relies on its multitargeting potential which is a new tendency in managing many challenging illnesses. In cancer for example, immunomodulatory, antioxidant, anti-inflammatory and antimutagenic effects could be of unequaled importance and need further intensive pharmacological and clinical investigation. This is even more important in some cancers such as prostate hyperplasia where all these qualities are added to the androgenic effect [18] of BP to possibly make it an adjuvant or even treatment in this disease. In neurodegeneration also, micronutrient richness, anti-inflammatory, antioxidant, enzyme, and excitotoxicity inhibition qualities of BP provide the same type of advantages. This potential was developed by us in a previous paper and will not be part of the present chapter.

15.2.1 Immune Response Modulation

The immunomodulatory effect of BP is frequently reported in the literature, but evidence, at least at experimental level, is still relatively sparse as we will see through numerous examples in this section. BP was reported to both exert immunostimulant actions, alleviate autoimmune processes [30] and induce an immunomodulatory effect to inhibit allergic reactions [31]. We will discuss antiallergic effects in a separate section. Likewise, pollen has widely been reported as anti-infective agent [29, 32, 33], an effect which is not only linked to its immunomodulatory role. Despite the low abundance of experimental studies and the scarcity of clinical ones, BP appears to present an interesting immunomodulatory potential on immune cells, immune response mediators and on gut microbiota diversity and performance. For instance, supplementing BP to broilers diet resulted in many major immune-stimulating effects, such as increasing leukocyte viability, T lymphocyte proliferation and antibody production (Immunoglobulins A and M, but not G were increased in response to a tested vaccine), decreasing heterophil/lymphocyte ratio, and enhancing growth and reproductive performance [34]. Another study in broilers, reported that BP feeding resulted, depending on its ratio in the diet, in a linear increase of IgM but not IgG, as well as in an increase of thymus weight [35]. In horses, either trained and aged ones, BP supplementation did not alter plasmatic IgG levels [36]. In contrast, ethanolic extract of carbonized *Typhae* spp. pollen (not bee-collected) showed significant increase of total Ig-G levels in experimentally immunized rats [37]. This difference remains to be elucidated as it may originate from vastly different factors. Studies in other animals such as livestock also reported that BP supplementation resulted in an important enhancement of immune response, and was even proposed as a good substitute to antibiotics [32]. Some other studies in rodents reported immune-protective effects of BP which manifested in enhancing immune cell proliferation and profiles, and in circulating levels of immunoglobulins and total proteins [35].

BP polysaccharides seem to be markedly involved in its immunomodulatory effects as it may emanate from experimental studies. These type of compounds extracted from *Pinus massoniana*, *Crataegus pinnatifida*, *Nelumbo nucifera*, and *Phleum pratense* exert an immunomodulatory effect through different mechanisms, such as promoting spleen lymphocyte proliferation, nitric oxide production, macrophage activation, immunoglobulin secretin, and the liberation of pro-inflammatory cytokines [38].

Likewise, a polysaccharide fraction of Chinese wolfberry BP (likely *Lycium chinense* although the botanical name was not given in the publication) exerted a marked modulation of immune response toward a murine macrophage cell line [38]. The most active fraction tested in this study manifested a potent cytotoxic activity on the tested cell line, markedly increased nitric oxide (3.5-fold), TNF-α levels (2.1-fold), IL-6, and IL-1β levels. Curiously, the authors noted that these immune-stimulating effects were not dose-dependent, and they proposed some possible explanatory mechanisms. It is worth noting that the most active polysaccharide fraction contained the highest concentration of zinc element. We could then deduce that some activities related to the pollen-driven immunomodulation may not depend on the amount of administered pollen, and that micronutrients could play a potential role in such activities as it is the case for known secondary metabolites.

Another important mechanism by which BP may improve immune system performance is through positively acting on intestinal microbiota which is known to play a key role in body immunity and inflammatory signaling. As an example, a water-soluble polysaccharide from *Fagopyrum esculentum* BP was reported to attenuate ceftriaxone-induced gut dysbiosis in mice after being administered for 3 weeks [39]. Among this attenuation traits, growth retardation, atrophy of immune organs including thymus and spleen, increased gut permeability, and alteration of intestinal barrier integrity were induced by ceftriaxone and were all restored after polysaccharide administration. Moreover, pollen polysaccharide increased microbial diversity and richness and improved community structure in gut microbiota of studied animals. Furthermore, secretory Immunoglobulin A was markedly upregulated after polysaccharide administration, an effect that was concomitant with modulate abundances of these immunoglobulin secretion-related bacteria (e.g., *Proteobacteria*) and inflammation-related bacteria (e.g., *Enterococcus*). Inflammation markers such as intestinal IL-6 and IL-1β and serum lipopolysaccharides were restored to normal levels in pollen-treated animals.

BP was also experimentally verified to mitigate immunosuppressive effect of doxorubicin in rodents, as well as concomitant-chemotherapy induced impairment of hematopoiesis [40]. This study focused on spleen and bone marrow and showed that pollen extract also exerted a potent inhibitory effect of cellular apoptosis in both tissues (pollen was unfortunately commercial without defining its botanical origin).

An appealing study on zebrafish reported recently that BP supplementation to these animals resulted in vertical transmission of immune modulatory effect to offspring [41]. This study reported that the offspring of breeders fed with BP had longer survival upon virus exposure (spring viremia of carp virus was used) and a higher neutrophil recruitment to experimentally induced tail wounds. The protective effect was more pronounced with longer periods of pollen feeding. Interestingly, unlike antiviral protection, the enhancement of immune response was not noted upon an experimental bacterial infection. This intriguing observation may give us interesting insights about possible transmission of immunity enhancement during pregnancy in at risk situations, since the high safety of BP may be predictable.

Another important observation is that multifloral BP may be more potent as immunostimulant than mono-floral pollen. As an evocative example, honeybees living on mass agricultural cultures of unique botanical species show a marked weakness in their immune system and an overall disease susceptibility and poor health [42]. Moreover, fermented pollen was reported to exert more marked immunomodulatory effects than raw BP (*Apilactobacillus kunkeei* and *Hanseniaspora uvarum* strains fermentation showed more potential than spontaneous fermentation [43].

Additionally, as an example appealing carefulness in designing, interpreting experimental and translational studies, a study in bumblebees reported that no difference in immune response measures was noted after feeding these bees with

Helianthus annuus BP although the pollen-fed bees manifested a significantly better resistance to induced infection with the gut pathogen *Crithidia bombi* which usually infect bumblebees [44]. The authors of this study affirmed that the involved mechanisms in this remain unknown but presumed that the observed reduction in infection could rely on unmeasured immunity traits or on other unknown parameters. The performances of these bees toward infection were additionally boosted when sunflower pollen was consumed with a mix of wildflower pollens. Furthermore, in rats fed with multifloral BP, immunohistochemistry staining in response to experimental antibodies did not show significant impact on Interleukin-12 (one of the key cytokines in immune response initiation) levels in liver tissues [45].

These observations, show that some classical "easy" explanations frequently advanced by certain authors, and linking immune response modulation by pollen to phenolic compounds and to other bioactivities, such as antioxidant and anti-inflammatory ones, may not be always true. Specific compounds in a specific combination which is present in a specific natural product may determine the effects, whether the host organism where the study is conducted may also determine its response which will be regulated by a complex interplay of intrinsic and environmental factors, as well as by diverse experimental conditions and protocols. This discussion demonstrates, once again, that BP bioprospection is still at a very early stage but doesn't rule out possible beneficial contribution of BP as preventive and complementary adjuvant in immune-weakened persons.

15.2.2 Anti-allergic Effect

BP may stand in a bit "unenviable" position regarding its effects in allergies. Indeed, it may evidently carry potential allergens originating from plant pollens, but have also been shown to exert marked anti-allergic effects in several *in vitro*, animal and human studies, and was proposed as a possible preventive and therapeutic tools in allergic conditions such as asthma, bronchitis, and rhinitis [29, 46, 47].

Numerous mechanisms have been described for antiallergic activity. In IgE-sensitized cell cultures, BP was found to inhibit mastocyte degranulation through preventing the binding of these globulins to mast cells, and to inhibit TNF liberation and other signaling pathways by these cells [35]. In ovalbumin-sensitized animal models, BP was also reported to decrease IgE and IgG1 production and to inhibit eosinophil activation and immune cell migration to pulmonary airways [35]. Other more generic mechanisms were assigned. It was for example found to downregulate the expression of β-hexosaminidase (an enzyme involved in mast cell degranulation and other biological processes [48]) and to inhibit tyrosine phosphorylation in proteins, a posttranslational modification which is implicated in mastocyte degranulation [47], among myriad other key cellular functions [49].

BP phenolic compounds such as some flavonoids (e.g., kaempferol, myricetin, luteolin and quercetin) are classically known for their antiallergic activities especially in IgE-mediated allergies, but other compounds such as unsaturated fatty acids (e.g., malonic and α-linolenic acids) have been linked to its antiallergic effect [31, 46].

Despite the rareness of BP allergies and some clear experimental insights about its antiallergic potential, studies to figure out the latter remain very scarce and highly fuzzy. More studies are first needed at experimental level which allow the proposition of possible mechanisms and compounds involved in modulating allergic reactions, through directly impairing these reactions or through gathering anti-inflammatory and antioxidant mechanisms to exert this impairment. This entails differentiating effects on cellular and humoral immunity as well as on innate and acquired, Ig-E- and Non-Ig-E- mediated immune responses. Of course, *in vivo* pre-clinical and clinical studies will be subsequently needed before drawing any practically and clinically relevant inferences.

15.2.3 Anti-infective Potential

The anti-infective potential of numerous phenolic and other components of BP is undeniably and largely known. However, studies done with the whole product for anti-infective potential are more recent. Nevertheless, many factors such as botanical and geographical origins, foraging bee species, collection period and timing, environmental conditions, processing manipulations, and extracting protocols, will produce some bias that should be analyzed carefully. Accordingly, bioactivities including anti-infectious ones will largely differ.

Efficacy of BP as anti-viral agent was reported *in vitro* to be good against *Herpes Simplex Virus* 1 and 2, and moderate against *Influenza* virus (H1N1, H3N2, and H5N1 strains) [50, 51]. BP, among other bee products, has been proposed by numerous researchers as an adjuvant product in managing recently emerged coronavirus disease, whereas numerous phytochemicals (phenolics and others) found are actively studied for such indication [33, 51].

A study of a wild multifloral BP showed that hydroethanolic extracts manifested marked inhibitory action on *Candida albicans*, and other clinically isolated strains (*C. famata*, *glabrata*, *guillermondii*, *krusei*, and *lusitaniae*) [52]. Antifungal activity on *Zygosacharomyces* and *Aspergillus* spp. was also reported [30]. Water and hydroalcoholic extracts of hemp pollen have also shown interesting antifungal properties against some human-infecting pathogens including some strains belonging to *Trycophyton* and *Arthroderma* genus, in addition to other bacterial pathogens such as *Staphylococcus aureus*, *Pseudomonas aeruginosa*, and *Escherichia coli* [53].

BP showed bactericidal, bacteriostatic, and antifungal activities [50]. *In vitro* antibacterial effect was verified against *Clostridioides perfringens* (a pathogen ten-fold higher in autist patients) [54]. Ethanolic extract of miscellaneous BP samples manifested good antibacterial activity against *Streptococcus pyogenes* but did not have significant antibacterial effect on *S. aureus*, *P. aeruginosa*, and *E. coli* [51]. Methanol extract of *Castanea sativa* displayed antibacterial effect against *S. aureus*,

Enterococcus faecalis, *Micrococcus luteus*, *Bacillus cereus*, *Enterococcus*, *E. coli*, and *Klebsiella pneumoniae*, with the effect on Gram-positive germs being higher than Gram-negative ones [51].

A study of ethanolic extracts of botanically diversified BP samples reported that they were markedly active against many food-borne pathogens including *Bacillus cereus*, *S. aureus*, *E. coli*, *P. aeruginosa*, *Salmonella typhimurium*, *Yersinia enterocolitica* [55]. This study also found that tested extracts were more active on Gram-positive strains. These results were in coordination with the findings of some previous studies reviewed by the authors, albeit minimum inhibitory concentrations were not the same, and extraction solvents were different between compared studies (included water, methanol and water). This repeatedly occurring observation in experimental studies is still not fully understood and needs to be elucidated.

A comparative study of methanol/water (80:20) extracts of BP, honey, bee bread, and beeswax found that the first had important antimicrobial activity against different strains of *Staphylococcus aureus*, *Enterococcus faecalis*, *Escherichia coli*, *Listeria monocytogenes*, and *Salmonella typhimurium* [56]. The antimicrobial effect was much weaker on the last two species, but BP potential was only exceeded by bee bread whether honey was much less active and beeswax inactive. The authors of this study reviewed many other previous findings from other studies that reported marked inhibitory effect of BP on pathogen bacteria such as *Salmonella enterica*, *L, monocytogenes*, and *S. aureus*, and other less or non-pathogen bacteria such as *E. coli*. Recurrent observations of the superiority of BP antibacterial action compared to honey and its inferiority compared to bee bread, and of high concentrations of pollen needed against some strains, were reported in the review, but we could not draw universal conclusions from the cited number of samples and studies.

Butanol extract of *Cistus criticus* BP showed good antibacterial activity against *S. aureus*, *S. epidermidis*, *P. aeruginosa*, *E. coli*, *Enterobacter cloacae*, and *Klebsiella pneumoniae*, while cyclohexane and dichloromethane extracts of the same pollen did not show antibacterial activity on these strains [51].

Numerous antimicrobial phytochemicals that are present in *Apis mellifera* BP are also found in pollen collected by other bee species, especially those that are widely studied such as stingless bees. Many phytocompounds (e.g., glycosides of apigenin, kaempferol, luteolin, myricetin, naringenin, quercetin, phenolic acids) present in both are widely known for their antibacterial potential against some pathogens including some very frequent ones in humans such as *Helicobacter pylori*, *S. aureus* and *E. coli* [57]. For redaction briefing reasons we did not give more details which can be found in cited references.

Moreover, BP has sometimes shown interesting antiparasitic potential. Rhusflavone, a biflavonoid that was recently isolated from mono-floral pollen of *Attalea funifera*, manifested high leishmanicidal activity against both promastigote (infective form) and amastigote (replicative form) forms of *Leishmania amazonensis* in a dose-dependent manner. It's important to underline that even if rhusflavone was less effective than Amphotericin B, its hemolytic toxicity of was much lesser [58]. Accordingly, a very recent meta-analysis compared the concentrations of diverse phytochemicals (mainly focused on phenolics including flavonoids, viz.

apigenin, kaempferol, luteolin, and quercetin, and chlorogenic and rosmarinic acids) in dozens of nectar and pollen types. Interestedly revealed that the concentration in which many of these phytochemicals is present in some samples exceed those that inhibit growth of *Leishmania* cell cultures [59]. This result gives a good example of how could a drug-free and abundant natural product fight one of the most disability-causing parasites in the world.

An observation that could be of great interest is possible biotechnological exploitation of BP as a medium to produce natural anti-infectious agents. We have repeatedly seen that fermented BP may present enhanced phytochemical and bioactive profiles. We have also evoked that BP bears some probiotic strains mainly originating from bee digestive microbiota. As an important illustrative example, *Streptomyces* strains are among the most versatile and successfully used microorganisms in biotechnological drug production [60]. These reminders elucidate a potential exploitation of BP in natural drug discovery. A recent study reported that *Streptomyces* strains isolated from the pollen stores of honey bees produced Piceamycin, a known antibiotic used efficiently against the fatal bee brood pathogen *Paenibacillus larvae* [61]. This very recent study focused on a beehive-attacking pathogen, but similar works may be planned against human pathogens.

From all previously cited examples we realize that BP is a very rich pool of bioactive compounds that may be still untapped or rarely exploited in drug discovery related to pathogen microbes that affect humans. Infectious and parasitic diseases continue to affect billions of people around the world, and numerous impediments continue to potentiate the global burden of these diseases. Some socioeconomic and political factors may obviously stand behind some of these impediments such as access to diagnosis and available efficient drugs, but many other aspects of hindering rely on frequent resistance of infectious organisms and on the scarcity of therapeutic tools even in rich countries. In this context BP may present an opportunity both for drug discovery and for large scale preventive, nutritional-based measures, or for environmental actions such as the case we have just seen for *Leishmania.* We have seen examples of different botanical species, extraction methods, and bee species, (the same observation is valid for geographical locations albeit not detailed here), and for all known kind of infectants including bacteria, virus, fungi, and parasites. Moreover, as we are detailing in many parts of this work, BP has the quality of gathering other complementary bioactivities that invaluably help in infection fighting such as nutritional richness, immune response modulation, microbiota enhancement, and anti-inflammatory and antioxidant potentials. A lot of engaged, standardized, and targeted work is therefore needed, but prospected benefits worth the effort.

15.2.4 Musculoskeletal Diseases

Considering the evidence that we gathered during previous sections and chapters, BP seems to be a promising candidate in managing many musculoskeletal diseases, especially those entailing oxidative, inflammatory and/or nutritional imbalances.

15.2.4.1 Rheumatic Diseases

Bee venom was frequently studied in arthritis and showed good results including in humans [30, 62]. However, BP is still very poorly studied in this widespread and disabling kind of disease. We have seen that anti-inflammatory and antioxidant effects of BP are widely verified. The immunomodulatory effect of BP is also evident albeit less studied. These three mechanisms are classically known to intervene in many rheumatic diseases and lead to the confirmed disease stage along chronic evolution and worsening. Likewise, the antiaging effect through other mechanisms such as telomere shortening [62], and its verified ability to mitigate these three disease pillars, BP could be helpful in managing these disorders at different stages of risk and onset. A study, already cited in this chapter, that used an experimental pouch inflammation, which induce a model similar to joint inflammation in humans, and reported marked inhibitory effect of inflammation and angiogenesis by BP [63].

Numerous inflammatory processes such as pro-inflammatory cytokine increase, autophagic processes, oxidative stress aggravation, autoimmune imbalances, inflammatory enzymes, gut liberation of inflammation triggers, dietary insufficiency of vitamin and mineral intake, etc., are alleviated or corrected by BP as we have amply detailed. These elemental theoretical and experimental arguments may constitute a rough outline for promising potential of BP in rheumatic diseases.

15.2.4.2 Osteoporosis

Theoretically, BP could supply a good and diversified source for bone metabolism and structure through anabolic and loss prevention roles, due to its richness of needed elements for such functions. However, the very few experimental studies conducted so far, reported very conflictual results.

After BP supplementation, one study reported higher calcium concentration in rat femoral bone in all tested dosages (5 and 10 mg/100 g body weight), one reported no changes in rats in all tested dosages (0.5 and 1.0% of feed), a third reported no changes in broilers, while a fourth reported a decreased bone mineralization, length, weight, and wall thickness in quails at 1% of BP in feed, and a fifth study reported reduced bone weight in rats at 0.75% of BP [64]. Trying to solve this puzzle, a very recent study compared the effect of BP supplementation by including BP at 0.5 and 0.75% of rat feed. After 3 months of supplementation, the authors noted a slight, but not significant, increase in body and femoral bone weight with 0.5% of pollen and curiously a decrease in cortical bone thickness and femoral bone weight as well as an increase in body weight with 0.75% pollen supplementation [64]. Moreover, 0.75% pollen-fed group manifested a decreased plasma activity of alkaline phosphatase (enzyme involved in bone formation) and a reduced size of primary and secondary osteoblasts, which could indicate an impaired blood flow and some signs of bone toughness.

Another recent study in quails reported similar findings. BP supplementation, at 1% of feed by weight, negatively affected bone structure, regardless of quail sex, by reducing bone length, weight, and wall thickness, and slowing bone maturation [65]. However, this study noted that BP supplementation increased bone mineralization which resulted in enhancing homeostasis of trabecular bone in metaphysis. This enhancement differed in its entailed processes between sexes. In male quails, it resulted from increased trabecular thickness, while it resulted from reducing trabecular space in females. The authors suspected a hormonal interference, and a role of mineral and vitamin D content of pollen, but further studies are needed to elucidate these effects, and to verify their universality or variability.

Considering these very intriguing observations, more experimental work, preferably with unified experimental protocol and more significant sampling, should be undertaken to figure out possible mechanisms by which BP could alter bone structure and function in a dose-dependent manner.

15.2.4.3 Sarcopenia

Sarcopenia is a widespread disorder affecting millions of people worldwide. A recent meta-analysis found that incidence rate in older people may range from 10% to 27%, considering that data variability could emanate from some factors including the absence of a clear definition and diagnostic standards to this pandemic disease [66].

Two studies in rodents have evaluated many aspects of BP (both used monofloral) potential to prevent, preserve and restore muscle alterations either in animals submitted to chronic swimming exercise or in malnourished ones [67]. In the formers, BP supplementation resulted in markedly reducing oxidative stress in gastrocnemius muscles, downregulating myostatin expression, and improving exercise-induced impairment of mitochondrial complexes activity. In the latter, BP led to visceral and subcutaneous adipose tissue restoration, plantaris and gastrocnemius muscle mass increase, cytokine level normalization, muscle protein synthesis improvement, mitochondrial complex IV activity increase, and mammalian target of rapamycin activation which could mean an autophagy modulation and anabolic promoting effect.

Since BP is a very well-balanced source of dietary proteins and other nutrients including those involved in muscle metabolism and detoxification such as vitamins and oligo-elements, this impact has a good support. Anti-inflammatory and antioxidant bioactivities may be added to these qualities to make BP a valuable source in people risking such condition, either by early detected disorders or by lifestyle and environmental factors such as sedentary life and nutritional imbalances or deficiencies. Unfortunately, studies of BP in sarcopenia are extremely rare, a fact that is strikingly faced by these encouraging elements in favor of a leading role of BP in preventing and managing this disease "abnormally neglected" wile "constituting a silent killer" which is certainly going to strongly spread in the future.

15.2.5 Dermatological Disorders

Throughout this chapter, we have highlighted some bioactive properties of BP that are of prime importance in skin protection and disorder repairing. This includes obviously anti-infective, anti-inflammatory, antioxidant, immunomodulating, and antiallergic potentials. These qualities make BP not only a valuable product in preventing and managing dermatological disorders, but also a precious source for dermo-cosmetic uses.

According to those different qualities, BP has been reported to promote granulation tissue proliferation in wound healing, prevent infections in new tissues, accelerate overall tissue regeneration, suppress oedemas, strengthen skin capillary network, promote skin cell metabolism and proliferation, and modulate sebum secretion [68–70]. We have already seen that numerous tested BP manifested marked inhibitory activities against bacterial and fungal pathogens affecting human skin. All these observations taken together assert that BP is undoubtedly a multitargeting potential drug in numerous skin disorders.

BP feeding to zebrafish models of melanoma did not result in a protective effect against tumor cell proliferation, and, although it significantly modified animal intestinal microbiota, gene transcript levels of proinflammatory mediators in the intestine didn't differ between pollen-fed and control animals [71]. This bring conflictual results with some previous experimental results [29] that reported a BP activity against melanoma. We have already seen that BP presents antineoplastic potential and we will shortly see that it is an excellent tyrosinase inhibitor. Studies on BP in melanoma remain extremely scarce and such result discrepancies are perfectly normal before sufficient studies sampling and diversity.

Extracellular vesicles became recently discovered and investigated in bee products entailing honeybee hypopharyngeal gland secretions (viz. BP, honey and royal jelly) [72]. These vesicles showed appealing importance especially in dermatological applications where they may contribute to some healing properties of these products such as anti-microbial and pro-regenerative effects [70, 72]. Preliminary experimental evidence shows mainly that these recent findings may present good incentive to investigate the role of these vesicles in wound healing, and some works are successfully undertaken to formulate novel products for wound healing exploiting the antibacterial, immunomodulatory, anti-inflammatory, and repairing effects of BP and other bee products, and enhancing bioavailability of bioactive compounds (e.g., collagen hydrogels, nanovesicle carriers) [70, 72, 73].

A considerable amount of data corroborate that impaired antioxidative defense and/or abnormal accumulation of reactive oxidant species [74–76], as well as inflammation especially in its and generalized chronic phase [76–78], are among foremost players in a large array of dermatological disorders and age-related frailties.

Polyphenols, the leading known bioactive compounds of BP, are a family of chemicals which is extensively investigated and widely known for promising and diversified potential in preventing and suppressing many dermatological disorders as well as an unvaluable pool of actives for cosmetic indications [76, 79–81]. Most of this evidence originates from polyphenol studies at molecular levels. Some molecules contained in BP such as luteolin [82], quercetin [83, 84], resveratrol [85], apigenin [86], kaempferol [87], and many other phytochemicals, are widely studied and show promising prospects against many dermatological disorders, while manifesting also a great potential in cosmetology. Studies of integrative preparations of natural products including these compounds as they are present in nature at least in combinations that are chemically more known, as it is the case for BP, are still lacking. Further studies must investigate total and multiphasic extracts of BP to mimic natural complementary and synergic environments in which phytochemicals and other natural compounds exist.

Cumulating evidence suggests that this product may be efficient at least in minor skin disorders such as benign infections and inflammatory conditions. More potential may also emanate from balanced BP composition such as supplying micronutrients that are classically known to be involved in many dermatological disorders including vitamins B, D and E, bio-elements such as selenium and zinc, a balanced unsaturated fatty acid ratio, proteins, etc. Many commercial formulations with such claims exist on the market either for disorders and wound reparations or for cosmetical purposes [68, 69]. Many other products, containing a combination of BP and other natural products, exist in commerce or have been studied [29, 70]. However, given the absence of comparative investigations in these products, it is very difficult to draw bee pollen-specific conclusions about activity.

15.3 Final Remarks

In this chapter a resume of the relevant research made with bee pollen was presented and discussed. Among the various bioactivities, screened is clear the pharmaceutical potential of this crude material for further investigation in drug discovery. Despite the amount of data already available for a high percentage of the possibilities, robust scientific evidence for these uses is still lacking.

References

1. Ahmad RS, Hussain MB, Saeed F, Waheed M, Tufail T (2017) Phytochemistry, metabolism, and ethnomedical scenario of honey: a concurrent review. Int J Food Prop 20(1):S254–S269. https://doi.org/10.1080/10942912.2017.1295257
2. Hadi A, Rafie N, Arab A (2021) Bee products consumption and cardiovascular diseases risk factors: a systematic review of interventional studies. Int J Food Prop 24(1):115–128. https://doi.org/10.1080/10942912.2020.1867568
3. El-Seedi HR et al (2020) Honeybee products: an updated review of neurological actions. Trends Food Sci Technol 101(May):17–27. https://doi.org/10.1016/j.tifs.2020.04.026

4. Aylanc V, Falcão SI, Ertosun S, Vilas-Boas M (2021) From the hive to the table: Nutrition value, digestibility and bioavailability of the dietary phytochemicals present in the bee pollen and bee bread. *Trends Food Sci. Technol.* 109(2020):464–481. https://doi.org/10.1016/j.tifs.2021.01.042
5. Izquierdo-Gascón M, Rubio-Gil Á (2022) Theoretical approach to Api-tourism routes as a paradigm of sustainable and regenerative rural development. J Apic Res:1–16. https://doi.org/10.1080/00218839.2022.2079285
6. Prendergast KS, Garcia JE, Howard SR, Ren ZX, McFarlane SJ, Dyer AG (2021) Bee representations in human art and culture through the ages. Art Percept 10(1):1–62. https://doi.org/10.1163/22134913-bja10031
7. Khalifa SAM et al (2020) Recent insights into chemical and pharmacological studies of bee bread. *Trends Food Sci. Technol.* 97(October 2019):300–316. https://doi.org/10.1016/j.tifs.2019.08.021
8. Camacho-Bernal GI et al (2021) Addition of bee products in diverse food sources: functional and physicochemical properties. Appl Sci 11(17):8156. https://doi.org/10.3390/app11178156
9. Tuoheti T, Rasheed HA, Meng L, Dong MS (2020) High hydrostatic pressure enhances the anti-proliferative properties of lotus bee pollen on the human prostate cancer PC-3 cells via increased metabolites. J Ethnopharmacol 261(1):113057. https://doi.org/10.1016/j.jep.2020.113057
10. Xi X et al (2018) The potential of using bee pollen in cosmetics: a review. J Oleo Sci 67(9):1071–1082. https://doi.org/10.5650/jos.ess18048
11. Lu P et al (2022) NMR and HPLC profiling of bee pollen products from different countries. Food Chem Mol Sci 5(July). https://doi.org/10.1016/j.fochms.2022.100119
12. Campos MG, Frigerio C, Bobiş O, Urcan AC, Gomes NGM (2021) Infrared irradiation drying impact on bee pollen: case study on the phenolic composition of eucalyptus globulus labill and Salix atrocinerea brot. Pollens. PRO 9(5):1–13. https://doi.org/10.3390/pr9050890
13. Habryka C, Socha R, Juszczak L (2021) Effect of bee pollen addition on the polyphenol content, antioxidant activity, and quality parameters of honey. Antioxidants 10(5):810. https://doi.org/10.3390/antiox10050810
14. Mărgăoan R et al (2019) Bee collected pollen and bee bread: bioactive constituents and health benefits. Antioxidants 8(12):1–33. https://doi.org/10.3390/antiox8120568
15. Mărgăoan R, Özkök A, Keskin Ş, Mayda N, Urcan AC, Cornea-Cipcigan M (2021) Bee collected pollen as a value-added product rich in bioactive compounds and unsaturated fatty acids: a comparative study from Turkey and Romania. LWT 149(June):111925. https://doi.org/10.1016/j.lwt.2021.111925
16. Taha EKA, Al-Kahtani S (2020) Macro- and trace elements content in honeybee pollen loads in relation to the harvest season. Saudi J Biol Sci 27(7):1797–1800. https://doi.org/10.1016/j.sjbs.2020.05.019
17. Al-Kahtani SN et al (2021) Effect of harvest season on the nutritional value of bee pollen protein. PLoS One 15(12 December):1–11. https://doi.org/10.1371/journal.pone.0241393
18. Baky MH, Abouelela MB, Wang K, Farag MA (2023) Bee pollen and bread as a super-food: a comparative review of their metabolome composition and quality assessment in the context of best recovery conditions. Molecules 28(2):715. https://doi.org/10.3390/molecules28020715
19. Rojo S, Escuredo O, Rodríguez-Flores MS, Seijo MC (2023) Botanical origin of Galician bee pollen (Northwest Spain) for the characterization of phenolic content and antioxidant activity. Foods 12(2):294. https://doi.org/10.3390/foods12020294
20. Xue F, Li C (2023) Effects of ultrasound assisted cell wall disruption on physicochemical properties of camellia bee pollen protein isolates. *Ultrason. Sonochem* 92(November 2022):106249. https://doi.org/10.1016/j.ultsonch.2022.106249
21. Alshallash K et al (2023) Bee pollen as a functional product – chemical constituents and nutritional properties. J Ecol Eng 24(2):173–183. https://doi.org/10.12911/22998993/156611

22. Végh R, Csóka M, Stefanovits-Bányai É, Juhász R, Sipos L (2023) Biscuits enriched with Monofloral bee pollens: nutritional properties, techno-functional parameters, sensory profile, and consumer preference. Foods 12(18):1–21. https://doi.org/10.3390/foods12010018
23. Hanafy NAN, Salim EI, Mahfouz ME, Eltonouby EA, Hamed IH (2022) Fabrication and characterization of bee pollen extract nanoparticles: their potential in combination therapy against human A549 lung cancer cells. Food Hydrocoll Heal 3(December):2023. https://doi.org/10.1016/j.fhfh.2022.100110
24. Chelucci E, Chiellini C, Cavallero A, Gabriele M (2023) Bio-functional activities of Tuscan bee pollen. Antioxidants 12(1):115. https://doi.org/10.3390/antiox12010115
25. Zhou E, Xue X, Xu H, Zhao L, Wu L, Li Q (2023) Effects of covalent conjugation with quercetin and its glycosides on the structure and allergenicity of Bra c p from bee pollen. *Food Chem* 406(November 2022):135075. https://doi.org/10.1016/j.foodchem.2022.135075
26. Qiao J et al (2023) Phenolamide and flavonoid glycoside profiles of 20 types of monofloral bee pollen. Food Chem 405(PA):134800. https://doi.org/10.1016/j.foodchem.2022.134800
27. Oroian M, Dranca F, Ursachi F (2022) Characterization of Romanian bee pollen—an important nutritional source. Foods 11(17):2633. https://doi.org/10.3390/foods11172633
28. Han S et al (2022) Lotus bee pollen extract inhibits isoproterenol-induced hypertrophy via JAK2/STAT3 signaling pathway in rat H9c2 cells. Antioxidants 12(1):88. https://doi.org/10.3390/antiox12010088
29. Algethami JS et al (2022) Bee pollen: clinical trials and patent applications. Nutrients 14(14):2858. https://doi.org/10.3390/nu14142858
30. Weis WA et al (2022) An overview about apitherapy and its clinical applications. Phytomedicine Plus 2(2):100239. https://doi.org/10.1016/j.phyplu.2022.100239
31. Durazzo A et al (2021) Bee products: a representation of biodiversity, sustainability, and health. Life 11(9):1–32. https://doi.org/10.3390/life11090970
32. Didaras NA, Karatasou K, Dimitriou TG, Amoutzias GD, Mossialos D (2020) Antimicrobial activity of bee-collected pollen and beebread: state of the art and future perspectives. Antibiotics 9(11):811. https://doi.org/10.3390/antibiotics9110811
33. Lima WG, Brito JCM, da Cruz Nizer WS (2021) Bee products as a source of promising therapeutic and chemoprophylaxis strategies against COVID-19 (SARS-CoV-2). Phyther Res 35(2):743–750. https://doi.org/10.1002/ptr.6872
34. Al-Kahtani SN, Alaqil AA, Abbas AO (2022) Modulation of antioxidant defense, immune response, and growth performance by inclusion of Propolis and bee pollen into broiler diets. Animals 12(13):1–13. https://doi.org/10.3390/ani12131658
35. Khalifa SAM et al (2021) Bee pollen: current status and therapeutic potential. Nutrients 13(6):1876. https://doi.org/10.3390/nu13061876
36. Kędzierski W et al (2020) Bee pollen supplementation to aged horses influences several blood parameters. J Equine Vet 90:103024. https://doi.org/10.1016/j.jevs.2020.103024
37. Gao M et al (Apr. 2021) Effects of carbonized process on quality control, chemical composition and pharmacology of Typhae Pollen: A review. *J. Ethnopharmacol.* 270(June 2020):113774. https://doi.org/10.1016/j.jep.2020.113774
38. Zhou W et al (2020) Antioxidant and immunomodulatory activities in vitro of polysaccharides from bee collected pollen of Chinese wolfberry. Int J Biol Macromol 163:190–199. https://doi.org/10.1016/j.ijbiomac.2020.06.244
39. Zhu L et al (2020) A polysaccharide from: Fagopyrum esculentum Moench bee pollen alleviates microbiota dysbiosis to improve intestinal barrier function in antibiotic-treated mice. Food Funct 11(12):10519–10533. https://doi.org/10.1039/d0fo01948h
40. Shaldoum F, El-kott AF, Ouda MMA, Abd-Ella EM (2021) Immunomodulatory effects of bee pollen on doxorubicin-induced bone marrow/spleen immunosuppression in rat. J Food Biochem 45(6):1–14. https://doi.org/10.1111/jfbc.13747
41. Di Chiacchio IM et al (2021) Bee pollen as a dietary supplement for fish: effect on the reproductive performance of zebrafish and the immunological response of their offspring. Fish Shellfish Immunol 119(July):300–307. https://doi.org/10.1016/j.fsi.2021.10.012

42. Bryś MS, Skowronek P, Strachecka A (2021) Pollen diet—properties and impact on a bee Colony. Insects 12(9):798. https://doi.org/10.3390/insects12090798
43. Filannino P et al (2021) Nutrients bioaccessibility and anti-inflammatory features of fermented bee pollen: a comprehensive investigation. *Front. Microbiol.* 12(February):622091. https://doi.org/10.3389/fmicb.2021.622091
44. Fowler AE, Sadd BM, Bassingthwaite T, Irwin RE, Adler LS (1853) Consuming sunflower pollen reduced pathogen infection but did not alter measures of immunity in bumblebees. Philos Trans R Soc B Biol Sci 377:2022. https://doi.org/10.1098/rstb.2021.0160
45. Jarosz PM, Jasielski PP, Zarobkiewicz MK, Sławiński MA, Wawryk-Gawda E, Jodłowska-Jędrych B (2022) Changes in histological structure, Interleukin 12, smooth muscle actin and nitric oxide synthase 1. and 3. Expression in the liver of running and non-running wistar rats supplemented with bee pollen or whey protein. *Foods* 11(8):1131. https://doi.org/10.3390/foods11081131
46. Abdelnour SA, Abd El-Hack ME, Alagawany M, Farag MR, Elnesr SS (2019) Beneficial impacts of bee pollen in animal production, reproduction and health. J Anim Physiol Anim Nutr (Berl) 103(2):477–484. https://doi.org/10.1111/jpn.13049
47. Giampieri F et al (2022) Bee products: an emblematic example of underutilized sources of bioactive compounds. J Agric Food Chem 70(23):6833–6848. https://doi.org/10.1021/acs.jafc.1c05822
48. Pingitore V et al (2022) Discovery of human hexosaminidase inhibitors by in situ screening of a library of mono- and divalent pyrrolidine iminosugars. Bioorg Chem 120(January):105650. https://doi.org/10.1016/j.bioorg.2022.105650
49. Taddei ML, Pardella E, Pranzini E, Raugei G, Paoli P (2020) Role of tyrosine phosphorylation in modulating cancer cell metabolism. Biochim Biophys Acta – Rev Cancer 1874(2):188442. https://doi.org/10.1016/j.bbcan.2020.188442
50. Asma ST et al (2022) General nutritional profile of bee products and their potential antiviral properties against mammalian viruses. Nutrients 14(17):3579. https://doi.org/10.3390/nu14173579
51. Nainu F et al (2021) Pharmaceutical prospects of bee products: special focus on anticancer, antibacterial, antiviral, and antiparasitic properties. Antibiotics 10(7):822. https://doi.org/10.3390/antibiotics10070822
52. Ilie CI et al (2022) Bee pollen extracts: chemical composition, antioxidant properties, and effect on the growth of selected probiotic and pathogenic bacteria. Antioxidants 11(5):959. https://doi.org/10.3390/antiox11050959
53. Acquaviva A et al (2022) Phytochemical and biological investigations on the pollen from industrial hemp male inflorescences. Food Res Int 161(August):111883. https://doi.org/10.1016/j.foodres.2022.111883
54. Alfawaz HA, El-Ansary A, Al-Ayadhi L, Bhat RS, Hassan WM (2022) Protective effects of bee pollen on multiple propionic acid-induced biochemical autistic features in a rat model. Meta 12(7):571. https://doi.org/10.3390/metabo12070571
55. Gercek YC, Celik S, Bayram S (2021) Screening of plant pollen sources, polyphenolic compounds, fatty acids and antioxidant/antimicrobial activity from bee pollen. Molecules 27(1):117. https://doi.org/10.3390/molecules27010117
56. Sawicki T, Starowicz M, Kłębukowska L, Hanus P (2022) The profile of polyphenolic compounds, contents of Total Phenolics and flavonoids, and antioxidant and antimicrobial properties of bee products. Molecules 27(4):1301. https://doi.org/10.3390/molecules27041301
57. Al-Hatamleh MAI, Boer JC, Wilson KL, Plebanski M, Mohamud R, Mustafa MZ (2020) Antioxidant-based medicinal properties of stingless bee products: recent progress and future directions. Biomol Ther 10(6):1–28. https://doi.org/10.3390/biom10060923
58. Gomes ANP et al (2021) Chemical composition of bee pollen and Leishmanicidal activity of Rhusflavone. Rev Bras 31(2):176–183. https://doi.org/10.1007/s43450-021-00130-z
59. Palmer-Young EC, Schwarz RS, Chen Y, Evans JD (2022) Can floral nectars reduce transmission of Leishmania? PLoS Negl Trop Dis 16(5):1–13. https://doi.org/10.1371/journal.pntd.0010373

60. Barbuto Ferraiuolo S, Cammarota M, Schiraldi C, Restaino OF (2021) Streptomycetes as platform for biotechnological production processes of drugs. Appl Microbiol Biotechnol 105(2):551–568. https://doi.org/10.1007/s00253-020-11064-2
61. Grubbs KJ et al (2021) Pollen Streptomyces produce antibiotic that inhibits the honey bee pathogen Paenibacillus larvae. Front Microbiol 12(February):632637. https://doi.org/10.3389/fmicb.2021.632637
62. Jagua-Gualdrón A, Peña-Latorre JA, Fernadez-Bernal RE (2020) Apitherapy for osteoarthritis: perspectives from basic research. Complement Med Res 27(3):184–191. https://doi.org/10.1159/000505015
63. Eteraf-Oskouei T, Shafiee-Khamneh A, Heshmati-Afshar F, Delazar A (2020) Anti-inflammatory and anti-angiogenesis effect of bee pollen methanolic extract using air pouch model of inflammation. Res Pharm Sci 15(1):66–75. https://doi.org/10.4103/1735-5362.278716
64. Martiniakova M, Bobonova I, Toman R, Galik B, Bauerova M, Omelka R (2021) Dose-dependent impact of bee pollen supplementation on macroscopic and microscopic structure of femoral bone in rats. Animals 11(5):1265. https://doi.org/10.3390/ani11051265
65. Tomaszewska E et al (2020) The effect of bee pollen on bone biomechanical strength and trabecular bone histomorphometry in tibia of young Japanese quail (Coturnix japonica). PLoS One 15(3):1–15. https://doi.org/10.1371/journal.pone.0230240
66. Petermann-Rocha F et al (2022) Global prevalence of sarcopenia and severe sarcopenia: a systematic review and meta-analysis. J Cachexia Sarcopenia Muscle 13(1):86–99. https://doi.org/10.1002/jcsm.12783
67. Ali AM, Kunugi H (2020) Apitherapy for age-related skeletal muscle dysfunction (Sarcopenia): a review on the effects of Royal Jelly, Propolis, and Bee Pollen. *Foods* 9(10):1362. https://doi.org/10.3390/foods9101362
68. Kurek-Górecka A, Górecki M, Rzepecka-Stojko A, Balwierz R, Stojko J (2020) Bee products in dermatology and skin care. Molecules 25(3):556. https://doi.org/10.3390/molecules25030556
69. Dumitru CD, Neacsu IA, Grumezescu AM, Andronescu E (2022) Bee-derived products: chemical composition and applications in skin tissue engineering. Pharmaceutics 14(4):1–29. https://doi.org/10.3390/pharmaceutics14040750
70. Peršurić Ž, Pavelić SK (2021) Bioactives from bee products and accompanying extracellular vesicles as novel bioactive components for wound healing. Molecules 26(12):3770. https://doi.org/10.3390/molecules26123770
71. Di Chiacchio IM et al (2022) Bee pollen in zebrafish diet affects intestinal microbiota composition and skin cutaneous melanoma development. Sci Rep 12(1):1–18. https://doi.org/10.1038/s41598-022-14245-3
72. Schuh CMAP, Aguayo S, Zavala G, Khoury M (2019) Exosome-like vesicles in Apis mellifera bee pollen, honey and royal jelly contribute to their antibacterial and pro-regenerative activity. J Exp Biol 222(20):jeb208702. https://doi.org/10.1242/jeb.208702
73. Chen X et al (2021) Identification of anti-inflammatory vesicle-like nanoparticles in honey. J Extracell Vesicles 10(4):e12069. https://doi.org/10.1002/jev2.12069
74. Jaganjac M, Tisma VS, Zarkovic N (2021) Short overview of some assays for the measurement of antioxidant activity of natural products and their relevance in dermatology. Molecules 26(17):5301. https://doi.org/10.3390/molecules26175301
75. Nakai K, Tsuruta D (2021) What are reactive oxygen species, free radicals, and oxidative stress in skin diseases? Int J Mol Sci 22(19):10799. https://doi.org/10.3390/ijms221910799
76. Liu H-M et al (2023) Possible mechanisms of oxidative stress-induced skin cellular senescence, inflammation, and cancer and the therapeutic potential of plant polyphenols. Int J Mol Sci 24(4):3755. https://doi.org/10.3390/ijms24043755
77. Pilkington SM, Bulfone-Paus S, Griffiths CEM, Watson REB (2021) Inflammaging and the skin. J Invest Dermatol 141(4):1087–1095. https://doi.org/10.1016/j.jid.2020.11.006
78. Orsmond A, Bereza-Malcolm L, Lynch T, March L, Xue M (2021) Skin barrier dysregulation in psoriasis. Int J Mol Sci 22(19):10841. https://doi.org/10.3390/ijms221910841

79. Bharadvaja N, Gautam S, Singh H (2023) Natural polyphenols: a promising bioactive compounds for skin care and cosmetics. Mol Biol Rep 50(2):1817–1828. https://doi.org/10.1007/s11033-022-08156-9
80. Schilrreff P, Alexiev U (2022) Chronic inflammation in non-healing skin wounds and promising natural bioactive compounds treatment. Int J Mol Sci 23(9):4928. https://doi.org/10.3390/ijms23094928
81. Tabolacci C et al (2023) Phytochemicals as immunomodulatory agents in melanoma. Int J Mol Sci 24(3):1–38. https://doi.org/10.3390/ijms24032657
82. Gendrisch F, Esser PR, Schempp CM, Wölfle U (2021) Luteolin as a modulator of skin aging and inflammation. Biofactors 47(2):170–180. https://doi.org/10.1002/biof.1699
83. Beken B, Serttas R, Yazicioglu M, Turkekul K, Erdogan S (2020) Quercetin improves inflammation, oxidative stress, and impaired wound healing in atopic dermatitis model of human keratinocytes. Pediatr Allergy Immunol Pulmonol 33(2):69–79. https://doi.org/10.1089/ped.2019.1137
84. Elmowafy M et al (2021) Olive oil/pluronic oleogels for skin delivery of quercetin: in vitro characterization and ex vivo skin permeability. *Polymers (Basel)* 13(11):1808. https://doi.org/10.3390/polym13111808
85. Lin MH, Hung CF, Sung HC, Yang SC, Yu HP, Fang JY (2021) The bioactivities of resveratrol and its naturally occurring derivatives on skin. J Food Drug Anal 29(1):15–38. https://doi.org/10.38212/2224-6614.1151
86. Yoon JH, Kim MY, Cho JY (2023) Apigenin: a therapeutic agent for treatment of skin inflammatory diseases and cancer. Int J Mol Sci 24(2):1498. https://doi.org/10.3390/ijms24021498
87. Kim J et al (2022) Kaempferol tetrasaccharides restore skin atrophy via PDK1 inhibition in human skin cells and tissues: bench and clinical studies. Biomed Pharmacother 156:113864. https://doi.org/10.1016/j.biopha.2022.113864

Index

A
Added value, 211
Aflatoxins, 181–183
Alkaloids, 63, 179, 187–188
Amino acid score (AAS), 20, 29
Anti-allergic activity, 86, 322, 324
Anti-infective activity, 322, 330
Antioxidants, 18, 31, 52, 54, 86, 95, 98, 104, 127–129, 131, 133–142, 148, 155, 158, 160, 161, 165, 167, 169, 199–201, 203, 204, 206–208, 211–214, 221, 232, 252, 254–256, 258, 260, 262, 264, 266, 268, 271, 292, 321, 324, 325, 327–330

B
Bacteria, 38, 64, 72, 140, 180, 184, 230–238, 240, 243, 258, 259, 309, 323, 326, 327
Bee bread, 30, 39, 53, 73, 86–92, 94, 185, 230, 232, 235, 253, 258, 259, 300, 312, 326
Bee-collected pollen (BCP), 30, 41, 56, 73–80, 104–119, 121, 128, 131, 132, 139, 142, 181, 182, 186, 188, 230–232, 234, 235, 237–242, 293, 302, 309, 310
Beekeeping, 184, 186, 214, 251, 288
Bee pollen, 13, 18, 51, 71, 136, 179, 199, 232, 251, 277, 319
Betaines, 198, 213–216, 221
Bioaccessibility of elements, 95, 96
Bioactive compounds, 18, 51–54, 64, 104, 121, 198, 199, 208, 211, 213, 221, 251, 252, 256, 259, 261, 262, 264, 294, 320, 327, 330
Biogenic elements, 85, 86
Bioindicators, 96, 97
Botanical markers, 139, 140, 204, 214
Botanical origins, 22, 25, 32, 42, 57, 58, 60, 61, 63, 74, 75, 79, 80, 93, 94, 97, 98, 104–107, 109–112, 115, 117, 132, 137, 139–141, 147–149, 162, 166, 200, 206, 209, 210, 212–217, 219, 220, 252, 255, 292, 294, 295, 297–298, 308, 312, 323

C
Carbohydrates, 18, 21, 31, 37, 51–65, 71, 85, 86, 98, 198, 216, 251, 252, 259, 266, 268, 277, 291–293, 295, 298, 299, 301, 320
Carotenoids, 37, 51, 52, 54, 71, 72, 86, 98, 104, 138, 147–161, 166, 251, 252, 254–256, 258, 260, 266, 269, 271, 291, 292, 301, 320, 321
Chemometric analysis, 97–98
Chemotaxonomy, 128, 141
Composition, 10, 52, 104, 128, 165, 198, 251, 288, 320
Contaminant elements, 87
Corbiculars, 230

D
Determination method, 56, 198
Dietary fibres, 51, 52, 54–56, 60–62, 64, 65
Digestibility, 18–21, 37–39, 41, 42, 63, 64, 128, 131, 204, 252, 254–256, 258, 261, 267, 292
Dispersibility, 295, 297

N. Ecem Bayram et al. (eds.), *Pollen Chemistry & Biotechnology*,
https://doi.org/10.1007/978-3-031-47563-4

E
Emulsifying properties, 38, 267, 295–297, 301, 312, 313

F
Fatty acids, 37, 54, 55, 60, 62, 71, 72, 74–80, 98, 132, 147, 161, 179, 198, 220, 243, 251, 252, 261, 262, 266, 291, 294, 300, 320, 324–325
Fermentation, 31, 37, 38, 42, 53, 55, 63, 64, 73, 80, 128, 131, 142, 209, 231, 232, 236, 240, 243, 253, 258–264, 271, 300–302, 323
Floral pollen, 17, 22–30, 39, 41, 74, 75, 79, 105, 297
Flower pollen, 30, 56, 104, 209, 230
Foaming properties, 31, 294, 295, 297, 313
Food safety, 62, 179, 184, 187, 288
Fumonisins, 183
Functional food components, 104, 121
Functional foods, 18, 31, 63, 64, 87, 95, 104, 121, 127, 128, 142, 147, 270, 288, 299, 313

G
Geographical origins, 18, 51, 52, 54, 56–59, 61, 71, 74, 79, 87, 92–94, 97, 98, 106, 121, 127, 137, 139, 140, 142, 198, 207, 251, 291–293, 295, 297, 302, 313, 325
Glucosinolates (GSLs), 37, 104, 198, 204–208, 221

H
Health effects, 147, 155, 165–166
Heavy metals, 167, 188
Honey bees, 1, 11, 17, 21, 24, 25, 29, 30, 32, 40, 42, 51, 277
Human nutrition, 30, 52–54, 62–64, 85, 95, 98, 119–121, 179, 186, 251–253, 259, 278

I
Immune response modulation, 322–324, 327

K
Kaempferol, 54, 117, 129, 132, 134, 135, 137–141, 215, 261, 324, 326, 327, 331

L
Lactic acid bacteria (LAB), 38, 53, 230–232, 236–237, 243, 258–264, 271
Lignin, 55, 216–220, 294
Lipids, 18, 22, 40, 51, 52, 54, 55, 62, 71–80, 85, 86, 121, 138, 157, 159, 162, 198, 201, 251, 252, 277, 291–294, 296, 299–301, 312, 313, 320

M
Mechanical device, 280
Microscopic fungi (MF), 218, 230, 232, 233, 238–242
Musculoskeletal disorders, 327–331

N
Nitrogen content, 22, 39, 40
Nitrogen-to-protein conversion factor, 40
Nutritional value of pollen, 24, 213

O
Ochratoxin (OTA), 181–183
Organic acids, 198, 208–210, 221, 259, 264, 320

P
Pesticide contamination, 184–186
Phenolamides, 104, 198–204, 221
Phenolic acids, 86, 103–121, 129, 132, 139, 252, 321, 326
Phenolic compounds, 51, 54, 60, 72, 98, 103–105, 108, 119, 121, 129, 198, 203, 207–208, 211–213, 251, 252, 255–256, 258, 271, 291, 292, 300, 320, 321, 324
Pollen, 1, 17, 51, 71, 104, 127, 147, 179, 198, 229, 251, 277, 319
Pollen anatomy, 1–13
Pollen bioactivities, 320
Pollen collection, 53–54, 80, 180, 285, 286
Pollen hydrophilic properties, 292–294
Pollenkitt, 37, 72, 73
Pollen lipophilic properties, 292–294
Pollen morphology, 1–13
Pollen traps, 39, 53, 62, 86, 179, 230, 232, 278–288
Polyphenols, 86, 128, 129, 131, 137, 139, 141, 142, 147, 211, 255, 261, 262, 266, 320, 330, 331

Q
Quercetin, 54, 129, 132–141, 254, 256, 261, 268, 291, 292, 324, 326, 327, 331

R
Recommended intake, 95, 165
Rutin, 132, 134–141, 256, 261

S
Sample treatment, 199–203, 205, 207–211, 219
Sensory properties, 292, 294, 295, 299, 301–312
Skin care agent, 155, 166, 199, 330, 331
Sporopollenin, 6, 37, 52, 56, 60, 72, 81, 95, 104, 105, 198, 216–221, 265, 294, 299
Stilbenes, 198, 211–213, 221
Sugars, 1, 18, 30, 51, 52, 54–60, 62, 64, 104, 121, 131, 252, 259, 266, 269, 292, 295, 301

T
Toxic compounds, 184

V
Viruses, 233, 242–243, 287, 323, 325, 327
Volatile compounds, 73, 74, 301, 321
Volatiles, 1, 37, 71, 73–74, 81, 261

W
Wettability, 31, 295, 297

Y
Yeast, 38, 231, 232, 238–240, 258, 259, 261–264, 269, 271, 300

Z
Zearalenone (ZEA), 180, 183

Printed in the USA
CPSIA information can be obtained
at www.ICGtesting.com
LVHW010018230124
769684LV00005B/72